Fortschritte der Chemie organischer Naturstoffe

Progress in the Chemistry of Organic Natural Products

46

Founded by L. Zechmeister
Edited by W. Herz, H. Grisebach, G. W. Kirby, and Ch. Tamm

Authors:
E. Fujita, S. Johne, R. Kasai,
M. Node, O. Tanaka

Springer-Verlag
Wien New York 1984

Dr. W. HERZ, Professor of Chemistry, Department of Chemistry,
The Florida State University, Tallahassee, Florida, U.S.A.

Prof. Dr. H. GRISEBACH, Biologisches Institut II, Lehrstuhl für Biochemie der Pflanzen,
Albert-Ludwigs-Universität, Freiburg i. Br., Federal Republic of Germany

G. W. KIRBY, Sc. D., Regius Professor of Chemistry, Chemistry Department,
The University, Glasgow, Scotland

Prof. Dr. CH. TAMM, Institut für Organische Chemie der Universität Basel,
Basel, Switzerland

With 7 Figures

© 1984 by Springer-Verlag/Wien
Softcover reprint of the hardcover 1st edition 1984
Library of Congress Catalog Card Number AC 39-1015

ISSN 0071-7886

ISBN-13: 978-3-7091-8761-6   e-ISBN-13: 978-3-7091-8759-3
DOI: 10.1007/978-3-7091-8759-3

Contents

List of Contributors

FUJITA, Prof. E., Institute for Chemical Research, Kyoto University, Uji, Kyoto-fu 611, Japan.

JOHNE, Prof. Dr. S., Institute of Plant Biochemistry, Academy of Sciences of the German Democratic Republic, DDR-4000 Halle (Saale), German Democratic Republic.

KASAI, Dr. R., Institute of Pharmaceutical Sciences, Hiroshima University School of Medicine, Kasumi 1-2-3, Minami-ku, Hiroshima 734, Japan.

NODE, Dr. M., Institute for Chemical Research, Kyoto University, Uji, Kyoto-fu 611, Japan.

TANAKA, Prof. O., Institute of Pharmaceutical Sciences, Hiroshima University School of Medicine, Kasumi 1-2-3, Minami-ku, Hiroshima 734, Japan.

Saponins of Ginseng and Related Plants

By O. TANAKA and R. KASAI, Institute of Pharmaceutical Sciences,
Hiroshima University School of Medicine, Kasumi, Hiroshima, Japan

With 1 Figure

Contents

Acknowledgements

We are greatly indebted to Emeritus Professor SHOJI SHIBATA of University of Tokyo for his valuable advice and encouragement. Thanks are also due to Mr. T. MORITA for his help in preparation of the manuscript.

I. Introduction

Ginseng, the famous plant drug, has been used as an expensive traditional medicine in oriental countries for more than 5000 years. The source plant of this drug is *Panax ginseng* C. A. Meyer (Araliaceae), a herb with fleshy roots which grows wild in cool and shady forests extending from Korea and North Eastern China to Far Eastern Siberia. Because the wild plant is relatively rare, it has been cultivated in Korea, China and Japan. After four or six years, the carrot-like roots of *P. ginseng* are steamed and dried to prepare "Red Ginseng", while the peeled roots dried without steaming are designated as "White Ginseng". In our chemical studies of this plant, White Ginseng was used unless otherwise stated. Dried lateral roots and roots dried without peeling are also used for the preparation of Ginseng prescription. Two closely related plants, American Ginseng (*P. quinquefolium* L., produced in the U.S.A. and Canada) and Sanchi-Ginseng (= Tienchi, roots of *P. notoginseng* (Burk.) F. H. Chen, cultivated in Yunnan, China) are also used for similar medicinal purposes.

As regards the chemical constituents of *Panax* spp., the presence of saponins in American Ginseng was reported as early as 1854 (*1*). Since the beginning of this century, a number of Japanese, European and Korean chemists have concerned themselves with isolation and structure elucidation of Ginseng saponins; these include ASAHINA (*2*), KONDO (*3*), KOTAKE (*4*), WAGNER-JAUREGG (*5*), HÖRHAMMER (*6*) and their coworkers. However, even the basic skeleton of the major sapogenin was not characterized until 1960.

Since 1958, SHIBATA and his successors have conducted chemical studies on the saponins of the crude drug. After many twists and turns, it was determined that the major saponins of Ginseng were not oleanane oligoglycosides which are very common in nature, but that the genuine sapogenins were represented by triterpenes of the dammarane type. This was the first example of the occurrence of dammarane saponins in nature. The complications encountered in the isolation and structure determination were mainly due to an unexpected acid catalyzed epimerization of the tertiary hydroxyl group on C-20 of the carbon skeleton which was followed by cyclization of the side chain. This undesirable reaction accompanied acid hydrolysis of the saponins to the sapogenins.

Studies on the chemical constituents of Ginseng and its relatives have also shown that the characteristic dammarane saponins are found in most other *Panax* species as well.

II. Structure and Chemistry of Dammarane Sapogenins

1. Panaxadiol, an Artifact

A sapogenin named panaxadiol (1) was isolated from the crude Ginseng saponin mixture by hydrolysis with dilute mineral acid in boiling aqueous ethanol (*7*). The skeleton of (1) was found to be that of a dammarane type triterpene by conversion of its monoacetate (2) into isotirucallenol (3) *via* monoketone (4) and panaxanol (5). The presence of a hydroxyl group at C-12 was confirmed by formation of the α,β-unsaturated ketone (6) upon acid treatment of (4); presence of the trimethyltetrahydropyran ring was proved by MS [base peak: m/z 127, representing fragment (7) in Chart 1] and NMR spectroscopy in 1963. While the tentative stereochemistry originally (*8*)

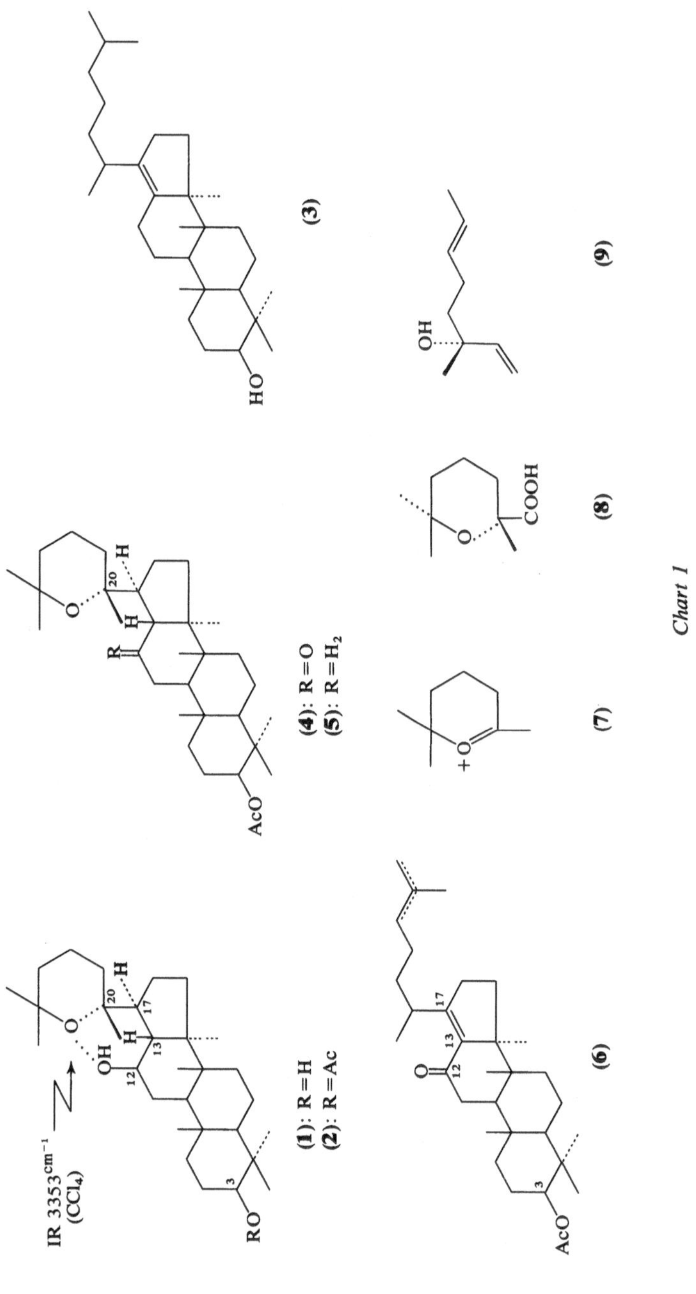

O. TANAKA and R. KASAI:

Chart 1

(3)

(9)

(8)

(7)

(4): R=O
(5): R=H₂

(6)

(1): R=H
(2): R=Ac

IR 3353 cm⁻¹ (CCl₄)

proposed for (1) turned out to be erroneous, the correct configurations of (1) at C-13 and C-17 (see Chart 1) were finally established (9, 10, 11) as a result of work on the genuine aglycone (see Chapter II-2). Moreover, stepwise degradation of (1) afforded (−)-cinenic acid (8) derived from R-(−)-linalool (9), thus elucidating the configuration of C-20 of (1) as R (12). It should be noted that the IR spectrum of (1) indicates strong hydrogen bonding between the 12β-hydroxyl group and the oxygen of the tetrahydrofuran ring and that the hydroxyl group is sterically hindered and therefore resistant toward acetylation.

2. Acid Catalyzed Reactions of Dammarane Type Triterpenes and the Genuine Sapogenin of Ginseng

In 1930, KOTAKE (4) reported that partial hydrolysis of the crude Ginseng saponins afforded a prosapogenin named α-panaxin. He also reported that hydrolysis of α-panaxin with conc. HCl at room temperature afforded an unidentified chlorine-containing sapogenin (4). The subsequent repetition of his experiment by SHIBATA's group showed that the same chloro compound (10) could be obtained directly by hydrolysis of the crude saponins with conc. HCl, while (1) was recovered unchanged after acid treatment (9). This observation led to the suspicion that (1) might not be a genuine sapogenin.

Dehydrochlorination of (10) with potassium t-butoxide (see Chart 2) yielded a compound (11), whose catalytic hydrogenation afforded a dihydro derivative (12). Treatment of (11) with dilute mineral acid in boiling aqueous ethanol gave (1) (9). The structure of (11), including the stereochemistry at C-13 and C-17 was shown (10, 11) (see Chart 2) to be 12β-hydroxydammarenediol-I by chemical correlation with betulafolienetriol (13) [a substance isomeric with (11) at C-3 and C-20, which is a constituent of the fresh leaves of white birch (13, 14)]. That the absolute configuration at C-20 of dammarenediol-I (14) and its homologous such as (11) is 20(R) and that of dammarenediol-II (15) and its homologues such as (13) is 20(S) was subsequently shown by two different groups (12, 15, 16).

Treatment of (11) with conc. HCl at room temperature yielded (10). Furthermore, catalytic hydrogenation of the saponin mixture in ethanol-acetic acid followed by hydrolysis with dilute mineral acid did not yield (1) but afforded the dihydro derivative (12). This evidence indicated that (1) and (10) must be artifacts formed during the process of acid hydrolysis; thus, at that stage, the genuine sapogenin seemed to be substance (11) which was named protopanaxadiol (9, 10, 11). However, as will be described later, investigation of the acid-catalyzed isomerization of dammarane type triterpenes and subsequent study of the mild hydrolysis of the purified

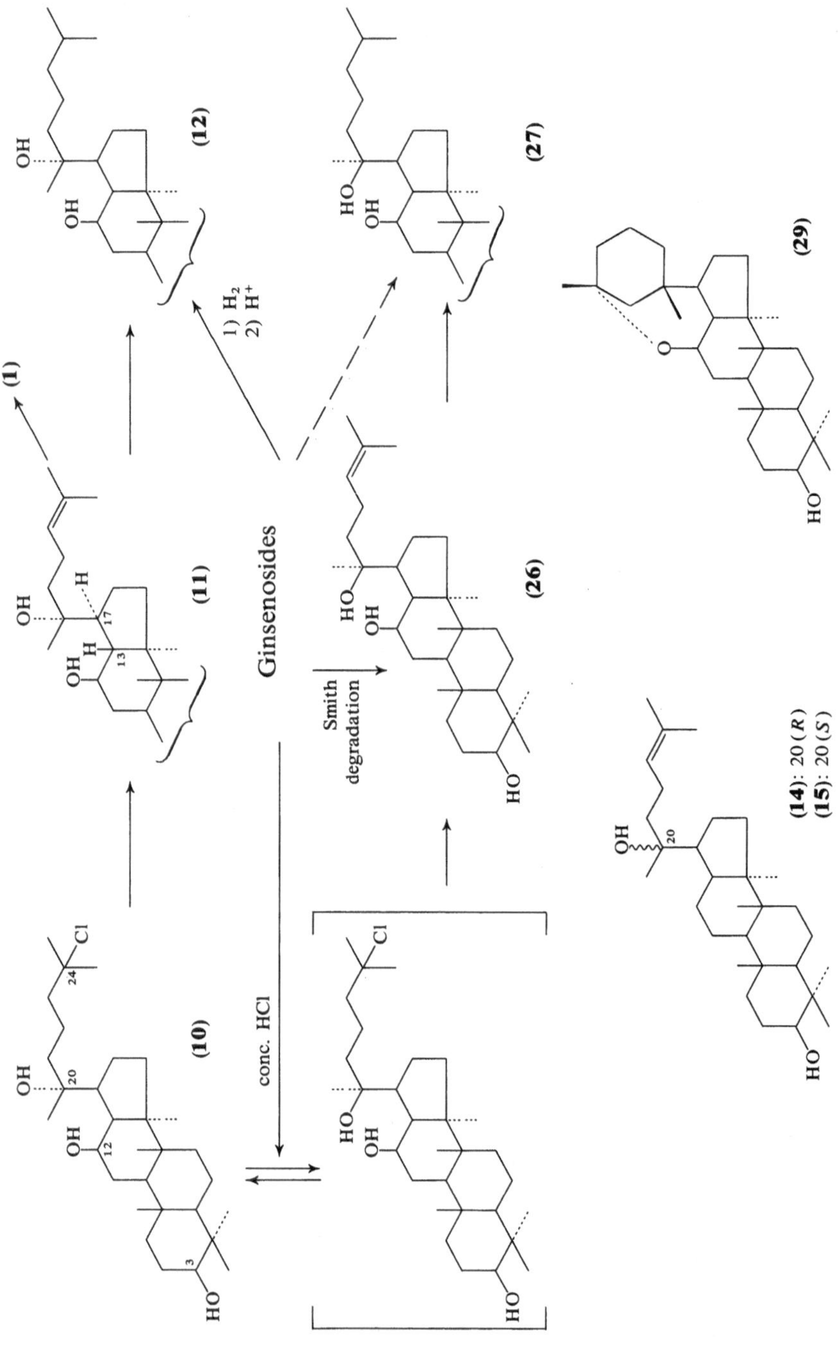

(12)

(27)

(29)

(1)

(11)

1) H₂
2) H⁺

Ginsenosides

17

13

H

H

(10)

Smith
degradation

(26)

conc. HCl

Cl

24

OH

20

OH

12

3

HO

Cl

HO
OH

HO

OH

20

(14): 20 (R)
(15): 20 (S)

HO

Chart 2

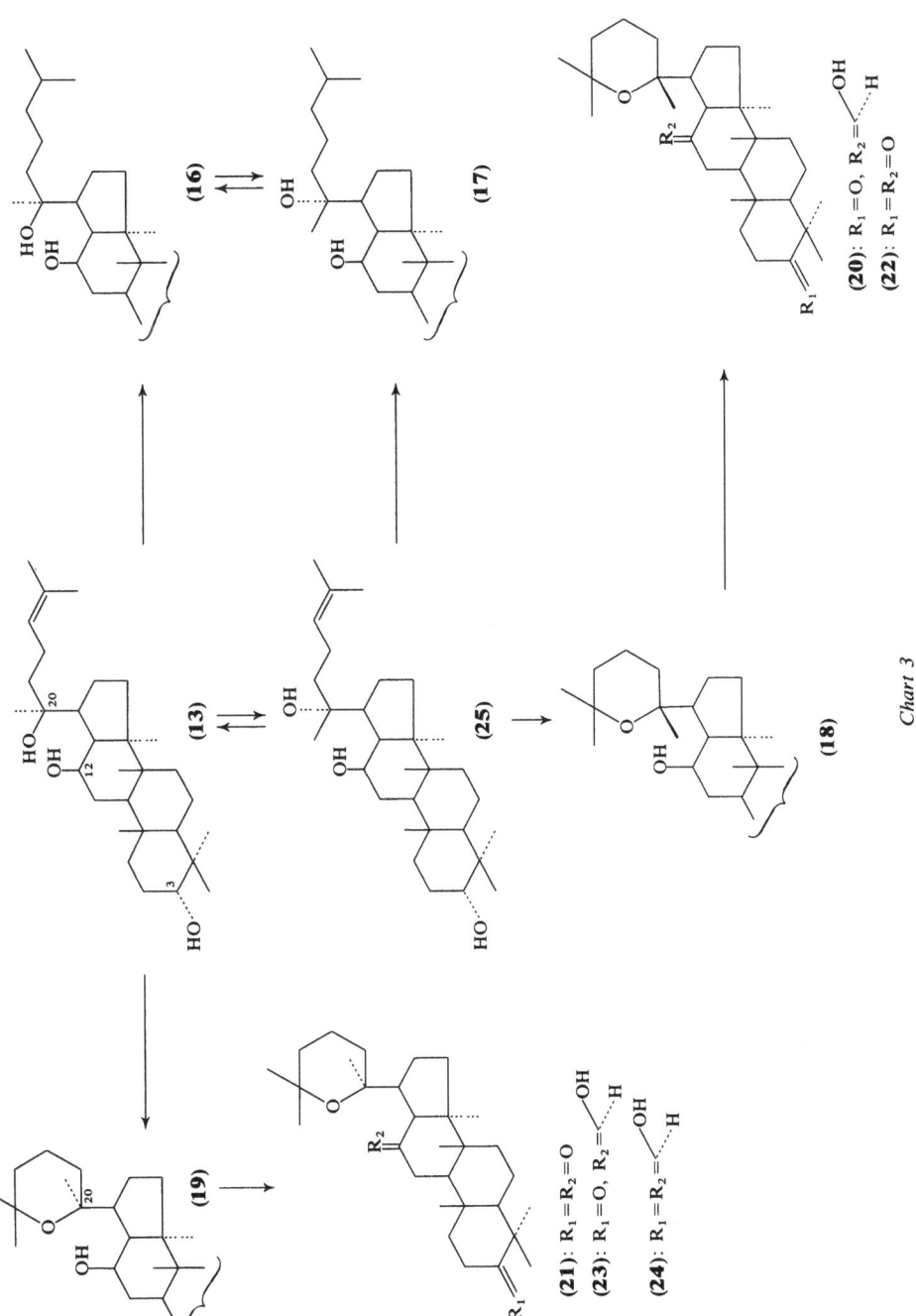

Chart 3

saponins led to the conclusion that the genuine sapogenin was not (11) but its C-20 epimer, 20(S)-protopanaxadiol or 12β-hydroxydammarenediol-II, the 3-epimer of (13). Hence, (11) was designated as 20(R)-protopanaxadiol in order to avoid confusion in the literature.

Because Ginseng is a very expensive drug, Shibata's group decided to use (13) as a model compound in its chemical investigations of the Ginseng sapogenin which are outlined in Chart 3. On catalytic hydrogenation, (13) afforded betulafolianetriol (16). Acid treatment of (16) under the conditions previously used for preparation of (1) from the crude saponin afforded an equilibrium mixture of (16) and its 20-epimer (17) with (17) in slight excess. The structure of (17) was substantiated by chromic acid oxidation which led to a 3,12-diketone identical with the diketone prepared from (12) (*16, 17*). Similar acid treatment of (13) afforded two compounds (18) and (19) in approximately equal yields. The presence of a trimethyltetrahydropyran ring in both compounds was demonstrated by MS and NMR spectroscopy. Mild oxidation of (18) afforded a ketone (20) which was identical with the 3-ketone derived from (1). Consequently, (18) could be formulated as 3-*epi*-panaxadiol. A diketone (21) prepared from (19) exhibited an ORD curve which was very similar to that of a 3,12-diketone (22) derived from (1). It follows that (19) must be an 20-epimer of (1) (*16, 17*). NaBH$_4$ reduction of the 3-ketone (23) derived from (19) afforded 20-*epi*-panaxadiol (24) (*18*).

Epimerization and cyclization of the side chain proceeded more slowly under the influence of *p*-toluenesulfonic acid in chloroform at room temperature (*16, 17*). In the early stages, only (19) was formed from (13), while (18) began to appear almost as soon as the C-20 epimer (25). The structure of the latter was confirmed by catalytic hydrogenation of (25) to (17).

The presence of a β-hydroxyl or methoxyl group on C-12 was shown to be essential for this acid catalyzed reaction, since compounds without such a group such as (14) underwent dehydration of the tertiary hydroxyl on C-20 in preference to epimerization and cyclization of the side chain (*16, 17*).

It followed that the same epimerization probably occurred during acid hydrolysis of the Ginseng saponins. The chloro compound (10) was obtained from the ether insoluble fraction of the crude hydrolysate of the saponin mixture with conc. HCl. The ether soluble fraction of the hydrolysate was subjected to dehydrochlorination to give an 20-epimer of (11) named 20(S)-protopanaxadiol (26) (Chart 2), the structure of which was elucidated by correlating it with (13) as follows. The dihydro derivative (27) of (26) was oxidized to give a diketone (28) which was identical with the diketone derived from (13) *via* (16) (*16, 17, 19*). Crystalline substances (1) and (12) of the 20(R)-series were also readily obtained from the less soluble fractions of the hydrolysates of the saponin mixture and from the hydrogenated saponin mixture, respectively, while the corresponding 20(S)

epimers, (24) and (27) were expected to be present in the more soluble part of the hydrolysate. However, isolation of the 20(S) epimers in the pure state from the respective mother liquor could not be accomplished as the chromatographic behaviour of the c-20 epimers of the 3β-hydroxy series did not differ appreciably. In contrast, members of the various c-20 epimeric pairs of the 3α-hydroxy series, (13) and (25), (16) and (17) and (18) and (19) were easily separated from each other by TLC and column chromatography. Thus, if the 3α-hydroxy compound (13) had not been used as a model, the acid catalyzed epimerization would not have been discovered and the genuine sapogenin might even now be believed to be (11).

In the field of carbohydrate chemistry, the so-called "Smith degradation" (20) has been used for regiospecific cleavage of oligo- or polysaccharide units. This reaction involves periodate oxidation followed by $NaBH_4$ reduction and subsequent very mild acid treatment. As is mentioned in the following Chapter, TLC of the glycoside fraction of Ginseng extract demonstrated the presence of a number of saponins, which were named ginsenosides. When these ginsenosides-Rb_1 (Rb_1), -Rb_2 (Rb_2) and -Rc (Rc) were isolated by preparative TLC and subjected to mild hydrolysis, Smith degradation afforded only the genuine sapogenin (26) common to all of these saponins (19, 21).

As a result of this work, it was discovered that substance (11) undergoes another type of acid catalyzed ring closure on treatment with BF_3 etherate at room temperature (22, 23). This affords a crystalline compound named isodehydropanaxadiol (29) in 28% yield, whose structure was deduced by mass and NMR spectroscopy. Compound (29) was also isolated from the crude hydrolysate of the saponin mixture with dilute mineral acid as one of the minor products.

III. Structures of Ginseng Saponins (1)

In Fig. 1, TLC patterns of the saponin fraction and each saponin of Ginseng are illustrated. Up to now, eighteen saponins have been isolated from White Ginseng; these are ginsenosides-Ro*, -Ra_1*, -Ra_2*, -Ra_3*, -Rb_1, -Rb_2, -Rb_3*, -Rc, -Rd, -Re, -Rf, -Rg_1, -Rg_2*, -Rg_3*, -Rh_1*, 20-gluco-ginsenoside-Rf* (glc-Rf), notoginsenoside-R_1* (NG-R_1, Chapter VIII-2) and quinquenoside-R_1* (Q-R_1, Chapter VIII-1) (* denotes a minor saponin). Of these saponins, Ro is a saponin based on oleanolic acid and is identical with chikusetsusaponin-V (C-V, Chapter VIII-3) which is the major saponin of the rhizomes of *Panax japonicus, P. japonicus* var. *major, P. pseudo-ginseng* subsp. *himalaicus* (Himalayan *Panax*) and is also obtained from the roots of *P. quinquefolium* (American Ginseng). Its structure is shown in Table III. All the others are dammarane saponins, one of the minor saponins, Q-R$_1$, having been first isolated from American Ginseng (Chapter VIII-1). The present Chapter deals with the structure elucidation of the major saponins, Rb$_1$, Rb$_2$, Rc, Rd, Re, Rf and Rg$_1$ and some of the minor saponins, Rg$_2$, Rb$_3$, glc-Rf and Rg$_3$. Chemical studies on the other minor saponins will be described in Chapter V.

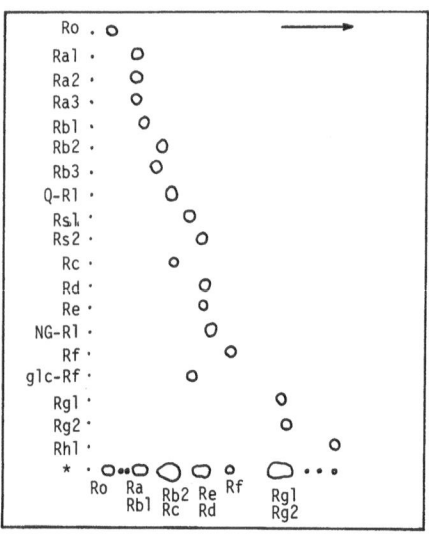
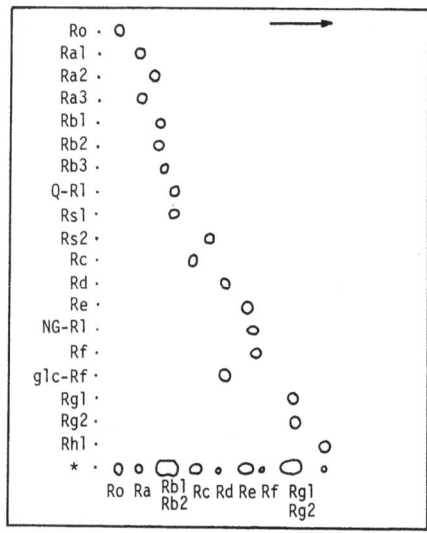

CHCl$_3$-MeOH-H$_2$0(14:6:1)   BuOH-AcOEt-H$_2$0(4:1:2, upper layer)
Plate:Silica Gel 100 F254(Merck)
* :Crude Saponin Fraction of the *P.ginseng* Roots.

Fig. 1. Thin Layer Chromatograms of the Saponins of *P. ginseng* Roots

1. Structure of the Prosapogenin Common to Ginsenosides-Rb$_1$, -Rb$_2$, -Rb$_3$, -Rc and -Rd

Ginsenosides-Rb$_1$, -Rb$_2$ and -Rc purified by preparative TLC were hydrolyzed with 50% aqueous acetic acid at 70°. From the methanol-insoluble fraction of each reaction mixture, a common prosapogenin (30) was obtained in crystalline form (24) which was identical with KOTAKE's α-panaxin (4) mentioned in Chapter II. Later, (30) was also obtained from Rd by the same treatment.

The structure of (30) was elucidated as follows (19, 24). Mineral acid hydrolysis of (30) yielded (1) and glucose, while Smith degradation of (30) afforded an aglycone which was identical not with the genuine sapogenin (26) but with its 20 epimer (11). The IR spectrum (in CCl$_4$) of an octaacetate of (30) exhibited a concentration independent -OH band at 3542 cm^{-1} which was characteristic of the intramolecular hydrogen bond

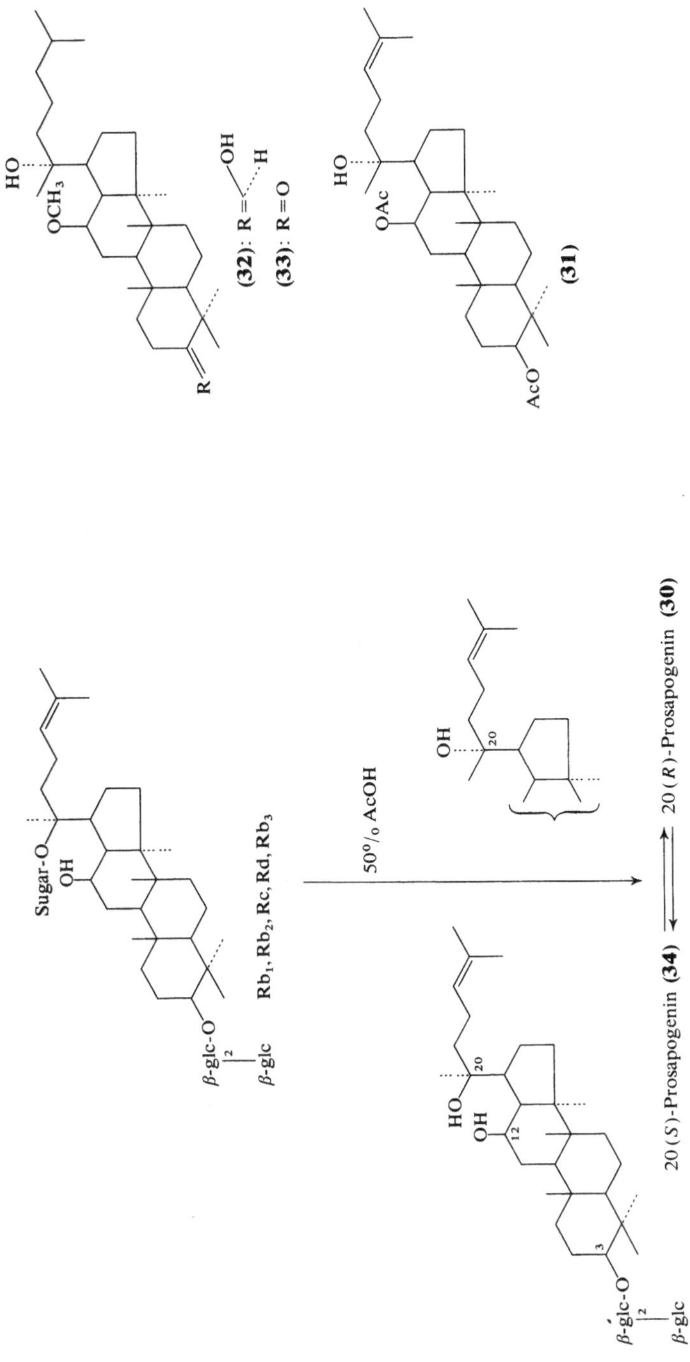

(32): R =⟨OH ·····H

(33): R = O

(31)

Rb₁, Rb₂, Rc, Rd, Rb₃

50% AcOH

20 (S)-Prosapogenin (34) ⇌ 20 (R)-Prosapogenin (30)

Chart 4

between the 12-o-acetyl group and the unacetylated 20-hydroxyl group of
3,12-di-o-acetyl-20(R)-protopanaxadiol (31). This indicated that both the
12- and 20-hydroxyl groups of (30) must be free and that a glycosyl linkage
should be located at the 3-hydroxyl group.

This was substantiated by the following evidence. Methylation of (30)
by HAKOMORI's procedure (25) followed by catalytic hydrogenation gave an
octamethyl ether of a dihydroprosapogenin which afforded a monomethyl
ether (32) after hydrolysis with dilute mineral acid. On chromic acid
oxidation, (32) gave a monoketone (33) whose ORD curve exhibited a
positive Cotton effect characteristic of a 3-ketone of the dammarane series,
and different from the negative Cotton effect typical of a 12-ketone (11).
The IR spectrum of (33) indicated that the intramolecular hydrogen bond
between the 20-hydroxyl and the 12-methoxyl groups was not affected by
the oxidation. The NMR spectrum of the octamethyl ether of the
dihydroprosapogenin exhibited doublets due to two anomeric protons at
δ 4.17 (J = 8 Hz) and 4.57 (J = 8 Hz) and on methanolysis afforded methyl
2,3,4,6-tetra-o-methyl-glucopyranoside and methyl 3,4,6-tri-o-methyl-
glucopyranoside, leading to the formulation of (30) as 3-o-β-sophorosyl-
20(R)-protopanaxadiol. Prosapogenin (30) was recently isolated from
Ginseng in 0.0002% yield and named ginsenoside-Rg₃ (26).

Absence of -OH absorption from the IR spectra of Rb₁, Rb₂, Rc and Rd
acetates indicated the presence of a o-glycosyl linkage at c-20, because
the 20 tertiary hydroxyl group of dammaranes resists acetylation and a
hydrogen bonded -OH band should be observed in consequence even after
acetylation if the 20-hydroxyl group were free. In parallel with the reported
facile hydrolysis of t-butyl glycosides (27, 28), it was concluded that the
c-20-glycosyl linkage of these saponins was hydrolyzed even on heating
with aqueous acetic acid to give an equilibrated mixture of prosapogenin
c-20-epimers, leading eventually to isolation of the 20(R)-epimer (30) from
the less soluble fraction of the hydrolysates (18). The 20(S)-epimer (34)
which is more soluble in organic solvents than (30), was isolated from the
methanol-soluble fraction. This selective hydrolysis of the 20-glycosyl
linkage has been used effectively not only in structural studies of a variety of
dammarane saponins but also in work studying the metabolism of
dammarane saponins in animals (Chapter XII).

2. Ginsenosides-Rb₁, -Rb₂, -Rb₃, -Rc and -Rd, Saponins of 20(S)-Protopanaxadiol

On the aforementioned evidence, Rb₁, Rb₂, Rc and Rd can be
formulated as 20-o-glycosides of 3-o-β-sophorosyl-20(S)-protopanaxadiol.
The isolation of Ginseng saponins on a preparative scale was accomplished

as follows (*29, 30, 31, 32*). An aqueous suspension of the methanolic extract was washed with ether and then extracted with 1-butanol saturated with water; the butanol layer containing the crude saponin fraction was subjected to column chromatography on silica gel [solvent, chloroform: methanol: water (lower layer) or 1-butanol: ethyl acetate: water (upper layer)] to give saponins, Ro (0.02% yield), Rb_1 (0.47%), Rb_2 (0.21%), Rb_3 (0.005%), Rc (0.26%), Rd (0.15%), Re (0.15%), Rf (0.05%), Rg_1 (0.21%), Rg_2 (0.01%) and glc-Rf (0.005%).

On dilute mineral acid hydrolysis, Rb_1 and Rd afforded glucose and (**1**), while Rb_2 and Rc yielded glucose and arabinose along with (**1**). On mild hydrolysis with hot aqueous acetic acid Rd gave (**30**), (**34**) and glucose, while Rb_1 afforded gentiobiose and Rb_2 and Rc gave bioses which consisted of glucose and arabinose. Methylation of these saponins followed by methanolysis and subsequent analysis of the resulting fully or partially methylated methyl monosaccharides by TLC and GLC established the structures of these four saponins as shown in Table I. The configuration assigned to the anomeric carbons was based on the coupling constants of the anomeric proton signals and on molecular rotation differences (*30*).

Ginsenoside-Rb_3, one of the minor saponins of Ginseng afforded (**26**) on Smith degradation and yielded (**30**), (**34**) and a biose (glucose-xylose) on partial hydrolysis. The structure of this saponin was established in the way described in the previous paragraph and is shown in Table I (*31*).

Table I. *Saponins of 20 (S)-protopanaxadiol*

Name	R_1	R_2	Source
Ra_1	-glc^2- glc	-glc^6- ara(pyr)^4- xyl	P. ginseng roots* (52)
Ra_2	-glc^2- glc	-glc^6- ara(fur)^2- xyl	P. ginseng roots* (52)
Ra_3	-glc^2- glc	-glc^6- glc^3- xyl	P. ginseng roots* (54)
Rb_1	-glc^2- glc	-glc^6- glc	P. ginseng roots (30), roots* (56), flower buds (55), leaves (55)

Table I *(continued)*

Name	R_1	R_2	Source
			P. notoginseng roots (*85, 88*), corms (*92*), flower buds (*119*), seeds (*118*), leaves (*118*) *P. quinquefolium* roots (*85*) *P. pseudo-ginseng* subsp. *himalaicus* rhizomes (*107*) *P. japonicus* var. *major* leaves (*120*)
Rb$_2$	-glc^2-glc	-glc^6-ara(pyr)	*P. ginseng* roots (*30*), roots* (*56*), fruits (*55*), flower buds (*55*), leaves (*55*) *P. notoginseng* corms (*92*), flower buds (*119*), roots (*85*) *P. quinquefolium* roots (*85*), leaves (*116*)
Rb$_3$	-glc^2-glc	-glc^6-xyl	*P. ginseng* roots (*31*), roots* (*56*) *P. notoginseng* seeds (*118*), leaves (*118*) *P. quinquefolium* roots (*86*), leaves (*117*) *P. pseudo-ginseng* subsp. *himalaicus* leaves (*111*) *P. japonicus* var. *major* leaves (*120*)
Rc	-glc^2-glc	-glc^6-ara(fur)	*P. ginseng* roots (*30*), roots* (*56*), fruits (*55*), flower buds (*55*), leaves (*55*) *P. notoginseng* roots (*85*), flower buds (*119*), seeds (*118*), leaves (*118*) *P. japonicus* var. *major* leaves (*120*) *P. quinquefolium* roots (*86*)
Rd	-glc^2-glc	-glc	*P. ginseng* roots (*30*), roots* (*56*), fruits (*55*), flower buds (*110*), leaves (*55, 109*) *P. notoginseng* roots (*85*), corms (*92*), seeds (*118*), flower buds (*119*) *P. quinquefolium* roots (*85*), leaves (*116*) *P. pseudo-ginseng* subsp. *himalaicus* leaves (*111*) *P. japonicus* var. *major* leaves (*120*), rhizomes (*105*)

Table I *(continued)*

Name	R_1	R_2	Source
			P. japonicus (Chinese) rhizomes (*104*)
Rg_3	$-glc^2-glc$	-H (20S and 20R)	*P. ginseng* roots* (*56*)
F2	-glc	-glc	*P. ginseng* leaves (*109*)
			P. notoginseng flower buds (*119*)
			P. quinquefolium roots (*86*)
Rs1	$-glc^2-glc^6-Ac$	$-glc^6-ara(pyr)$	*P. ginseng* roots* (*56*)
Rs2	$-glc^2-glc^6-Ac$	$-glc^6-ara(fur)$	*P. ginseng* roots* (*56*)
$Q-R_1$	$-glc^2-glc^6-Ac$	$-glc^6-glc$	*P. ginseng* roots* (*56*)
			P. quinquefolium roots (*86*)
Gy-XVII	-glc	$-glc^6-glc$	*P. notoginseng* roots (*92*)
			P. quinquefolium roots (*86*)
			Gynostemma pentaphyllum (Cucurbitaceae) (*87*)
Gy-IX	-glc	$-glc^6-xyl$	*P. notoginseng* seeds (*118*), leaves (*118*)
			Gynostemma pentaphyllum (*87*)
NG-R4	$-glc^2-glc$	$-glc^6-glc^6-xyl$	*P. notoginseng* roots (*92*)
NG-Fa	$-glc^2-glc^2-xyl$	$-glc^6-glc$	*P. notoginseng* leaves (*118*), seeds (*118*)
NG-Fc	$-glc^2-glc^2-xyl$	$-glc^6-xyl$	*P. notoginseng* leaves (118), seeds (*118*)
NG-Fe	-glc	$-glc^6-ara(fur)$	*P. notoginseng* leaves (*118*)
$PG-F_8$	$-glc^{6\!\diagup Ac}_{2\diagdown glc}$	$-glc^6-xyl$	*P. pseudo-ginseng* subsp. *himalaicus* leaves (*111*)
C-Ia	$-glc^6-xyl$	-H	*P. japonicus* rhizomes (*103*)
C-III	$-glc^{6\diagup xyl}_{2\diagdown glc}$	-H	*P. japonicus* rhizomes (*99, 101*)

-glc: β-D-glucopyranosyl, -xyl: β-D-xylopyranosyl, -ara(pyr): α-L-arabinopyranosyl, -ara(fur): α-L-arabinofuranosyl, Ac: acetyl, roots*: Red-Ginseng.

Q-: Quinquenoside, Gy-: Gypenoside, NG-: Notoginsenoside, PG-: Pseudo-ginseno-side, C-: Chikusetsusaponin.

O. TANAKA and R. KASAI:

Table II. *Saponins of 20 (S)-protopanaxatriol*

Name	R_1	R_2	R_3	Source
Re	-glc²-rha	-glc	-H	*P. ginseng*
				roots (*32*), roots* (*56*), fruits (*55*),
				flower buds (*110*), leaves (*55, 109*)
				P. notoginseng
				roots (*85*), corms (*92*)
				P. quinquefolium
				roots (*85*), leaves (*116*)
				P. trifolius
				roots (*65*)
				P. japonicus
				leaves (*121*)
				P. japonicus (Chinese)
				rhizomes (*104*)
				P. pseudo-ginseng subsp. *himalaicus*
				leaves (111)
Rf	-glc²-glc	-H	-H	*P. ginseng*
				roots (*32*), roots* (*56*)
				P. trifolius
				roots (*65*)
glc-Rf	-glc²-glc	-glc	-H	*P. ginseng*
				roots (*31*), roots* (*56*)
				P. notoginseng
				roots (*92*)
				P. japonicus var. *major*
				rhizomes (*105*)
Rg_1	-glc	-glc	-H	*P. ginseng*
				roots (*29*), roots* (*56*), fruits (*55*),
				flower buds (*110*), leaves (*55, 109*)
				P. notoginseng
				roots (*85*), corms (*92*)
				P. quinquefolium
				roots (*85*), leaves (*116, 117*)
				P. japonicus (Chinese)
				rhizomes (*104*)
				P. zingiberensis
				rhizomes (*106*)

Table II *(continued)*

Name	R₁	R₂	R₃	Source
Rg₂	-glc²-rha	-H	-H	*P. ginseng* roots *(32)*, roots* *(56)* *P. notoginseng* roots *(92)* *P. quinquefolium* roots *(86)* *P. japonicus* rhizomes *(103)* *P. japonicus* (Chinese) rhizomes *(104)* *P. trifolius* roots *(65)*
Rh₁	-glc	-H	-H	*P. ginseng* roots *(55)* *P. notoginseng* roots (92)
F1	-H	-glc	-H	*P. ginseng* leaves *(55)* *P. japonicus* leaves *(121)*
F3	-H	-glc⁶-ara(pyr)	-H	*P. ginseng* flower buds *(55)*, leaves *(109)*
NG-R1	-glc²-xyl	-glc	-H	*P. ginseng* roots* *(56)* *P. notoginseng* roots *(48)*, corms *(92)*
NG-R2	-glc²-xyl	-H	-H	*P. notoginseng* roots *(48)* *P. japonicus* (Chinese) rhizomes *(104)* *P. japonicus* var. *major* rhizomes *(105)*
NG-R3	-glc	-glc⁶-glc	-H	*P. notoginseng* roots *(92)*
NG-R6	-glc	-glc⁶-α-glc	-H	*P. notoginseng* roots *(92)*
C-L5	-H	-glc⁶-ara(pyr)⁴-xyl	-H	*P. japonicus* leaves *(121)*
C-L10	-H	-H	-glc	*P. japonicus* leaves *(121)*

-glc: β-D-glucopyranosyl, -α-glc: α-D-glucopyranosyl, -xyl: β-D-xylopyranosyl, -ara(pyr): α-L-arabinopyranosyl, roots*: Red-Ginseng.
NG-: Notoginsenoside, C-: Chikusetsusaponin.

Table III. *Saponins of Oleanolic Acid*

Name	R_1	R_2	Source
Ro(C-V)	-glcUA^2-glc	-glc	*P. ginseng* roots (*30*), roots* (*56*) *P. quinquefolium* roots (*85*) *P. pseudo-ginseng* subsp. *himalaicus* rhizomes (*107*) *P. japonicus* rhizomes (*99, 102*) *P. japonicus* (Chinese) rhizomes (*104*) *P. japonicus* var. *major* rhizomes (*105*) *P. zingiberensis* rhizomes (*106*) *P. trifolius* roots (*65*)
C-IV	-glcUA^4-ara(fur)	-glc	*P. japonicus* rhizomes (*99, 100*) *P. pseudo-ginseng* subsp. *himalaicus* rhizomes (*108*) *P. japonicus* (Chinese) rhizomes (*104*) *P. zingiberensis* rhizomes (*106*)
C-IVa	-glcUA	-glc	*P. japonicus* rhizomes (*103*) *P. pseudo-ginseng* subsp. *himalaicus* rhizomes (*107*) *P. japonicus* var. *major* rhizomes (*105*) *P. japonicus* (Chinese) rhizomes (*104*) *P. zingiberensis* rhizomes (*106*)
C-Ib	-glcUA$_4^6$$\diagdown$$^{glc}_{ara(fur)}$	-H	*P. japonicus* rhizomes (*103*)
Z-R$_1$	-glcUA^2-glc	-H	*P. zingiberensis* rhizomes (*106*)

-glcUA: β-D-glucuronic acid, -glc: β-D-glucopyranosyl, -ara(fur): α-L-arabinopyranosyl.
C-: chikusetsusaponin, Z-: zingibroside.

3. Panaxatriol and Ginsenosides-Rf, -Rg$_1$, -Rg$_2$ and 20-Gluco-ginsenoside-Rf (glc-Rf), Saponins of 20(S)-Protopanaxatriol

Ginsenoside-Rg$_1$ obtained in 0.21% yield was purified through its crystalline decaacetate (29). On dilute mineral acid hydrolysis, Rg$_1$ yielded glucose and a crystalline compound named panaxatriol (35) which seemed to be an artifact formed from the genuine sapogenin during the process of the acid hydrolysis. The structure of (35) was established as 6α-hydroxypanaxadiol by comparison of its IR, NMR and mass spectra with the spectra of (1), zeorin (37), a lichen substance, and their derivatives (33).

Smith degradation of Rg$_1$ gave a genuine sapogenin (36) which afforded (35) on acid treatment with dilute mineral acid. Hydrolysis of Rg$_1$ with conc. HCl at room temperature and dehydrochlorination of the resulting mixture of chlorides (38) and (39) with potassium t-butoxide in dimethylsulfoxide followed by chromatography on silica gel gave the pair of c-20 epimers (36) and (40) which were catalytically reduced to the respective dihydro compounds (41) and (42). Treatment of (42) with dilute mineral acid gave an equilibrium mixture of (41) and (42). The c-20 configuration assigned to these epimers was based on the observation that in 20-hydroxy-12-keto dammaranes with partial structures 43(S) and 43(R), the 20(S) series exhibits partial and the 20(R) series exhibits complete hydrogen bonding between the 12-keto and the 20-hydroxyl group as demonstrated by IR spectroscopy. This can be attributed to the difference in conformational stability of the two hydrogen-bonded forms (see Chart 6). On chromic acid oxidation, (41) and (42) afforded the triketones (44) and (45) whose IR spectra indicated that the chirality of (44) at c-20 should be S and that of (45) R. Consequently, the genuine sapogenin (36) of Rg$_1$ can be designated as 20(S)-protopanaxatriol.

Catalytic hydrogenation of Rg$_1$ followed by repeated methylation afforded a dihydro-decamethyl ether (46) whose IR spectrum showed no -OH absorption, indicating the presence of a glycosyl linkage on the 20-hydroxyl group. Acid hydrolysis of (46) gave a dimethyl ether (47) whose oxidation afforded a ketone (48). The ORD curve of this substance (Chart 7) displayed the negative Cotton effect characteristic of triterpenoid 6-ketones such as zeorinone (49), but differed from that of 3- and 12-ketones of triterpenes of this type. This permitted formulation of (47) as 3,12-o-dimethyl-dihydroprotopanaxatriol, although the configuration of c-20 of (47) and (48) has not yet been determined. The locations of the glycosidic linkages were thus shown to be at c-6 and c-20.

On methylation followed by methanolysis, Rg$_1$ yielded only methyl 2,3,4,6-tetra-o-methyl-glucopyranoside. As the coupling constants of both anomeric proton signals of (46) indicated β-configuration, structure of Rg$_1$ was established as 6,20-di-o-β-D-glucopyranosyl-20(S)-protopanaxatriol

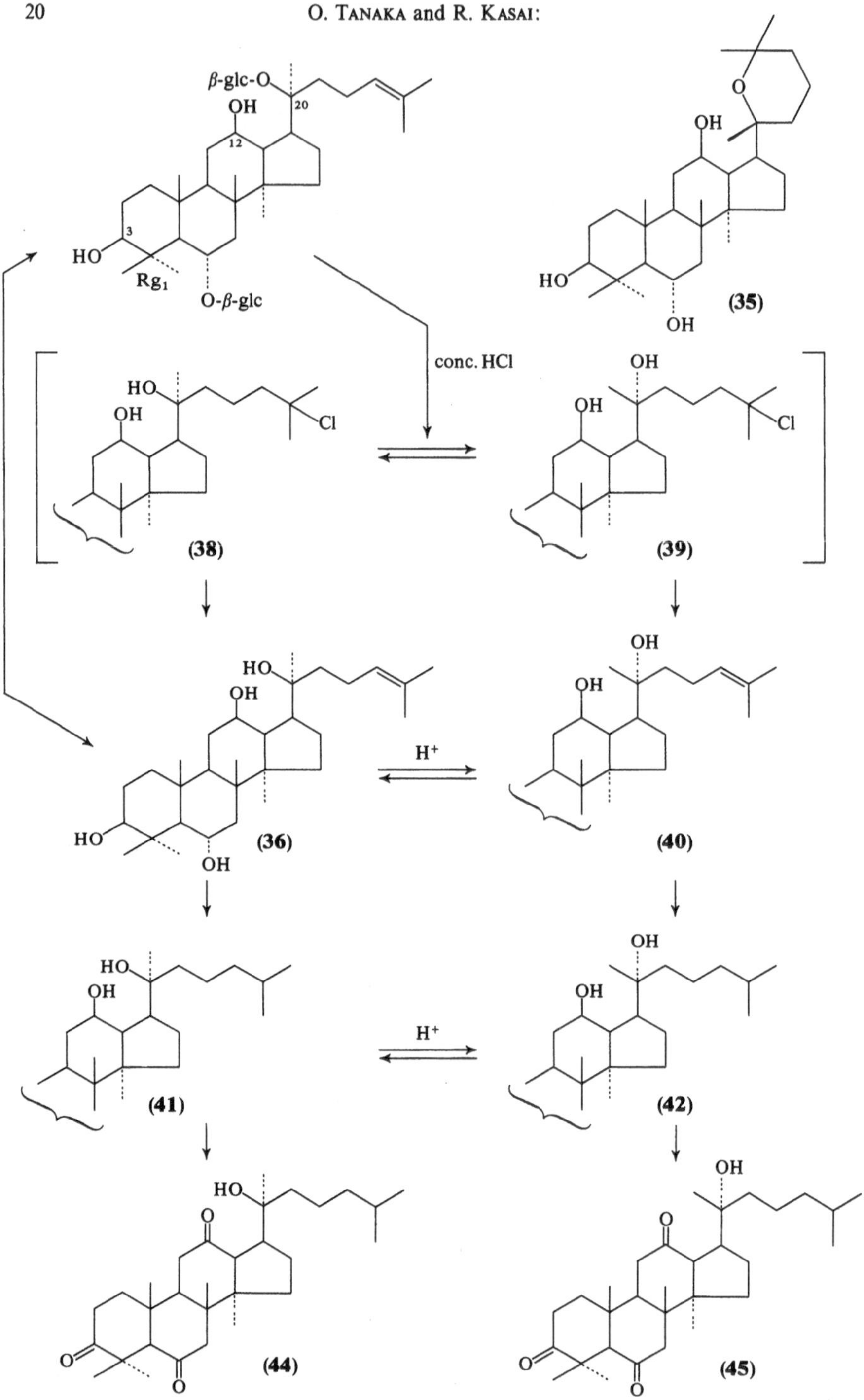
β-glc-O
OH
20
12
3
HO
Rg1
O-β-glc
O
OH
HO
OH
(35)
conc. HCl
HO
OH
Cl
(38)
OH
OH
Cl
(39)
HO
OH
HO
OH
(36)
H+
OH
OH
(40)
HO
OH
(41)
H+
OH
OH
(42)
HO
O
O
O
(44)
OH
O
O
O
(45)

Chart 5

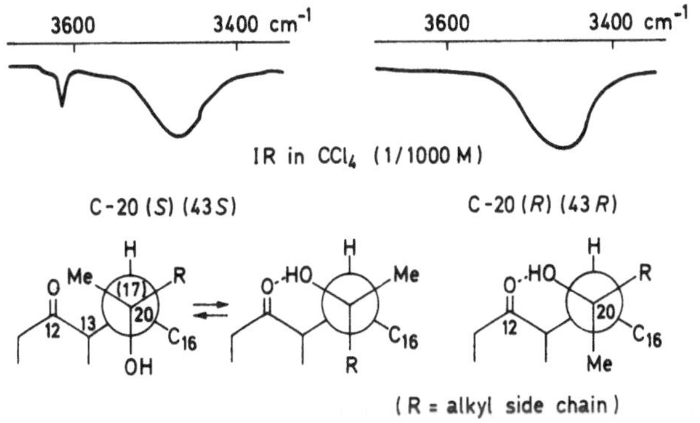

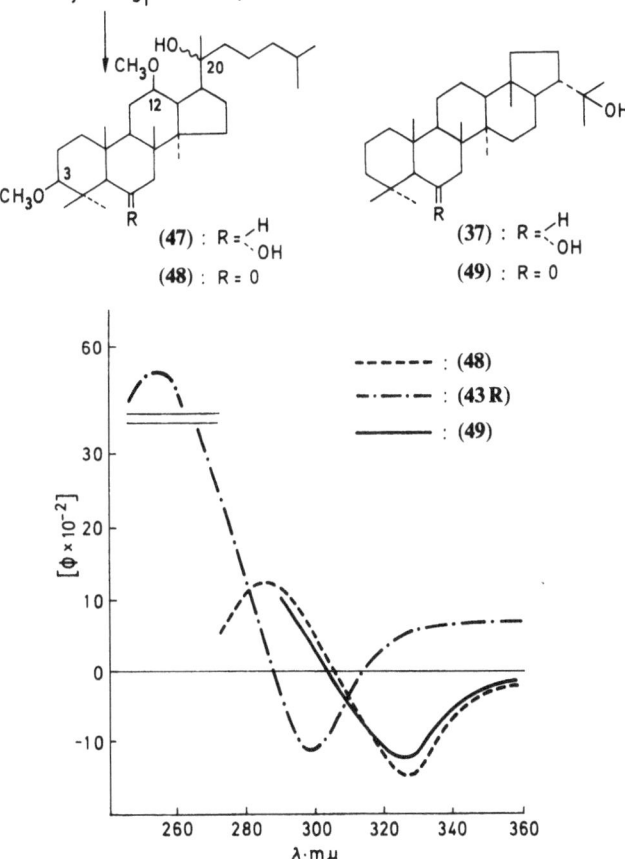

$$3600 \qquad 3400\ cm^{-1} \qquad 3600 \qquad 3400\ cm^{-1}$$

IR in CCl$_4$ (1/1000 M)

C-20 (S) (43 S)   C-20 (R) (43 R)

(R = alkyl side chain)

Chart 6

dihydro-Rg$_1$-decamethylether (46)

(47) : R = $<^H_{OH}$ (37) : R = $<^H_{OH}$

(48) : R = O   (49) : R = O

-------- : (48)

-·-·-· : (43 R)

——— : (49)

$[\phi \times 10^{-2}]$

$\lambda \cdot m\mu$

Optical Rotatory Dispersion Curve

Chart 7

(29). This formulation of Rg_1 was further confirmed by the mass spectrum of its decaacetate (Chapter IV-4) (34), which disproved another structure proposed for the genuine sapogenin by ELYAKOV et al. (35).

The structures of Re, Rf and Rg_2 were established by SANADA et al. (32). Smith degradation of these saponins yielded their common genuine sapogenin (36). On heating with dilute mineral acid, Re and Rg_2 afforded (35), rhamnose and glucose, while Rf gave (35) and glucose. The IR spectra of methylated Rf and Rg_2 showed a strong -OH band near 3380 cm^{-1} due to hydrogen bonding between the 20-hydroxyl and the 12-methoxy group, indicating the absence of a glycosidic linkage on the C-20-hydroxyl, while the methyl ether of Re did not exhibit any -OH absorption. Treatment of Re with 50% aqueous acetic acid at 70° yielded glucose and a prosapogenin which was identical with Rg_2 or with its 20 epimer or represented a mixture of both epimers, indicating that Re must be represented by 20-O-glucosyl-Rg_2. Dilute mineral acid hydrolysis of the methyl ethers of hydrogenated Re, Rf and Rg_2 afforded (47), proving the location of a glycosidic linkage at their 6-hydroxyl groups. Structures of these three saponins were finally established by methanolysis of their methyl ethers as shown in Table II. Their anomeric configurations were shown by ^{1}H-NMR spectrometry and molecular rotation differences. In 1972, HAN and HAN reported isolation and structure study of saponin C which is identical with Re (36).

Later, a minor saponin of (36) named 20-gluco-ginsenoside-Rf (glc-Rf) was also isolated from Ginseng by SANADA et al. Its structure was established in a manner similar to that described above and is shown in Table II (31).

In subsequent studies, a number of other saponins of (36) have been isolated from various parts of Panax spp. In all of these, the 3-hydroxyl carries no sugar.

IV. Modern Techniques Used in Structure Determination

For rapid and unambiguous identification of glycosides, especially glycosides of an unstable aglycone such as the Ginseng saponins, the following enzymic and spectroscopic procedures have been developed.

1. Enzymatic Hydrolysis of Glycosides

As already mentioned, the acid-unstable genuine dammarane sapogenins can be obtained from the saponins by means of Smith degradation or its modifications. However, yields by this procedure are generally low and a relatively large amount of the glycosides is required for preparation and identification of a sapogenin.

YOSIOKA and coworkers reported that soil bacterial cultivation using a strain selected on a medium containing 6 g of a mixture of Rb_1, Rb_2 and Rc as sole carbon source afforded 42 mg of (26) and 410 mg of a glucoside (50) (37, 38). The structure of (50) was established as 20-O-β-D-glucosyl-20(S)-protopanaxadiol.

KOHDA and TANAKA have screened a variety of commercially available crude glycosidase preparations for their effectiveness in the hydrolysis of Ginseng saponins (39). Emulsin and crude preparations of cellulase and amylase exhibited very low hydrolytic activities, while crude preparations of pectinase, naringinase and hesperidinase, especially the last, had high activities. Thus, incubation of a mixture of Rb_1, Rb_2 and Rc with crude hesperidinase at pH 4.0 for 40 hr gave (50) in ca. 70% yield together with a small amount of (26) and similar incubation of Rg_1 afforded its genuine sapogenin (36) almost quantitatively. Hesperidinase produced by *Aspergillus niger* was discovered by OKADA et al. (40, 41); the crude preparation has been used in Japan to hydrolyze hesperidin (flavanol rhamnoglucoside) in an industrial process. Since purified hesperidinase does not hydrolyze the saponins, the activity must be due to some other enzymes present in the crude preparation. The crude enzyme is also capable of hydrolyzing efficiently many other plant glycosides such as stevioside (42) and is now an important reagent for studying the chemistry of natural glycosides derived from unstable aglycones.

2. ¹³C-NMR Spectroscopy of Dammarane Type Triterpenes

Full assignments of the signals in the ¹³C-NMR spectra of dammarane type triterpenes including (11), (26), and (36) have been published (43). These are shown in Table IV. While C-20 epimers in the dammarane series are difficult to distinguish by TLC, optical rotation, IR, mass and ¹H-NMR spectra, chemical shift differences of the C-17, C-21 and C-22 signals in the ¹³C-NMR spectra of 20-epimers of 12β-hydroxydammaranes, such as the genuine sapogenins of Ginseng saponins, are significant. These differences arise from the γ-gauche effect (44) associated with the conformation around the C-17, C-20 linkage which is fixed by strong hydrogen bonding (Chart 8) and promise to be of great utility for structure determination of Ginseng saponins. It is now possible to establish the structure of the basic sapogenin, including its configuration of C-20, by ¹³C-NMR spectroscopy, taking into consideration the glycosylation shifts (see next section) without the necessity of isolating it first by mild hydrolysis.

Table IV. ¹³C Chemical Shifts for Dammarane Triterpenes in Chloroform-d and Pyridine-d₅ (in parentheses)

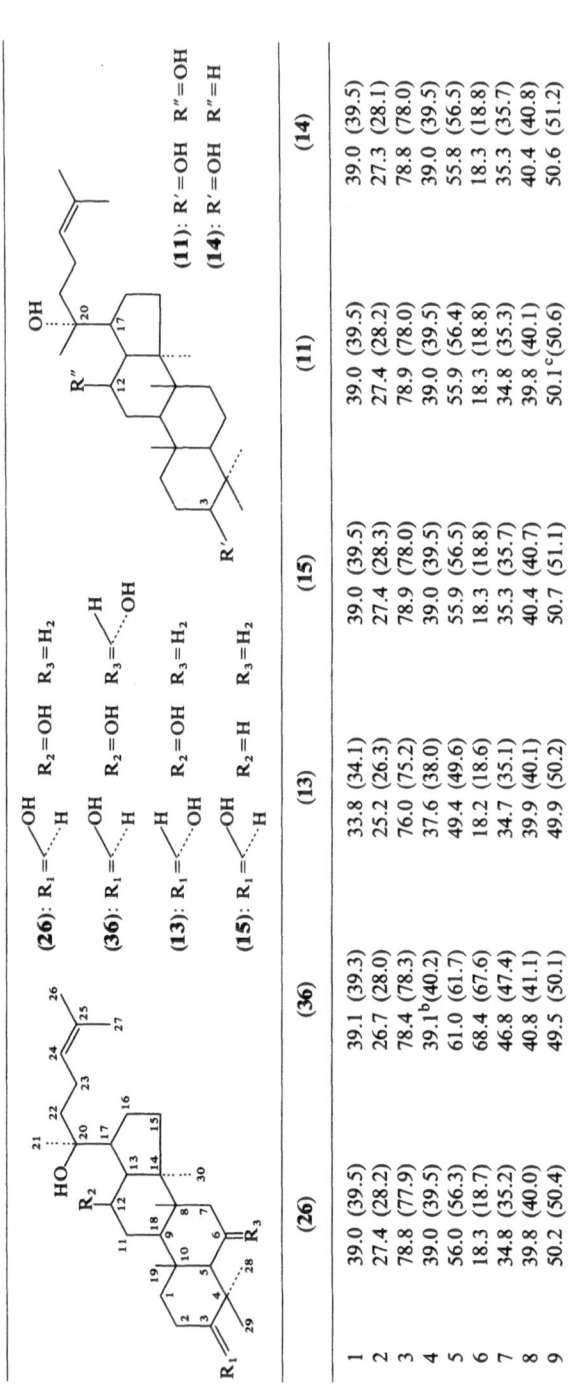

(26): $R_1 = $ ⟨OH / H⟩ $R_2 = OH$ $R_3 = H_2$

(36): $R_1 = $ ⟨OH / H⟩ $R_2 = OH$ $R_3 = $ ⟨H / OH⟩

(13): $R_1 = $ ⟨H / OH⟩ $R_2 = OH$ $R_3 = H_2$

(15): $R_1 = $ ⟨OH / H⟩ $R_2 = H$ $R_3 = H_2$

(11): R′=OH   R″=OH
(14): R′=OH   R″=H

	(26)	(36)	(13)	(15)	(11)	(14)
1	39.0 (39.5)	39.1 (39.3)	33.8 (34.1)	39.0 (39.5)	39.0 (39.5)	39.0 (39.5)
2	27.4 (28.2)	26.7 (28.0)	25.2 (26.3)	27.4 (28.3)	27.4 (28.2)	27.3 (28.1)
3	78.8 (77.9)	78.4 (78.3)	76.0 (75.2)	78.9 (78.0)	78.9 (78.0)	78.8 (78.0)
4	39.0 (39.5)	39.1ᵇ(40.2)	37.6 (38.0)	39.0 (39.5)	39.0 (39.5)	39.0 (39.5)
5	56.0 (56.3)	61.0 (61.7)	49.4 (49.6)	55.9 (56.5)	55.9 (56.4)	55.8 (56.5)
6	18.3 (18.7)	68.4 (67.6)	18.2 (18.6)	18.3 (18.8)	18.3 (18.8)	18.3 (18.8)
7	34.8 (35.2)	46.8 (47.4)	34.7 (35.1)	35.3 (35.7)	34.8 (35.3)	35.3 (35.7)
8	39.8 (40.0)	40.8 (41.1)	39.9 (40.1)	40.4 (40.7)	39.8 (40.1)	40.4 (40.8)
9	50.2 (50.4)	49.5 (50.1)	49.9 (50.2)	50.7 (51.1)	50.1ᶜ(50.6)	50.6 (51.2)

10	37.1 (37.3)	39.2ᵇ(39.3)	37.2 (37.5)	37.1 (37.4)	37.1 (37.4)	37.1 (37.4)
11	31.2 (32.0)	30.9 (31.9)	31.2ᵇ(31.8)	21.6 (21.9)	31.2ᵇ(32.2)	21.4 (21.9)
12	70.8 (70.9)	70.5 (70.9)	70.7 (70.9)	25.4 (25.8)	70.7 (70.8)	25.3 (25.8)
13	47.7 (48.5)	47.2 (48.1)	47.4 (48.3)	42.3 (42.6)	48.5 (49.2)	42.2 (42.6)
14	51.6 (51.6)	51.3 (51.6)	51.6 (51.6)	50.3 (50.6)	51.6 (51.7)	50.0 (50.4)
15	31.1 (31.3)	30.9 (31.3)	31.0ᵇ(31.2)	31.2 (31.7)	31.1ᵇ(31.5)	31.1 (31.5)
16	26.6 (26.8)	26.4 (26.8)	26.6 (26.9)	27.6 (28.1)	26.4 (26.6)	27.5 (28.1)
17	53.6 (54.7)	53.5 (54.6)	53.6 (54.5)	49.9 (50.3)	49.9ᶜ(50.6)	49.5 (49.9)
18	16.2ᵃ(16.2)ᵃ	17.2ᵃ(17.5)ᵃ	16.0ᵃ(16.3)ᵃ	16.2ᵃ(16.5)ᵃ	16.2ᵃ(16.3)ᵃ	16.2ᵃ(16.5)ᵃ
19	15.7ᵃ(15.8)ᵃ	17.2ᵃ(17.4)ᵃ	15.7ᵃ(15.8)ᵃ	15.5ᵃ(16.3)ᵃ	15.7ᵃ(15.9)ᵃ	15.5ᵃ(16.3)ᵃ
20	74.0 (72.9)	73.9 (72.9)	73.7 (72.9)	75.4 (74.0)	74.6 (72.9)	75.8 (74.4)
21	26.8 (26.9)	26.7 (26.9)	26.6 (26.9)	24.9 (25.3)	21.8 (22.7)	23.5 (24.5)
22	34.8 (35.8)	24.6 (35.7)	34.7 (35.7)	40.5 (41.9)	42.3 (43.2)	41.8 (42.9)
23	22.4 (22.9)	22.3 (22.9)	22.3 (22.8)	22.6 (23.3)	21.8 (22.7)	22.3 (23.0)
24	125.2 (126.2)	125.0 (126.2)	125.3 (126.2)	124.8 (126.0)	124.6 (126.0)	124.7 (126.0)
25	131.4 (130.6)	131.4 (130.6)	131.2 (130.5)	131.5 (130.6)	131.9 (130.6)	131.4 (130.6)
26	25.8 (25.8)	25.7 (25.8)	25.7 (25.8)	25.7 (26.1)	25.8 (25.9)	25.7 (25.8)
27	17.8 (17.6)	17.7 (17.7)	17.7 (17.6)	17.7 (17.7)	17.8 (17.7)	17.7 (17.7)
28	28.1 (28.6)	30.9 (31.9)	28.3 (29.3)	28.0 (28.7)	28.1 (28.7)	28.0 (28.7)
29	15.5ᵃ(16.4)ᵃ	15.5ᵃ(16.4)ᵃ	22.1 (22.4)	15.4ᵃ(15.8)ᵃ	15.4ᵃ(16.5)ᵃ	15.4ᵃ(15.8)ᵃ
30	16.9 (17.0)	16.9 (17.0)	16.9 (16.9)	16.5 (16.9)	17.2 (17.3)	16.4 (16.8)

a, b, c: Assignments may have to be interchanged in each vertical column.

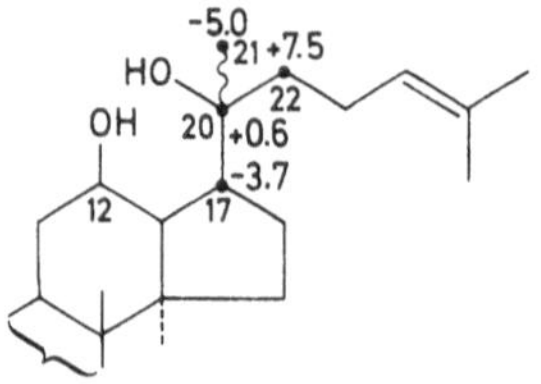

O. TANAKA and R. KASAI:

$\Delta \delta = \delta C$ of 20 (R) - epimer $- \delta C$ of 20 (S) - epimer

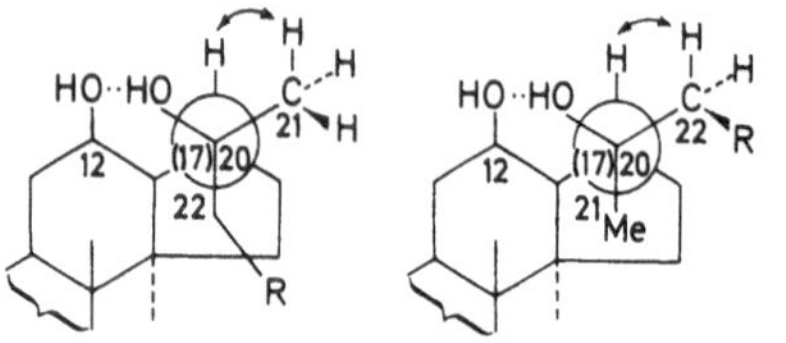

20 (S) 20 (R)

Chart 8

3. Glycosylation Shifts in ^{13}C-NMR Spectra

To expand the applicability of ^{13}C-NMR spectrometry to the chemistry of natural glycosides, α- and β-epimeric pairs of glucosides, mannosides, rhamnosides and arabinosides of a variety of alcohols were synthesized and the carbon resonance displacements of both sugar and aglycone moieties on glycoside formation, the so-called "glycosylation shifts", where studied (*45, 46, 47*). On glycosylation, the anomeric carbon of the sugar moiety experiences a decreasing amount of deshielding as the nature of the alcohol is varied from methyl to primary to secondary and then to tertiary; thus glycosylation of tertiary alcohols results in almost no shift of the carbon signal (Table V).

Table V. ^{13}C Chemical Shifts^a for Anomeric Carbon (36) (37) (38)

agly-cone	(free)	MeOH	n-PrOH	iso-PrOH	trans-4-t-butyl-cyclohexanol	t-BuOH
α-D-glc	94.1	101.3 (166 Hz)	100.2 (166 Hz)	98.4 (164 Hz)	98.5 (166 Hz)	94.7 (164 Hz)
α-D-man	95.7	102.6 (166 Hz)	101.2 (166 Hz)	99.2 (166 Hz)	99.5 (164 Hz)	96.1 (165 Hz)
α-L-rha	95.8	102.4 (168 Hz)	101.1 (166 Hz)	99.2 (166 Hz)	99.1 (168 Hz)	95.7 (164 Hz)
β-L-ara	94.6	102.1 (168 Hz)	100.9 (166 Hz)	99.0 (167 Hz)	99.0 (168 Hz)	95.1 (168 Hz)
β-D-glc	98.8	105.5 (156 Hz)	104.4 (152 Hz)	102.4 (154 Hz)	102.4 (156 Hz)	98.9 (151 Hz)
β-D-man	95.7	102.7 (156 Hz)	101.8 (155 Hz)	99.2 (154 Hz)	99.4 (155 Hz)	96.2 (153 Hz)
β-L-rha	94.6^b	102.6 (158 Hz)	101.3 (152 Hz)	99.5 (154 Hz)	99.3 (154 Hz)	95.8 (152 Hz)
α-L-ara	99.2	105.8 (158 Hz)	104.7 (156 Hz)	103.0 (157 Hz)	103.0 (158 Hz)	99.1 (157 Hz)

glc: glucopyranose, man: mannopyranose, rha: rhamnopyranose, ara: arabinopyranose. In parentheses: coupling constant $^1J_{C'-H'}$ of anomeric carbon.
^{a} In pyridine-d_5, ^{b} in D$_2$O.

References, pp. 65—76

Carbinyl carbon signals such as the α-carbons of aglycone alcohols are commonly displaced downfield by about 7 ppm on β-D-glucosylation, β-D-mannosylation, β-L-rhamnosylation and α-L-arabinosylation regardless of the class of alcohols. The paramagnetic shift is somewhat smaller for α-D-glucosylation and β-L-arabinosylation and even more so for α-D-manno-sylation and α-L-rhamnosylation (Table VI).

Signals of the carbons vicinal to the C – OH function (the β-carbons) are generally displaced upfield on glycosylation. In secondary alcohols, the degree of this shielding depends to a remarkable extent on the chirality of the carbinyl carbon of the aglycone as well as the chirality of the anomeric carbon of the sugar. As regards the β-carbons of secondary alcohols [for definition of β-(pro-R) and β'-(pro-S) carbons, see Chart 9], a β'-carbon is always more shielded (about 4 ppm) than a β-carbon (about 2 ppm) on β-D-glucosylation, β-D-mannosylation, α-L-rhamnosylation and α-L-ara-binosylation when the chirality of the anomeric carbon in the free sugar is R, while α-D-glucosylation, β-L-glucosylation, α-D-mannosylation, β-L-rhamnosylation and β-L-arabinosylation result in greater shielding of a β-carbon (about 4 ppm) than a β'-carbon (about 2 ppm) when the chirality of the anomeric carbon atom of the free sugar is S (see Table VII).

Table VI. Glycosylation Shift^a of Aglycone α-C (36) (37) (38)

	MeOH	n-PrOH	iso-PrOH	trans-4-t-butyl-cyclohexanol	t-BuOH
	$\Delta\delta_{\alpha\text{-}C}$	$\Delta\delta_{\alpha\text{-}C}$	$\Delta\delta_{\alpha\text{-}C}$	$\Delta\delta_{\alpha\text{-}C}$	$\Delta\delta_{\alpha\text{-}C}$
α-D-glc	+5.5	+5.7	+6.3	+6.6	+6.9
α-D-man	+5.0	+5.0	+5.1	+4.6	+6.7
α-L-rha	+5.0	+5.1	+5.2	+5.3	+6.4
β-L-ara	+5.9	+6.3	+6.3	+6.7	+6.7
β-D-glc	+7.3	+7.3	+7.6	+7.4	+7.5
β-D-man	+7.1	+7.1	+6.8	+6.8	+7.5
β-L-rha	+7.0	+7.0	+6.9	+6.9	+7.4
α-L-ara	+6.7	+7.0	+7.3	+7.2	+7.1

^{a} In Pyridine-d_5, Δδ: δα-C of glycoside − δα-C of aglycone.

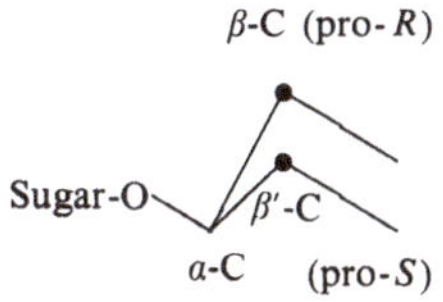

β-C (pro-R)

Sugar-O   β'-C

a-C   (pro-S)

Chart 9

Table VII. *Glycosylation Shift[a] of β-C and β'-C* (*36*) (*37*) (*38*)

		n-PrOH	iso-PrOH	trans-4-t-butyl cyclohexanol		t-BuOH	cholestanol	
	Chir.[c]	β-C[b]	β-C[b]	β'-C[b]	β-C[b]	β'-C[b]	β-C[b]	β-C[b]	β'-C[b]
α-D-glc	S	−3.5	−3.9	−2.0	−4.2	−2.4	−2.8		
α-D-man	S	−3.2	−3.2	−2.1	−4.3	−2.5	−2.7		
β-L-rha	S	−3.4	−3.6	−1.7	−3.9	−2.1	−2.8		
β-L-ara	S	−3.4	−3.9	−2.1	−4.2	−2.3	−2.8		
β-L-glc	S							−4.0	−2.3
β-D-glc	R	−3.3	−1.8	−3.6	−2.2	−3.9	−2.6		
β-D-man	R	−3.3	−1.9	−4.3	−2.2	−4.0	−2.8		
α-L-rha	R	−3.5	−2.1	−4.1	−2.6	−4.5	−3.0		
α-L-ara	R	−3.3	−1.8	−3.7	−2.1	−3.9	−2.6		

[a] In Pyridine-d_5,   [b] δβ-C or β'-C of glycoside − δβ-C or β'-C of aglycone,   [c] Chirality of anomeric carbon atom as a free form.

Table VIII. *^{13}C Chemical Shift[a] for Anomeric-C and Glycosylation Shift of α-C, β-C and β'-C of Glycosides of (51) and (52)*

		anomeric-C		$\Delta\delta_{\alpha\text{-}C}$[b]		$\Delta\delta_{\beta\text{-}C}$[b]	$\Delta\delta_{\beta'\text{-}C}$[c]	$\Delta\delta_{\beta\text{-}C}$[c]	$\Delta\delta_{\beta'\text{-}C}$[c]
	Chir.[d]	(51)	(52)	(51)	(52)	(51)		(52)	
α-D-glc	S	96.1	102.1	+4.7	+10.5	−5.9	−2.1	−1.3	−2.6
α-D-man	S	97.1	103.7	+3.3	+10.8	−6.3	−2.1	−1.5	−2.8
β-L-rha	S	98.4	103.1	+5.7	+9.9	−5.0	−2.1	−1.4	−2.0
β-L-ara	S	96.9	102.8	+5.0	+10.1	−5.7	−2.1	−1.2	−2.4
β-D-glc	R	105.9	101.5	+10.4	+6.4	−1.8	−1.1	−2.1	−4.9
β-D-man	R	103.6	98.4	+10.2	+5.8	−2.1	−1.5	−2.1	−5.0
α-L-rha	R	103.0	97.2	+10.6	+4.0	−2.9	−1.8	−2.4	−6.2
α-L-ara	R	106.3	101.4	+10.2	+5.9	−1.8	−1.3	−2.3	−5.3

[a] In pyridine-d_5,   [b] δα-C of glycoside − δα-C of aglycone,   [c] δβ-C or β'-C of glycoside − δβ-C or β'-C of aglycone,   [d] Chirality of anomeric carbon atom as a free form.

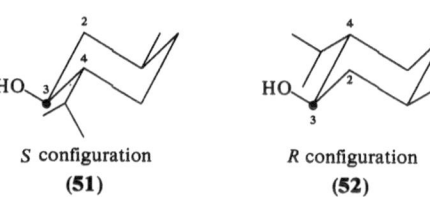

S configuration   R configuration
    (51)   (52)

Glycosylation Shifts.

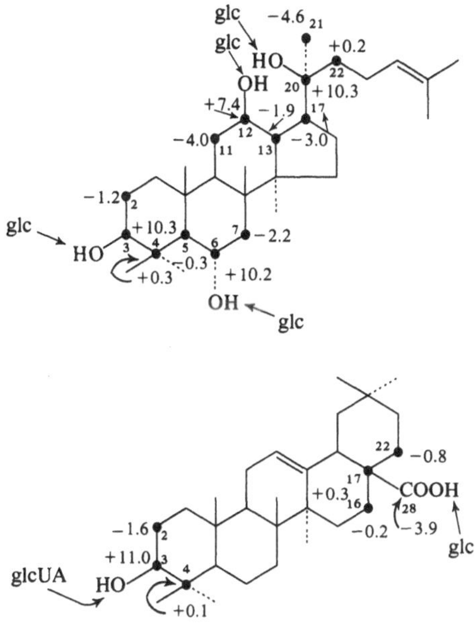

Chart 10

glc: β-D-glucopyranose
glcUA: β-D-glucuronic acid

Glycosylation shifts in relatively hindered alcohols such as d- and 1-menthol (**51**, **52**), and 3β-, 6α- and 12β-hydroxylated dammarane type triterpenes differ from those in the less hindered alcohols discussed previously owing to a change in the conformation of the glycosidic linkage. The magnitude of these anomalous shifts also depends on both the stereochemistries of the carbinyl and the anomeric carbon as shown in Chart 10 and Table VIII. Thus, in the β-D-glucoside, β-D-mannoside, α-L-rhamnoside and α-L-arabinoside of (**51**), where c-3 is S and the configuration of the anomeric carbon in the free sugar is R and in the α-D-glucoside, α-D-mannoside, β-L-rhamnoside and β-L-arabinoside of (**52**) where c-3 is R, the signals due to the anomeric carbons and c-3 are more deshielded than those in the less hindered alcohols. Contrariwise, the signals of the anomeric carbons and c-3 of the β-D-glucoside, β-D-mannoside, α-L-rhamnoside and α-L-arabinoside of (**52**) and the α-D-glucoside, α-D-mannoside, β-L-rhamnoside and β-L-arabinoside of (**51**) are somewhat less deshielded than those of the corresponding glycosides of less hindered alcohols. Similar characteristic shifts were observed for β-D-glucosides of 3$\beta(S)$-, 6$\alpha(S)$- and 12$\beta(R)$-hydroxydammaranes. Unusual glycosylation shifts were also exhibited by the carbons of the sugar moieties in 1,2-linked oligoglycosides attached to the 6α-hydroxyl of (**36**) (**48**).

Chemical shifts of the sugar carbons of some of the saponins are shown in Table IX.

Table IX. ^{13}C-NMR Chemical Shifts of Sugar Moieties (in Pyridine-d_5)

		Ra$_1$	Ra$_2$	Ra$_3$	Rb$_1$	Rb$_2$	Rb$_3$	Rc
3-glc	1	104.9	105.0	104.9	105.0	105.0	104.8	104.9
	2	83.0	83.1	83.5	82.9	83.0	83.4	83.1
	3	78.0[a]	77.3[a]	78.0[a]	77.2[a]	78.1[a]	77.7[a]	77.8[a]
	4	71.5	71.6	71.7	71.5	71.5	71.7	71.5
	5	78.0[a]	78.0[a]	78.0[a]	78.0[a]	78.1[a]	78.8[a]	77.8[a]
	6	62.7	62.7	62.7	62.6	62.7	62.8	62.6
glc	1	105.6	105.7	106.9	105.6	105.7	105.6	105.6
	2	76.8	76.9	77.0	76.7	76.9	76.7	76.8
	3	79.1[a]	79.1[a]	79.2[a]	78.8[a]	79.0[a]	78.1[a]	78.9[a]
	4	71.5	71.6	71.7	71.5	71.5	71.7	71.5
	5	78.0[a]	78.0[a]	78.0[a]	78.0[a]	78.7[a]	78.1[a]	78.0[a]
	6	62.7	62.7	62.7	62.6	62.7	62.8	62.6
20-glc	1	97.9	98.0	98.0	97.9	97.9	97.9	97.9
	2	74.7	74.8	74.8[c]	74.9	74.8	74.8	74.9
	3	79.1[a]	78.0[a]	78.0[a]	78.0[a]	78.7[a]	78.1[a]	78.0[a]
	4	71.5	71.6	71.7[b]	71.5	71.5	71.7	71.5
	5	76.8	76.9	77.0	76.7	76.6	76.6	76.3
	6	69.7	68.2	69.6	71.5	69.0	69.8	68.3
glc	1				104.9	105.0		
	2				74.1	74.9		
	3				87.5	78.0[a]		
	4				71.3[b]	71.5		
	5				78.0[a]	78.0[a]		
	6				62.4	62.6		
ara	1	104.9	108.0			104.5		109.9
	2	72.6	90.6			72.0		83.3
	3	73.7	79.1[a]			73.9		78.6
	4	78.0	85.3			68.5		85.8
	5	65.5	62.7			65.5		62.6
xyl	1	106.6	104.3	106.3			105.2	
	2	75.2	74.8	75.3[c]			74.2	
	3	78.0	78.0[a]	77.0			77.3[a]	
	4	70.8	70.9	70.8			70.8	
	5	67.0	67.2	67.3			66.4	

		Rd	Rg$_3$	F2	Rs$_1$	Rs$_2$	NG-R4
3-glc	1	105.0	105.3	106.9	104.8	104.8	105.0
	2	83.3	83.3	75.7	84.1	84.0	82.6
	3	78.1[a]	78.0	79.2[a]	77.9[a]	77.8[a]	77.7[a]
	4	71.6	71.5	71.6[b]	71.6[b]	71.8[b]	71.4[b]
	5	78.1[a]	78.0	78.2[a]	77.9[a]	77.8[a]	77.7[a]
	6	62.7	62.7	62.8	62.7	62.5	62.6
glc	1	105.9	106.0		106.1	105.9	105.2
	2	77.0	77.0		76.6	76.5	76.5

Table IX *(continued)*

		Rd	Rg[3]	F2	Rs[1]	Rs[2]	NG-R4
	3	79.1[a]	78.0		79.0[a]	78.7[a]	78.6[a]
	4	71.6	71.5		70.9	70.8	71.4[b]
	5	78.1[a]	78.0		75.2	75.2	77.7[a]
	6	62.7	62.7		64.7	64.5	62.6
20-glc	1	98.2			97.9	98.0	97.8
	2	75.0			74.8	74.9	74.7
	3	78.1[a]			78.4[a]	78.3[a]	77.7[a]
	4	71.6			71.3[b]	71.2[b]	71.4[b]
	5	78.1[a]			76.6	76.5	76.5
	6	62.7			69.1	68.3	71.2[b]
glc	1			98.2			105.0
	2			75.1			74.7
	3			78.7[a]			77.7[a]
	4			71.8[b]			71.4[b]
	5			78.2[a]			76.5
	6			62.8			69.6
ara	1				104.5	109.9	
	2				72.0	83.3	
	3				73.9	79.0	
	4				68.4	85.8	
	5				65.4	62.5	
xyl	1						105.2
	2						74.7
	3						77.7[a]
	4						71.0
	5						66.7
$\underline{C}H_3CO$					20.8	20.8	
$CH_3\underline{C}O$					170.8	170.8	

		NG-Fa	NG-Fc	NG-Fe	Gy-IX	Gy-XVII	Q-R1	PG-F8
3-glc	1	104.7	104.7	106.6	106.9	106.7	104.9[b]	104.9
	2	82.8	82.9	75.5	75.6	75.6	84.3	83.4
	3	77.7[a]	77.9[a]	78.9[a]	79.1[a]	79.0[a]	77.0[a]	78.0[a]
	4	71.6	71.6	71.8	71.4[b]	71.6	71.5	71.5[b]
	5	77.7[a]	77.9[a]	78.0[a]	78.2[a]	78.6[a]	78.2[a]	74.6[a]
	6	62.8	62.9	62.9	63.1	62.9	62.7	64.5
glc	1	103.1	103.2				106.1	105.9
	2	84.4	84.5				76.6	76.7[c]
	3	78.1[a]	78.1[a]				79.1[a]	79.1[a]
	4	71.1	71.1				71.5	71.8[b]
	5	77.7[a]	77.9[a]				75.2	78.0[a]
	6	62.8	62.9				64.7	63.0
xyl	1	106.3	106.4					
	2	75.7	75.9					

Table IX *(continued)*

		NG-Fa	NG-Fc	NG-Fe	Gy-IX	Gy-XVII	Q-R1	PG-F8
	3	79.1	78.5[b]					
	4	70.6	71.1					
	5	67.3	67.3					
20-glc	1	97.9	98.0	97.9	98.0	97.9	98.0	98.0
	2	75.0	75.4	74.9	74.7	75.1	74.9	74.9
	3	78.1[a]	78.1[a]	78.7[a]	78.6[a]	78.1[a]	78.2[a]	78.0[a]
	4	71.6	71.7	71.8	71.8[b]	71.6	71.5	71.8[b]
	5	76.9[a]	76.8[a]	76.3[a]	76.8[a]	76.8	76.6	76.9[c]
	6	70.6	70.2	68.3	70.1	71.6	71.5	70.0
glc	1	105.2				105.1	105.2[b]	
	2	74.7				74.6	74.9	
	3	78.1[a]				78.1[a]	78.2[a]	
	4	71.6				71.6	71.5	
	5	78.1[a]				78.1[a]	78.2[a]	
	6	62.8				62.7	62.7	
xyl	1		105.7		105.6			105.4
	2		74.7		74.7			74.6
	3		79.2[b]		77.8[a]			77.6
	4		70.7		71.0[b]			71.1
	5		66.9		66.8			66.7
ara	1			109.9				
(fur)	2			83.1				
	3			78.7[a]				
	4			85.5				
	5			62.5				
$\underline{C}H_3CO$							20.9	20.8
$CH_3\underline{C}O$							170.8	170.6

		Re	Rf	glc-Rf	Rg₁	Rg₂	Rh₁
6-glc	1	101.6	103.2	103.8	105.7	102.4	105.9
	2	79.1	79.5	79.1	75.3	79.0[a]	75.4
	3	78.0[a]	78.7[a]	78.1	80.0[a]	79.9[a]	80.0[a]
	4	72.1	71.7	71.6	71.6[b]	73.2	71.8
	5	78.0[a]	79.8	79.8	79.3[a]	78.8	79.5[a]
	6	62.9	62.9	62.9	62.9	63.7	63.1
glc	1		103.2	103.8			
	2		75.9	75.9			
	3		78.4[a]	78.1			
	4		72.3	72.3			
	5		79.8	79.8			
	6		63.3	63.3			
rham	1	101.6				102.4	
	2	72.1				72.8	
	3	72.1				72.8	

Table IX *(continued)*

	Re	Rf	glc-Rf	Rg$_1$	Rg$_2$	Rh$_1$
4	73.8				74.5	
5	69.3				69.9	
6	18.6				19.2	
20-glc 1	98.1		98.2	98.1		
2	75.0		75.1	74.9		
3	79.1		78.1	78.8[a]		
4	71.1		71.6	71.3[b]		
5	78.5[a]		79.8	77.8[a]		
6	62.7		62.9	62.6		

	F1	F3	NG-R1	NG-R2	NG-R3	NG-R6	C-L5	C-L10
6-glc 1			103.4	103.3	105.7	105.8	100.4	1 12-glc
2			79.4	79.6	75.1	75.4[b]	75.1	2
3			78.9[a]	78.6	79.7[a]	79.9[a]	78.3[a]	3
4			71.4	71.6	71.5	71.8	70.9	4
5			79.9	79.9	79.3[a]	79.1[a]	77.7[a]	5
6			62.6	62.7	62.6	62.7	62.8	6
xyl 1			104.5	104.6				
2			75.6	75.6				
3			78.9[a]	78.6				
4			71.4	71.6				
5			67.0	67.1				
20-glc 1	98.0	98.0	98.1		97.9	97.9	97.9	
2	74.9	74.9	75.0		74.6	75.2[b]	74.6	
3	78.9[a]	79.1	78.5[a]		78.0[a]	78.7[a]	79.1	
4	71.4	71.8	71.1		71.5	71.8	71.7	
5	78.0[a]	76.6	79.9		76.7	76.1	76.6	
6	62.7	69.1	62.6		71.5	68.0	69.7	
glc 1					105.0	100.3		
2					74.6	73.9		
3					78.0[a]	75.2[b]		
4					71.5	71.8		
5					78.0[a]	73.9		
6					62.6	62.7		
ara 1		104.5					104.3	
2		72.1					72.7	
3		74.0					73.7	
4		68.5					78.4	
5		65.5					65.5	
xyl 1							106.7	
2							75.2	
3							78.4	
4							70.8	
5							67.2	

Table IX (continued)

		C-L9a	C-L9bc	C-LT5	C-LT8	C-LN4	M7cd
3-glc	1			106.4	106.9	106.4	
	2			75.2[b]	75.7	75.4	
	3			78.3[a]	79.1[a]	78.5[a]	
	4			71.6	71.8	72.0	
	5			77.8[a]	78.4[a]	76.4[b]	
	6			62.7	63.0	69.8	
xyl	1					105.2	
	2					74.3	
	3					77.5	
	4					70.9	
	5					66.6	
12-glc	1	100.2	100.1				
	2	75.1	75.1				
	3	78.3[a]	78.3[a]				
	4	71.1	71.1				
	5	77.4[a]	77.3[a]				
	6	62.4	62.4				
20-glc	1			98.2	98.4	98.2	98.2
	2			75.2[b]	75.7	75.4	75.1
	3			77.8[a]	78.6[a]	78.3[a]	78.7[a]
	4			71.6	71.8	72.0	71.4
	5			76.3	77.9[a]	76.0[b]	78.2[a]
	6			70.3	63.0	69.8	62.8
glc	1			105.0			
	2			74.7[b]			
	3			77.8[a]			
	4			71.6			
	5			77.8[a]			
	6			62.7			
ara	1					104.2	
	2					72.0	
	3					73.8	
	4					68.3	
	5					65.4	

		PG-F11	M-R1	M-R2
6-glc	1	101.6	103.9	103.5
	2	79.1[a]	79.9	79.9
	3	78.1[a]	78.6	78.8
	4	72.1	71.7	71.3[a]
	5	78.1[a]	79.9	80.4
	6	63.0	63.0	63.0
glc	1		103.9	
	2		76.0	
	3		78.6	

References, pp. 65—76

Table IX *(continued)*

	PG-F11	M-R1	M-R2
4		72.4	
5		79.9	
6		63.5	
rham 1	101.6		
2	72.4		
3	72.1		
4	73.9		
5	69.2		
6	18.6		
xyl 1			104.9
2			75.9
3			78.8
4			71.7[a]
5			67.3

	Ro(C-V)	C-IV	C-IVa
3-glcUA 1	105.3	107.0	107.2
2	82.8	75.2	75.4
3	77.1[a]	76.6[a]	78.0
4	73.2	78.8[c]	73.4
5	77.9[a]	76.1[a]	78.0
6	172.6	172.4	172.6
glc 1	105.9		
2	77.9[a]		
3	78.2[a]		
4	71.7		
5	77.9[a]		
6	62.7		
ara 1		108.6	
2		82.4	
3		78.0[c]	
4		87.8	
5		62.6	
xyl 1			
2			
3			
4			
5			
28-glc 1	95.7	95.7	95.7
2	74.2	74.1	74.1
3	· 79.3[b]	79.2[b]	79.2[a]
4	71.2	71.1	71.2
5	78.9[b]	78.8[b]	78.8[a]
6	62.2	62.2	62.2

a, b, c: Assignments may have to be interchanged in each vertical column.

4. Mass Spectrometry (MS)

MS-analysis of glycosyl ions of acetylated or trimethylsilyl (TMSi) glycosides is effective in determining the structure of sugar moieties.

As mentioned in Chapter III-1, C-20 hydroxylated dammaranes and their glycosides are very unstable. Mass spectra of acetates (*34*) or TMSi ethers (*49*) of dammarane type saponins exhibited neither the molecular ion nor fragment ions with an intact oxygen function at C-20, but instead contain ions B and B′ (see Chart 11) which are diagnostic.

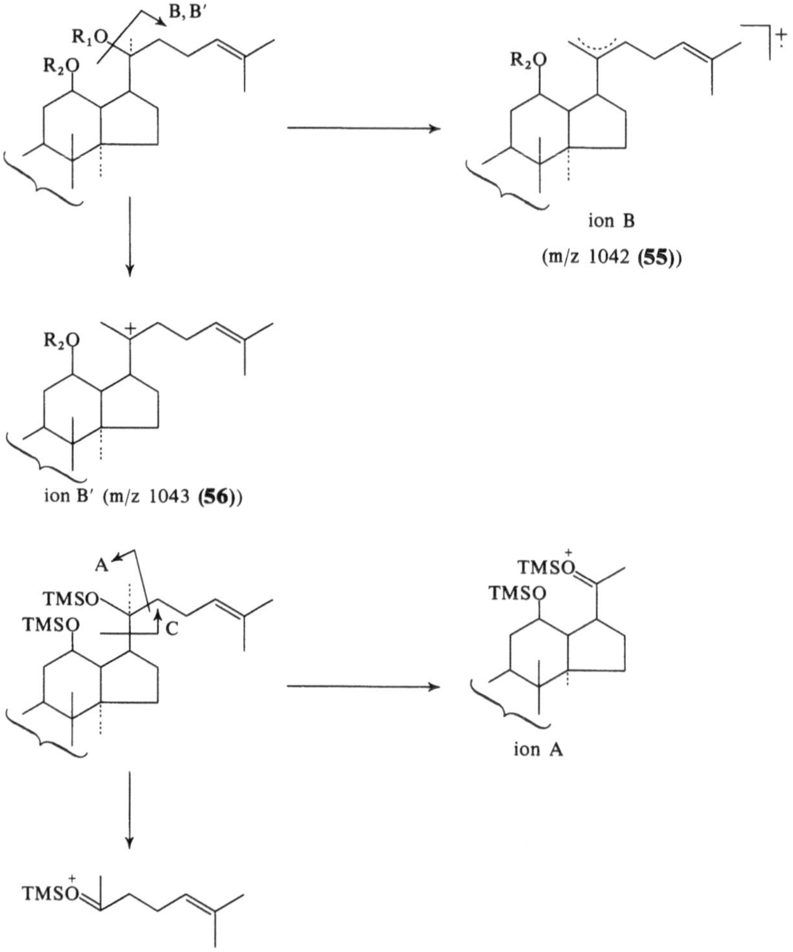
B, B′
R_1O
R_2O
ion B
(m/z 1042 (**55**))
ion B′ (m/z 1043 (**56**))
A
TMSO
TMSO
C
TMSO
TMSO
ion A
TMSO
ion C (m/z 199)

Chart 11

N-Trimethylsilylimidazole readily trimethylsilylates even the hindered 20-hydroxyl (49). Besides ions B and B', TMSi ethers of dammarane type triterpenes of their saponins having a free hydroxyl group at C-20 show the characteristic strong ions A and C (see Chart 11) which are also useful criteria for the absence of a glycosyl linkage at C-20.

KOMORI et al. (50) and recently, SCHULTEN and SOLDATI (51) reported that the field desorption (FD) mass spectra of dammarane saponins exhibited a fairly strong M + Na ion without the need for any prior derivatization. This appears to be of promise for determining the molecular weight and the homogeneity of saponins.

It should be emphasized that identification and criteria of purity of saponins based only on comparisons of TLC, optical rotation and IR spectrometry have frequently led to erroneous conclusions and that these must be substantiated by means of a combination of the procedures described in this chapter.

V. Structures of Ginseng Saponins (2)

By means of the new techniques described in the preceeding chapter as well as the modern separation procedures described in Chapter VI, isolation and structure elucidation of the minor saponins of Ginseng were accomplished recently.

1. Minor Saponins of Ginseng, Ginsenosides-Ra₁, -Ra₂, -Ra₃ and -Rh₁

Although TLC indicated the presence of a minor saponin named ginsenoside-Ra, little effort was devoted to its isolation and structure determination prior to 1981. BESSO et al. showed that the minor saponin actually was a mixture of three saponins by means of reverse phase TLC on octadecyldimethylsilyl silica gel (52). Reverse phase column chromatography of Ra permitted separation into three saponins in 0.03%, 0.02% and 0.005% yield which were named ginsenosides-Ra₁, Ra₂ and Ra₃.

The ^{13}C-NMR spectra of all three saponins exhibited signals due to the aglycone moiety at almost the same positions as those of Rb₁, Rb₂, Rc and Rd (Chapter IV-2,3, Chart 10), indicating that the minor saponins were glycosides of (26) with glycosyl linkages at both the 3- and the 20-hydroxyl groups (52). That the saponins contains five monosaccharide units was demonstrated by the presence of five anomeric carbon signals.

On hydrolysis with mineral acid, Ra₁ and Ra₂ afforded glucose, arabinose and xylose, while mild hydrolysis with aqueous acetic acid yielded oligosaccharides (53) from Ra₁ and (54) from Ra₂, along with the epimeric common prosapogenin pair (30) and (34) previously obtained from Rb_1, Rb_2, Rc, Rd etc. (Chapter III-1). The EI-MS of the acetates of both saponins showed a pair of B-type ions, m/z 1042 (55) and 1043 (56) (Chapter IV-4, Chart 11) and ions due to [terminal pentose $(Ac)_3$] (m/z 259), [hexose $(Ac)_4$] (m/z 331), [pentose-pentose $(Ac)_5$] (m/z 475) and [hexose-hexose $(Ac)_7$] (m/z 619).

The ^{13}C-NMR spectrum (Chapter IV-3, Table IX) of Ra_1 revealed that the signals of the sugar moiety consisted of those due to the 3-O-β-sophorosyl unit of (30) and those associated with the 20-O-β-D-xylopyrano-syl-(1→4)-α-L-arabinopyranosyl-(1→6)-β-D-glucopyranosyl group of chikusetsusaponin-L5 (C-L5), a saponin from the leaves of *P. japonicus* (Chapter IX-4, Table I). This led to the formulation of Ra_1 as 20(*S*)-protopanaxadiol 3-O-β-D-glucopyranosyl-(1→2)-β-D-glucopyranoside-20-O-β-D-xylopyranosyl-(1→4)-α-L-arabinopyranosyl-(1→6)-β-D-glucopyra-noside, i. e., xylo-ginsenoside-Rb_2. Permethylation of [(53): β-D-xylopyra-nosyl-(1→4)-α-L-arabinopyranosyl-(1→6)-D-glucopyranose] followed by methanolsyis and analysis of the resulting fully or partially methylated methyl glycosides was consistent with this formulation (52) (Table I). KOIZUMI, SHOJI *et al.* also isolated the same saponin and determined its structure by enzymic partial hydrolysis (53).

Comparison of the ^{13}C-NMR spectrum of the sugar moiety of Ra_2 with that of Rc (Chapter IV-3, Table IX) revealed the presence in the spectrum of Ra_2 of an additional set of signals assignable to a terminal β-xylopyranosyl unit (52). Further, on comparing Ra_2 with Rc, a signal at δ 109.9, characteristic of the anomeric carbon of α-arabinofuranosides was dis-placed upfield by 1.9 ppm and a resonance at δ 83.3 due to C-2 of the α-arabinofuranosyl unit was deshielded by 7.3 ppm, while the other signals of Rc remained almost unchanged. This evidence led to the formulation of Ra_2 as 20(*S*)-protopanaxadiol 3-O-β-D-glucopyranosyl-(1→2)-β-D-gluco-pyranoside-20-O-β-D-xylopyranosyl-(1→2)-α-L-arabinofuranosyl-(1→6)-β-D-glucopyranoside, i. e., xylo-ginsenoside-Rc (Table I). Permethylation of [(54): β-D-xylopyranosyl-(1→2)-α-L-arabinofuranosyl-(1→6)-D-glucopyranose] followed by methanolysis and analysis of the resulting fully or partially methylated methyl glycoside was in accord with this formulation.

The molecular weight of Ra_3 was determined by FD-MS (54). On partial hydrolysis, Ra_3 afforded (30) and a trisaccharide. GC-MS analysis of the hydrolysates of the permethyl ether of this trisaccharide as well as EI-MS (as the acetate) and ^{13}C-NMR spectrometry of Ra_3 led to the formulation of this saponin as xylo-Rb_1, whose β-D-xylopyranosyl unit was located at the 3-hydroxyl group of the terminal glucosyl unit of the 20-O-β-gentiobiosyl moiety of Rb_1 (44) (Table I).

Ra_1, Ra_2 and Ra_3 are the most polar dammarane saponins because of the presence of five monosaccharide units.

The least polar Ginseng saponin named ginsenoside-Rh_1 was isolated in 0.002% yield (55). On hydrolysis with crude hesperidinase (Chapter IV-1), Rh_1 yielded (36) and glucose. The structure of Rh_1 was established by comparing its ^{13}C-NMR spectrum with those of (36) and Rg_1. On going from (36) to Rh_1, the signal due to C-6 was deshielded by 10.4 ppm and the signal due to C-7 was displaced upfield by 2.2 ppm, while the other signals of the aglycone moiety of Rh_1 where observed at almost the same positions as those of (36) (Chapter IV-2, Table IV, Chart 10). As regards the sugar carbon signals, a set of resonances assignable to the 6-O-β-D-glucopyranosyl unit of Rg_1 was also found in the spectrum of Rh_1 at almost the same frequences. It follows that Rh_1 must be formulated as the 6-O-β-D-glucopyranosides of (36) (Table II).

2. Saponins of Red Ginseng

In Asian countries steamed Ginseng, the so-called Red Ginseng, has been more commonly used than White Ginseng. From Red Ginseng supplied by the Office of Monopoly, Republic of Korea, KASAI et al. isolated all of the saponins found in White Ginseng in similar yields (56). Four minor saponins were also isolated from Red Ginseng, two being identical with notoginsenoside-R_1 (NG-R_1, 0.007% yield), a saponin of P. notoginseng (Sanchi-Ginseng, Chapter VIII-2, Table II), and quinquenoside-R_1(Q-R_1, 0.015% yield) identical with monoacetylginsenoside-Rb_1 first isolated from American Ginseng (Chapter VIII-1, Table I), and found also in White Ginseng.

The remaining two minor saponins were new and designated as ginsenoside-Rs1, (Rs-1, 0.008% yield) and ginsenoside-Rs2, (Rs-2, 0.01% yield) (56). Both of these had an acetoxyl group and afforded Rb_2 and Rc on alkaline saponification, respectively. The EI-MS of the TMSi ethers of both saponins exhibited ions at m/z 799 [(hexose-hexose) $(TMSi)_6$Ac] and 421 [hexose $(TMSi)_3$Ac], indicating the presence of an acetoxyl group in the terminal glucosyl moiety of the sophorosyl group of the parent saponins, Rb_2 and Rc. The location of this acetoxyl group was elucidated by ^{13}C-NMR spectrometry by considering the acylation shift (57). On going from Rb_1 and Rc to Rs-1 and -2, the signals due to C-6 and C-5 of the glucosyl moiety were displaced downfield by 2.0 ppm and upfield by 3.5 ppm, respectively, while the frequences of the other carbon signals remained unchanged (see Chapter IV-3, Table IX). It follows that the acetyl group of both saponins must be located on the 6-hydroxyl group of the terminal glucosyl moiety of the 3-O-β-sophorosyl group of Rb_2 and Rc (Table I) (46). These two saponins have not yet been isolated from White Ginseng.

VI. Analysis of Ginseng Saponins

To permit chemical evaluation of Ginseng or preparations of Ginseng sold for medicinal or health food purposes, methods for separative analysis of the major Ginseng saponins have been investigated.

1. Quantitative Analysis of Panaxadiol and Panaxatriol

SAKAMOTO *et al.* reported developed condition suitable for hydrolysis of the saponins so as to permit their quantitative analysis as (1) and (35) (*18*). The 20-epimeric mixtures of (1) and (35) produced in this fashion were determined by GLC in the form of their TMSi ethers (*18*), by preparative TLC followed by colorimetry (*58, 59, 60, 61*) and by TLC densitometry in the form of their naphthoates (*62*). This procedure is still useful for analysis of complex Ginseng prescriptions where direct TLC or HLC separation cannot be employed without preliminary purification.

2. Separative Analysis of Ginsenosides

SANADA *et al.* developed a method for rapid separative analysis of saponins in Ginseng roots by means of TLC densitometry (*63*). They obtained excellent separation of the saponins by TLC on silica gel using the following solvent systems a) chloroform-methanol-water (65:35:10, lower phase); b) chloroform-methanol-ethyl acetate-water (2:2:4:1, lower phase); c) 1-butanol-ethyl acetate-water (4:1:5, upper phase) and d) chloroform-1-butanol-methanol-water (20:40:15:20, lower phase). After spraying with 10% H_2SO_4, the plate was heated at 140° C for a few minutes, immediately covered with a glass plate of the same size to prevent the spots from changing color and then subjected to densitometry. This procedure is useful not only for the analysis of Ginseng but also for quantitative analysis of other plant glycosides (*64*). Two dimensional TLC followed by TLC densitometry was reported by LEE *et al.* (*65*).

Separation of Ginseng saponins by reverse phase HLC has been accomplished on a column designed for carbohydrate analysis (Waters) using an acetonitrile-water system as solvent (detection: refraction index difference) (*66, 67*) on a μ-Bondapac-C_{18} column (Waters) using the same solvent system (detection: end-absorption at 203 nm) (*68, 69*). For more sensitive detection in the UV region, the Ginseng saponins were benzoylated and then subjected to HLC on a TSK-gel LS-410 column (Toyo Soda Co. Ltd., Japan) using an *n*-hexane-dichloromethane-acetonitrile as solvent system (detection: absorption at 254 nm) to accomplish high sensitive analysis (*70*). Recently, SCHULTEN and SOLDATI reported separation and identification of Ginseng saponins by HLC or TLC coupled with FD-MS and multiple internal reflection infrared spectrometry (*51*). To optimize conditions for the HLC separation of Ginseng saponins, the retention behavior of the sapogenins in silica gel HLC was investigated by KUNIHIRO *et al.* (*71*). Very recently, KAIZUKA *et al.* accomplished the quantitative analysis of Ginseng saponins by HLC on a column of silica gel using the solvent system chloroform-methanol-ethanol-water (*72*). Reverse phase HLC using highly porous synthetic polymers is useful for removal of non-saponin substances from the crude saponin fraction. These modern separative methods are also useful for separation of the saponins on a preparative scale.

For highly selective analysis of Ginseng saponins on an ultramicro scale, SANKAWA *et al.* developed a radioimmunoassay for Rg_1 using a Rg_1-serum albumin conjugate (*73*).

3. Localization of Saponins in Ginseng Roots

As Table X shows, lateral roots contain significantly more Ginseng saponins, especially saponins of (**26**), than commercial White Ginseng which is prepared from 4- or 6 year aged main roots after removal of lateral roots and periderm. The saponin content of 1- or 2 year aged roots is also somewhat higher than that of White Ginseng (*18, 69*).

Table X. *Distribution of Ginsenosides in Panax ginseng C. A. Meyer* (*53*)

	Rg_1	Re	Rf	Rg_2	Rb_1	Rc	Rb_2	Rd	Total
					% content				
Leaves	1.078	1.524	—	—	0.184	0.736	0.553	1.113	5.188
Leafstalks	0.327	0.141	—	—	—	0.190	—	0.107	0.765
Stem	0.292	0.070	—	—	—	—	0.397	—	0.759
Main root	0.379	0.153	0.092	0.023	0.342	0.190	0.131	0.038	1.348
Lateral roots	0.406	0.668	0.203	0.090	0.850	0.738	0.434	0.143	3.532
Root hairs	0.376	1.512	0.150	0.249	1.351	1.349	0.780	0.381	6.148

KUBO *et al.* (*74*) further investigated these differences in saponin content and found that the saponins of Ginseng roots are localized outside the cambium, i. e., not in the xylem or pith but in the periderm and cortex. They mentioned that the peeling process for the production of White Ginseng results in a remarkable loss of the biologically active saponins and is therefore undesirable from the pharmaceutical point of view.

OTSUKA *et al.* (*75*) reported a high saponin content in a Ginseng corm (a rhizome part on a head of the root) as part of a study on the separation of the saponins by means of droplet counter current chromatography (DCC).

4. Seasonal Variation of Ginseng Root Constituents

KIM *et al.* (*76*) found that the yield of methanolic extract from roots harvested in winter was more than two-fold greater than from roots collected in the summer and that this difference was mainly due to a remarkable increase in the sucrose content of the winter collection. On the other hand, the content of dammarane saponins in the roots is higher in the summer (Table XI). In other words, Ginseng extracts obtained from winter collections, the saponins and probably the other active principles are significantly diluted by an increase in the amount of sucrose, although the yield of total extract appear to be very high. This result shows that summer is the best season for harvesting roots that yield extracts with high potency.

Table XI. *Seasonal Variation of Contents of Mono- and Oligosaccharides and Dammarane-Saponins*

Date of harvest	Yield of MeOH extract (%)*	Contents of carbohydrates in extract (%)			Contents of saponins (%) in extract (in roots*)	
		sucrose	glucose	fructose	group A[a]	group B[b]
March/20	40.2	67.6	2.2	1.6	4.9 (1.9)	2.1 (0.9)
April/5	38.7	72.4	—	—	7.3 (2.8)	2.4 (0.9)
April/20	29.3	57.2	—	—	7.6 (2.2)	2.5 (0.7)
May/5	23.7	30.4	—	—	13.0 (3.1)	4.7 (1.1)
May/20	24.4	20.0	—	—	13.0 (3.2)	4.8 (1.2)
June/20	20.9	31.6	0.5	1.1	14.4 (3.0)	5.9 (1.2)
July/20	16.2	30.4	—	—	15.5 (2.5)	7.1 (1.2)
Aug./20	16.8	39.2	—	—	16.6 (2.8)	6.0 (1.0)
Sept./20	13.8	44.4	0.1	0.2	12.3 (1.7)	5.0 (0.7)
Oct./20	15.4	62.4	—	—	11.5 (1.8)	4.7 (0.7)
Nov./20	23.6	76.4	—	—	9.2 (2.2)	3.2 (0.7)
Dec./20	43.6	62.8	0.9	0.6	5.1 (2.2)	2.0 (0.9)
Jan./20	34.5	75.6	0.5	0.4	6.4 (2.2)	2.5 (0.8)
Feb./20	34.1	60.0	—	—	5.6 (1.9)	2.2 (0.8)

* Calculated from dry-weights of roots,   [a] Saponins of 20(*S*)-protopanaxadiol,
[b] Saponins of 20(*S*)-protopanaxatriol,   — not determined.

It was also found that keeping fresh roots harvested in the summer at 2° for one month resulted in a remarkable increase of extract. This is also due mainly to the formation of sucrose from starch (*77*).

References, pp. 65—76

VII. Production of Saponins by Tissue Culture

Field cultivation of Ginseng is laborious and requires more than four years and special precautions to avoid direct sunshine and infection. Also, because of soil depletion, Ginseng cannot be cultivated on the same plot for at least 10–15 years. Hence tissue culture of this medicinal plant has been attempted by several groups in Japan, Korea, U.S.A. and China as a means for producing biologically active principles on a manufacturing scale.

FURUYA et al. (78) who investigated Ginseng tissue culture under various conditions in a jar fermenter succeeded in large scale production of callus which contains 5 – 10 times as much Ginseng saponins as commerical Ginseng roots.

VIII. Saponins of Other *Panax* spp.

Several *Panax* species other than Ginseng occur in the Northern hemisphere from the Eastern Himalayas onward through China and Japan to North America. These *Panax* species can be divided into two groups, depending on the shape of the underground parts.

Group A contains carrot-like roots and consists of the following species.

1) *P. ginseng*
2) *P. quinquefolium* L. (American Ginseng, wild and cultivated in North America)
3) *P. trifolius* L. (ground-nut or dwarf Ginseng, wild in North America)
4) *P. notoginseng* (Burk.) F. H. Chen (Chinese name: Sanchi-Ginseng or Tienchi, cultivated in Yunnan, China but the locality of the wild specimen is still obscure.)
5) *P. pseudo-ginseng* Wall. subsp. *pseudo-ginseng* Hara (very rare in the Eastern Himalayas) (79).

Group B contains long horizontal rhizomes and is mainly distributed from the Eastern Himalayas to Japan through southern China. It consists of the following species, subspecies or varieties.

1) *P. japonicus* C. A. Meyer = *P. pseudo-ginseng* Wall. subsp. *japonicus* (Meyer) Hara (79) [Japanese name: Chikusetsu-ninjin or Tochiba-ninjin, Chinese name: Zhujie-shen, Zhao-shen or Zhujie-sanchi, wild throughout Japan and also in southern China (80)]
2) *P. japonicus* C. A. Meyer var. *major* (Burk.) C. Y. Wu *et* K. M. Feng [= *P. pseudo-ginseng* Wall. var. *major* (Burk.) Li; Chinese name: Zu-tzi-shen, wild in southern China (80)]
3) *P. zingiberensis* C. Y. Wu *et* K. M. Feng [wild in southern China (80)]

4) *P. pseudo-ginseng* Wall. subsp. *himalaicus* Hara and its varieties; var. *angustifolius* (Burk.) Li and var. *bipinnatifidus* (Seem.) Li (*79*) [Himalayan *Panax*, wild in the Eastern Himalayas).

There are several other *Panax* species; for example, *P. japonicus* C. A. Meyer var. *stipuleanatus* H. T. Tsai *et* K. M. Feng etc. is found in southern China from Yunnan to Tibet, and seems to be closely related to Himalayan *Panax* (*80*).

Chemical studies on most of these *Panax* spp. have only been carried out recently. The saponins of Group A species are mainly oligoglycosides of dammarane type triterpenes, while the rhizomes of Group B contain large amounts of the common oleanolic acid saponins (C-V etc. Table III) along with dammarane saponins which are characteristic of each species.

1. Saponins of American Ginseng and Dwarf Ginseng

American Ginseng has been used for similar medicinal purposes in the U.S.A. and China. Comparisons of the saponins of American Ginseng with those of Ginseng were carried out by SHIBATA, ANDO *et al.* (*81, 82*) and by OTSUKA *et al.* (*75*). KIM and STABA (*83*) were the first to isolate several saponins from this drug; subsequently LUI and STABA (*84*) analyzed the saponins of American Ginseng by two-dimensional TLC and reported that the saponin composition, especially the presence of Rg_1, is liable to variation.

SANADA *et al.* (*85*) identified the Ginseng saponins Ro (C-V) (0.07% yield), Rb_1 (1.57%), Rb_2 (0.02%), Rc (0.22%), Rd (0.77%) and Re (0.89%) in roots of American Ginseng cultivated in Japan. In addition to these saponins, BESSO *et al.* (*86*) isolated and identified Rg_1 (0.15% yield) and several minor saponins in commercial American Ginseng. The minor saponins were Rg_2 (0.008% yield), Rb_3 (0.03%), ginsenoside-F2 (F2, 0.018%, a saponin from Ginseng leaves, see Chapter IX-1, Table II), pseudo-ginsenoside-F_{11} (PG-F_{11}, 0.04%, an ocotillol type saponin from leaves of Himalayan and American *Panax*, see Chapter IX-2) and gypenoside-XVII [Gy-XVII, 0.03%, a saponin isolated from *Gynostemma pentaphyllum*, Cucurbitaceae, by TAKEMOTO *et al.* (*87*), Table I] along with a new saponin named quinquenoside-R_1 (Q-R_1) (0.01%) which was proved to be mono-o-acetyl-Rb_1. By means of MS and the acylation shift (*57*) in ^{13}C-NMR spectrometry, the acetyl group of Q-R_1 was shown to be located on the 6-hydroxyl of the terminal glucosyl moiety of the β-sophorosyl group of Rb_1 (*68*) (Table I). Q-R_1 was also isolated from Ginseng (see Chapter V-2).

Dwarf Ginseng (*P. trifolius*) is distributed from southern Canada to northern U.S.A. This plant is smaller than other *Panax* spp. and has a round tuberous root. LEE *et al.* (*65*) reported that the saponin content in roots of this plant was very low, the total content being only 0.0061%, and deduced by means of two-dimensional TLC the presence of C-V, Re, Rf, Rg_2 and several unidentified saponins.

2. Saponins of Chinese Sanchi-Ginseng

This well-known Chinese traditional crude drug which consists of the roots of *P. notoginseng* cultivated in Yunnan has been used as a tonic and a hemostatic. Recently, it has attracted much attention because of its action in preventing and curing coronary disease. Several Ginseng dammarane saponins, Rb_1 (1.62% yield), Rd (0.32%), Re (0.51%) and Rg_1 (2.07%) have been isolated from this drug, while the oleanolic acid saponin C-V (= ginsenoside-Ro) could not be detected (*85, 88, 89*).

Later, by means of reverse-phase HLC, ZHOU, TANIYASU *et al.* (*48*) showed that what in the earier studies appeared to be a homogeneous Re fraction was actually a mixture of Re and a new dammarane saponin named notoginsenoside-R1 (NG-R1, 0.16% yield). The structure of NG-R1 was established (*48*) as 20(*S*)-protopanaxatriol 6-o-β-D-xylopyranosyl-(1→2)-β-D-glucopyranoside-20-o-β-D-glucopyranoside (Table II) as a result of the anomalous chemical shifts exhibited by the 1,2-linked sugar moiety attached to the 6-hydroxyl of (**36**). This has been referred to in Chapter IV-3 (see also Table IX). Identification of the signals is based on selective deuteration of the sugar moiety (*90, 91*). Further work on the minor saponins of this drug led to isolation of several known dammarane saponins, Rh_1 (0.06% yield), Rg_2 (0.03%) (*48*), glc-Rf (0.005%) and Gy-XVII (0.036%) (*92*). Isolation and structure elucidation of several new minor dammarane saponins were also achieved; these include noto-ginsenosides-R2 (NG-R2 yield: 0.04%, Table II) (*48*) and -R3 (NG-R3 yield: 0.007%), -R4 (NG-R4, 0.028%) and -R6 (NG-R6, 0.002%) (*92*) (Tables I, II). NG-R6 was formulated as the 6-o-β-D-glucopyranoside-20-o-α-D-glucopyranosyl-(1→6)-β-D-glucopyranoside of (**36**) which is the first example of a naturally occurring saponin with an α-glucosyl linkage (isomaltoside). The α-anomeric configuration was substantiated by ^{1}H- and ^{13}C-NMR spectrometry.

Corms of this plant gave Rb_1 (5.2% yield), Rb_2 (0.12%), Rd (1.0%), Re (0.63%), Rg_1 (5.7%) and NG-R1 (1.1%) in significantly higher yields than the roots (*92*).

3. Saponins of *P. japonicus* Rhizomes (Japanese Chikusetsu-Ninjin and Chinese Zhujie-Shen)

Chikusetsu-ninjin, the rhizome of *P. japonicus* which grows wild throughout Japan has been used as a stomachic, expectorant and antipyretic in traditional medicine.

Several Japanese phytochemists have studied the saponins of this drug, with oleanolic acid (57) being identified as the sapogenin (*93, 94, 95, 96, 97, 98*). SHOJI and coworkers (*99, 100, 101, 102, 103*) who have studied the saponins extensively since 1968, isolated the following oleanolic acid saponins, chikusetsusaponins-IV (C-IV, 0.43% yield) and -V (C-V = ginsenoside-Ro, 5.35% yield) along with chikusetsusaponin-III [C-III, the dammarane saponin of (26), 1.17% yield] and several minor saponins, chikusetsusaponins-IVa (C-IVa), -Ib (C-Ib), both oleanolic acid saponins and -Ia (C-Ia, a dammarane saponin) (Tables I and III). The significant difference in saponin composition between Chikusetsu-ninjin and Ginseng root supports the distinction made between these two crude drugs in traditional oriental medicine. In contrast to Ginseng roots, Chikusetsu-ninjin has not been used for restoration of virility.

In Yunnan, Guizhou, Hunnan, Jiangxi, Zhejiang and Szechwan, at altitudes of 1500 – 2500 m, there grows a wild *Panax* named Zhjie-shen, Zhao-shen or Zhujie-sanchi which has been used as expectorant, analgesic and antitussive in Chinese traditional medicine (*80*). This plant has bamboo-like long horizontally creeping rhizomes and is considered as botanically identical with Japanese Chikusetsu-ninjin (*P. japonicus*). The external and internal structures of the two plants are so similar that they are difficult to distinguish morphologically (*104*).

A Chinese-Japanese cooperative chemical study (*104*) of the rhizomes of this plant from Zhaotong, Yunnan at 2100 m led to isolation of the oleanolic acid saponins, C-IV, C-IVa, and C-V in yields of 3.4, 2.8, and 3.1%. In this respect, then Zhujie-shen is very similar to Japanese Chikusetsu-ninjin. However, the composition of the dammarane saponin fraction of Zhujie-shen which consisted of Rd (0.04% yield), Re (0.12%), Rg₁ (0.15%), Rg₂ (0.05%), NG-R2 (0.02%, see Chapter VIII-2, Table II) and pseudo-ginsenoside-F₁₁ (PG-F₁₁, 0.04%, an ocotillol-type saponin from leaves of Himalayan *Panax* and also from leaves of American Ginseng, see Chapter IX-2), was significantly different from that of the Japanese plant.

4. Saponins of Rhizomes of *P. japonicus* var. *major* (Chinese Zu-Tzi-Shen)

This *Panax* species is distributed from Tibet to Yunnan at higher altitudes (2500 – 4500 m) than Zhujie-shen and is commercially available in China as a traditional drug. The internodes of its long creeping rhizome are elongated and slender in contrast to those of Chikusetsu-ninjin and Zhujie-shen which are short and thick (*80*).

The China-Japan cooperative research group (*105*) also investigated the saponin composition of rhizomes collected in Likiang, Yunnan and found that while the rhizomes of this plant also contain a fairly large amount of oleanolic acid saponins, i. e. C-IVa (0.19% yield) and C-V (0.95% yield) (Table III), the dammarane saponin composition is evidently different from those of Chikusetsu-ninjin and Zhujie-shen. Rd was isolated in a relatively good yield (0.67%) along with the minor saponins, glc-Rf (0.01%) and NG-R2 (0.03%) (see Table II). In addition to these known saponins two new dammarane saponins having the ocotillol-type side chain were isolated. Structures of these saponins which were named majonosides-R1 (M-R1, 0.07%) and -R2 (M-R2, 0.11%) were established without chemical degradation and mainly by MS and ^{13}C-NMR spectrometry by comparing their properties with those of PG-F$_{11}$ isolated from leaves of Himalayan *Panax*. The common aglycone of both M-R1 and -R2 is the c-24 epimer of PG-F$_{11}$ [see Chapter IX-2, Table XII and Chart 14, compound (**66**)].

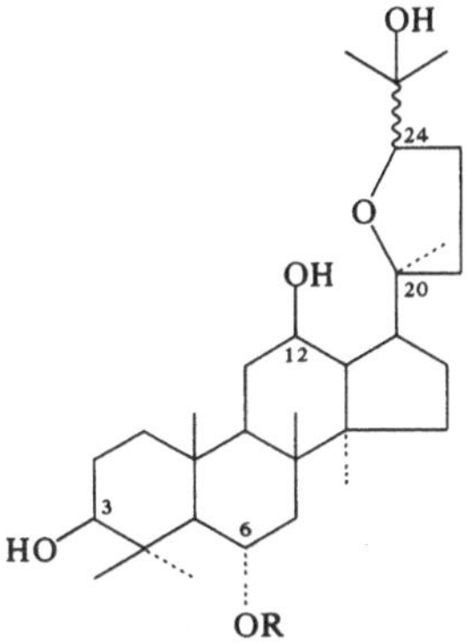

M-R1: R = -glc^2 – glc, 24(*S*)	*P. japonicus* var. *major* rhizomes (*105*)
M-R2: R = -glc^2 – xyl, 24(*S*)	*P. japonicus* var. *major* rhizomes (*105*)
PG-F$_{11}$: R = -glc^2 – rha, 24(*R*)	*P. pseudo-ginseng* subsp. *himalaicus* leaves (*111*)
	P. quinquefolium leaves (*116, 117*) roots (*85*)
	P. japonicus (Chinese) rhizomes (*104*)

Chart 12

5. Saponins of *P. zingiberensis* Rhizomes

This medicinal plant grows wild in the southern region of Yunnan at altitude of about 2000 m. From ginger-like rhizomes of this plant, YANG, ZHOU et al. (*106*) recently isolated the oleanolic acid saponins, C-IV (0.28% yield), -IVa (0.025%) and -V (2.1%) (see Table III) along with the dammarane saponins, Rg$_1$ (0.59%) and Rh$_1$ (trace) (see Table II). A new saponin named zingibroside-R$_1$ (Z-R$_1$) based on oleanolic acid was also isolated in 0.078% yield (see Table III).

6. Saponins of Rhizomes of Himalayan *Panax, P. pseudo-ginseng* subsp. *himalaicus*

From the rhizomes of Himalayan *Panax* collected near Khosa (1800 m) in western Bhutan, the oleanolic acid saponins, C-IV, -IVa and -V were isolated in yields of 0.3, 0.6 and 7.25%, respectively (*107, 108*). One of the major dammarane saponins of Ginseng, Rb_1 was also isolated in somewhat higher yield (1.05%) than from Ginseng roots.

Recently, *Acanthopanax senticosus* Harms (Araliaceae), so-called "Siberian-Ginseng", and *Pfaffia paniculata* (Matius) Kuntze (Acanthaceae), so-called "Brazilian-Ginseng", have been sold as health foods or for medicinal use similar to that of real Ginseng. It should be noted that the chemical constituents of these two plants are completely different from those of Ginseng and other *Panax* spp. and that dammarane saponins could not be detected in either of them.

IX. Saponins of Aerial Parts of *Panax* spp.

Leaves and flower buds of *Panax* spp. have been practically unutilized for medicinal purposes. To find better and cheaper sources of biologically active dammarane saponins and to find evidence for possible taxonomic relationships, the saponins of the aerial parts of this genus have also been investigated recently.

1. Saponins of *P. ginseng* Leaves and Flower Buds

Dammarane saponins which have been isolated from the leaves of this species include Rb_1, Rb_2, Rc, Rd [aglycone: (**26**)], Re and Rg_1 [aglycone: (**36**)] in yields of 0.1, 0.4, 0.2, 1.5, 1.5 and 1.5%, respectively, as well as three new saponins named ginsenosides-F1, -F2 and -F3 in yields of 0.4, 0.2 and 0.2% (*55, 109*). The aglycone of F1 and F3 is (**36**) and that of F2 is (**26**); their structures were established by YAHARA *et al.* (*109*) as shown in Tables I and II.

Flower buds of *P. ginseng* are usually removed before ripening for the better growth of roots. Rb_1, Rb_2, Rc, Rd, Re, Rg_1 and F3 were isolated from the flower buds in yields of 0.2, 0.2, 0.2, 0.5, 2.8, 0.2 and 0.03%, respectively, whereas the fruits gave Rb_2, Rc, Rd, Re and Rg_1 in yields of 0.2, 0.1, 0.1, 6.0 and 0.04% (*55, 110*). In addition to these saponins, a new dammarane saponin named ginsenoside-M7cd with an oxygenated side chain was isolated in 0.0039% yield from the flower buds as a minor saponin; its structure was elucidated by ^{13}C-NMR spectrometry as shown in Chart 13 (*55*).

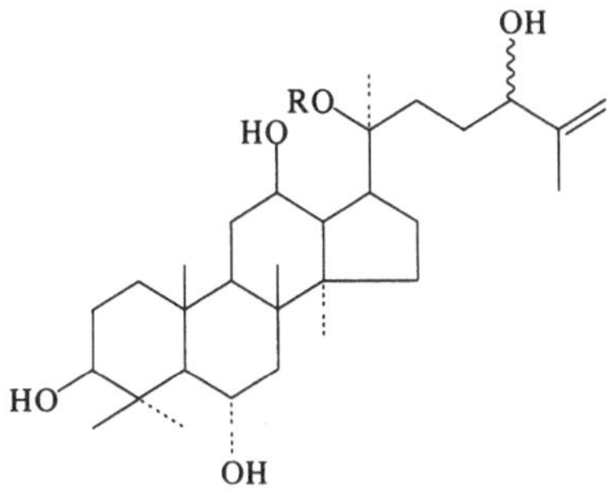

M7cd: R = -glc   P. ginseng
flower buds (55)

Chart 13

It is noteworthy that the yield of saponins based on (36) from the aerial parts is higher than the yield from the roots; the isolation of Re from the flower buds in such yield indicates the potential use of these parts as a source of this saponin. The total saponin content in the stems was found to be less than 0.5% (see Table X).

2. Saponins of Leaves of Himalayan and American *Panax* spp.

From dried leaves (85 g) of *P. pseudo-ginseng* subsp. *himalaicus* collected in Bhutan, three known dammarane saponins of Ginseng roots, Rb_3, Rd and Re were isolated in yields of 0.9, 0.1 and 0.1%. Further, two new saponins named pseudo-ginsenosides-F_8 (PG-F_8) and -F_{11} (PG-F_{11}) were obtained in yields of 0.1 and 0.4%, respectively (111). The IR, ^{1}H- and ^{13}C-NMR spectra of PG-F_8 demonstrated the presence of an acetoxyl group; mild alkaline hydrolysis yielded Rb_3, indicating that PG-F_8 is a mono-o-acetyl-Rb_3. The location of the acetoxyl group was revealed by the following evidence. The MS of the TMSi ether of PG-F_8 showed fragment peaks at m/z 799 [(glucose-glucose)-Ac(TMSi)$_6$], 729 [(xylose-glucose)-(TMSi)$_6$], 709 (799-TMSiOH), 451 [terminal glucose-(TMSi)$_4$] and 349 [terminal xylose-(TMSi)$_3$] (112). These ions disclosed that the O-acetyl group must be located on the inner glucosyl unit of the 3-o-β-sophorosyl moiety of Rb_3. As already mentioned in Chapter V-2, the ^{13}C-acylation shifts constitute the most powerful and substantial tool for determining the position of an O-acyl group (57). When the carbon resonances of the sophorosyl moieties of Rb_3 (Table IX) and PG-F_8 are compared, one of the two glucosyl-c-6 signals of Rb_3, that at δ 62.8, was shielded in PG-F_8 by +1.7 ppm, appearing at δ 64.5 and one of the two glucosyl-c-5 signals of Rb_3 (any two of three appearing at δ 77.7, 78.1 and 78.8) was displaced in PG-F_8 to δ 74.6, while the other signals remained practically constant. It follows that in PG-F_8, the acetyl group must acylate the 6-hydroxyl of the inner glucosyl unit of the β-sophorosyl moiety of Rb_3 (111) (Table I).

O. TANAKA and R. KASAI:

Table XII. ^{13}C Chemical Shifts of Ocotillol Type Triterpenes in Pyridine-d_5

(58): $R_1 = {}^{OH}_{H}$ $R_2 = OH$ $R_3 = OH$

(61): $R_1 = {}^{H}_{OH}$ $R_2 = H$ $R_3 = OH$

(63): $R_1 = O$ $R_2 = H$ $R_3 = H$

(65): $R_1 = {}^{OAc}_{H}$ $R_2 = H$ $R_3 = OH$

(66): $R' = {}^{OH}_{H}$ $R'' = OH$ $R''' = OH$

(62): $R' = {}^{H}_{OH}$ $R'' = H$ $R''' = OH$

(64): $R' = O$ $R'' = H$ $R''' = H$

	(58)	(66)	(61)	(62)	(63)	(64)	(65)
1	39.4	39.5	34.2	34.3	39.9	39.9	38.6
2	28.0	28.1	26.5	26.5	34.2	34.2	23.9
3	78.3	78.4	75.3	75.3	216.2	216.4	80.6
4	40.3	40.3	38.1	38.1	47.4	47.4	38.0
5	61.8	61.9	49.7	49.4	55.3	55.3	56.1
6	67.6	67.7	18.6	18.6	19.8	19.9	18.3
7	47.4	47.5	35.2	35.2	34.8	34.8	34.9

	1	2	3	4	5	6	7
8	41.0	41.2	40.2	40.2	40.4	40.5	39.9
9	50.4	50.2	50.6	50.5	50.0	50.2	50.6
10	39.4	39.3	37.7	37.7	36.9	37.2	37.1
11	32.3	32.2	31.7	32.5	22.2	22.4	32.2
12	71.1	70.8	71.1	70.7	26.0	26.1	71.0
13	48.3	49.1	48.4	49.4	43.3	43.3	48.3
14	52.0	52.2	52.2	52.3	50.0	50.2	52.1
15	31.7	32.6	32.3	32.2	31.7	31.7	31.6
16	25.4	25.8	25.5	25.7	27.4	27.4	25.4
17	49.3	49.5	49.9	49.9	50.0	50.2	49.4
18	17.7^a	17.8^a	16.5^a	16.6^a	16.0^a	16.1^a	16.6^a
19	17.4^a	17.2^a	15.6^a	15.7^a	15.1^a	15.3^a	16.4^a
20	86.6	87.0	86.7	87.1	86.2	86.4	86.6
21	26.9^b	26.9^b	26.9^b	26.9^b	23.3^b	26.3^b	26.9^b
22	32.8	32.6	32.8	32.5	36.2	35.4	32.7
23	28.6	28.7	28.7	28.6	26.8	26.8	28.6
24	85.6	88.4	85.6	88.3	84.1	87.4	85.5
25	70.2	70.0	70.1	70.0	71.1	70.4	70.3
26	27.1^b	26.6^b	27.3^b	26.5^b	26.0^b	26.8^b	27.1^b
27	27.6^b	29.0^b	27.6^b	29.0^b	26.8^b	26.8^b	27.6^b
28	31.8	31.9	29.4	29.4	26.8^b	27.1^b	28.0^b
29	16.4^a	16.5^a	22.4	22.4	21.1	21.1	15.4^a
30	18.2^a	18.1^a	18.2^a	18.0^a	16.4^a	16.5^a	18.3^a

a, b: Assignments may have to be interchanged in each vertical column.

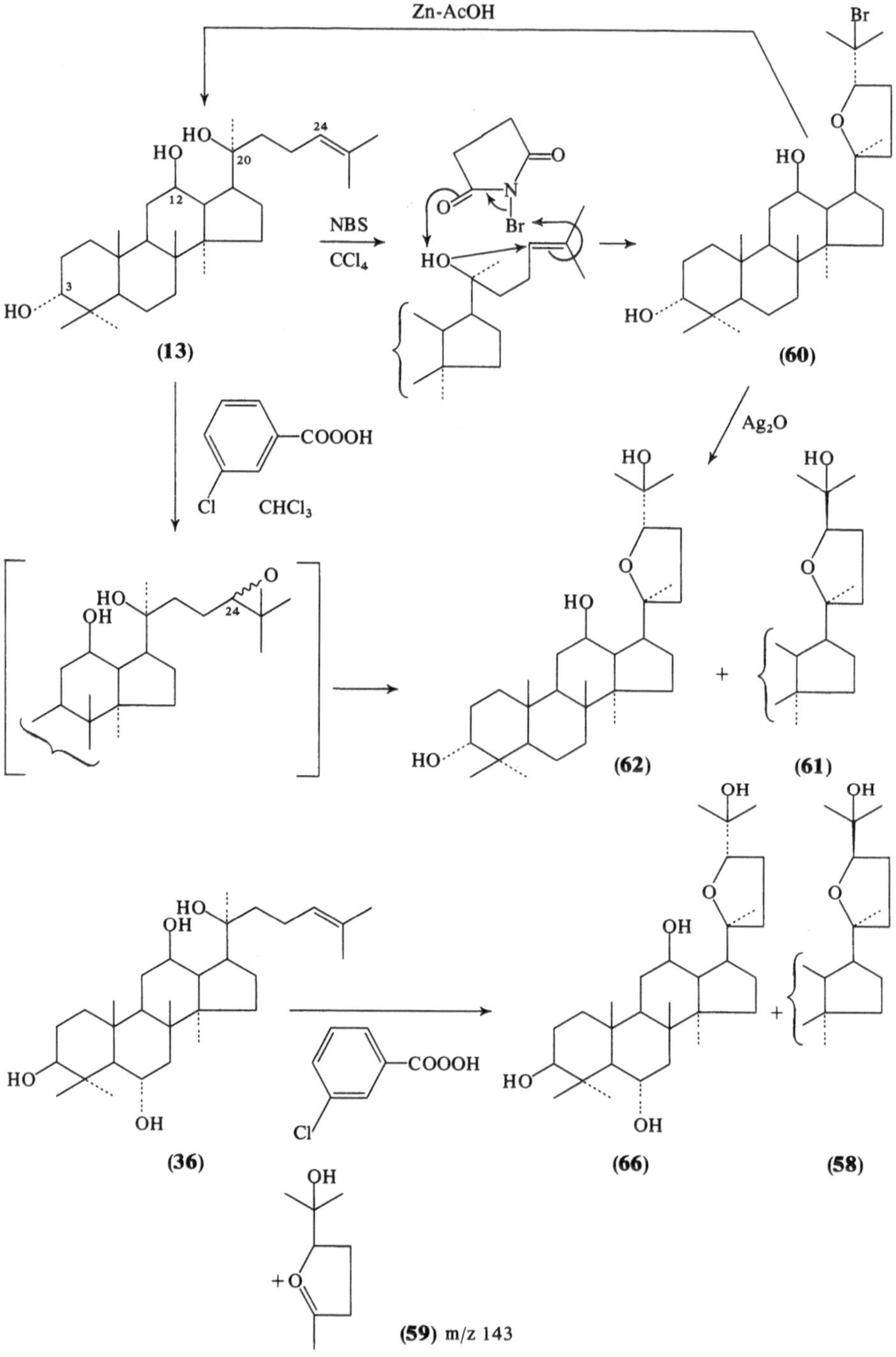

(59) m/z 143

Chart 14

Another new saponin, PG-F_{11}, on hydrolysis with crude hesperidinase (see Chapter IV-1) afforded glucose, rhamnose and an acid unstable aglycone (58) (*111*). The MS of (58) and an acetate of PG-F_{11} exhibited strong peak at m/z 143 (59) characteristic of ocotillol type triterpenes (*113*). In connection with the structure study of isodehydropanaxadiol (29) (see Chapter III), OHSAWA *et al.* (*22, 23*) earlier discovered a new cyclobromination of γ,δ-unsaturated alcohols under the influence of N-bromosuccinimide in CCl_4. Thus (13) afforded stereospecifically the bromo compound (60), which on reduction with zinc in acetic acid at room temperature, regenerated (13). The structure of (60) was elucidated by X-ray analysis. The chirality of c-24 of ocotillol type triterpenes including betulafolienetriol-oxide-I (61) and -II (62) which are constituents of white birch leaves, was established by correlation with (60) as shown in Chart 14 (*113*). Further investigation (*111*) disposed of a comment by RAO *et al.* (*114*) on this correlation.

Table XII lists signals in the ^{13}C-NMR spectra of 24 epimeric ocotillol type triterpenes such as ocotillone (63), cabralenone (64) and 3-o-acetylpyxinol (65) (*115*), as well as (61) and (62). Significant differences in the chemical shifts of c-20 and c-24 are found for compounds epimeric at c-24 (*111*). Comparison of the ^{13}C-NMR spectrum of (58) with the spectra of 20(S)-protopanaxatriol (36), (61), (62) and (65), showed that the ring skeleton of (58) must be identical with that of (36) and that its side chain can be represented by that of (61) and (65) with c-24-R, but not by that of epimer, (62).

The structure of (58) was further confirmed by its preparation from (36); peracid oxidation of the side chain double bond of (36) gave a mixture of very unstable epoxides epimeric at 24, which underwent spontaneously cyclization to (58) and its c-24 epimer (66). In the ^{13}C-NMR spectrum of the latter, c-24 and the other side chain carbons exhibited shifts essentially identical with those of (62), thus substantiating the assignment of chirality at c-24. The names 20(S)-protopanaxatriol-oxide I and II were proposed for (58) and (66).

As already mentioned in Chapter IV-3, the glycosylation shift rule has been used with great effect for determining the location of glycosyl linkages. On going from (58) to PG-F_{11}, the signal due to c-6 was deshielded and the signal assigned to c-7 was somewhat displayed upfield, while the other carbon resonances remained almost unchanged. This indicated that the location of the glycosyl linkage of PG-F_{11} should be limited to the 6α-OH of (58). Furthermore, comparison of the sugar carbon signals of PG-F_{11} with those of other saponins in the data-bank showed that those of PG-F_{11} were almost superimposable on those of the α-L-rhamnopyranosyl-$(1\rightarrow2)$-β-D-glucopyranosyl unit of Re, thus leading to the formulation of PG-F_{11} as the 6-o-α-L-rhamnopyranosyl-$(1\rightarrow2)$-β-D-glucopyranoside of (58), the first example of a naturally occurring ocotillol type glycoside (*111*) (Chart 12). PG-F_{11} was also isolated from the leaves (*116, 117*) and roots (*86*) of American Ginseng and from the rhizomes of Chinese *Panax japonicus* (Zhujie-shen) (*104*) (see Chapter VIII-1 and -3).

As mentioned in Chapter VIII-4, rhizomes of *P. japonicus* var. *major* yielded the ocotillol type saponins, majonosides-R1 and -R2, the aglycone of both of which was identical with 20(*S*)-protopanaxatriol-oxide II (**66**), the C-24 epimer of (**58**) (*105*).

Leaves and stems of American Ginseng furnished Rb_3, Rd, Re, Rg_1 and PG-F_{11} in yields of 0.1, 0.2, 0.1, 0.1 and 0.1%, respectively (*116, 117*). Since the saponin content of the stems should be very low (see Chapter VI-3, Table X), the leaves must produce these saponins in remarkably high yield. It is noteworthy that the saponin composition of American Ginseng leaves is very similar to that of Himalayan *Panax,* although the saponin composition of the underground parts of these species were significantly different (see Chapter VIII).

3. Saponins of Aerial Parts of Chinese *Panax* spp.

Leaves of *P. notoginseng* (Sanchi-Ginseng) cultivated in Yunnan gave seven dammarane saponins (*118*), three of which were identified as the known Ginseng saponins Rb_1 (0.03% yield), Rb_3 (0.71%) and Rc (0.39%). Another known saponin was identified as gypenoside-IX (Gy-IX, 0.03%), the saponin of (**26**) (*87*) isolated from *Gynostemma pentaphyllum* (Cucurbitaceae). The structures of three new dammarane saponins named notoginsenosides-Fa (NG-Fa, 0.01% yield), -Fc (NG-Fc, 0.05%) and -Fe (NG-Fe, 0.005%) were elucidated as shown in Table I by means of MS. ^{1}H and ^{13}C-NMR spectrometry as well as by partial hydrolysis.

From seeds of this plant Rb_1, Rd, Rc, Rb_3, NG-Fa, NG-Fc, and Gy-IX were isolated in yields of 0.001, 0.067, 0.42, 1.2, 0.087, 0.15 and 0.014%, respectively (*118*). The flower buds which are commercially available in China as a crude drug, gave Rb_1, Rb_2, Rc, Rd and F2 in yields of 0.4, 0.1, 1.0, 0.1 and 0.1%, respectively (*119*).

YANG *et al.* (*120*) studied the saponin composition of leaves of *P. japonicus* var. *major* collected in Yunnan, and isolated Rb_1, Rb_3, Rc and Rd in yields of 0.1, 0.4, 0.1 and 0.7%, respectively.

It has been reported that the pharmacological activities of saponins of 20(*S*)-protopanaxadiol (**26**) differ somewhat from those of 20(*S*)-protopanaxatriol (**36**); for example, saponins of (**26**) such as Rb_1 exhibit a sedative action while those of (**36**) such as Rg_1 stimulate the central nervous system (see Chapter XIII). It is therefore pharmacologically significant that the common sapogenin of the aerial parts of these Chinese *Panax* collections is represented exclusively by 20(S)-protopanaxadiol (**26**) and that no saponin based on 20(S)-protopanaxatriol (**36**) has as yet been detected. This is in contrast with the aerial parts of Ginseng, American Ginseng and Himalayan *Panax* which contain a large amount of saponins based on (**36**) along with a relatively small amount of saponins derived from (**26**).

4. Saponins of Leaves of *P. japonicus* (Japanese Chikusetsu-Ninjin)

Extensive study of the leaves of Chikusetsu-ninjin disclosed that their saponin composition depends significantly upon the localities in Japan from which they are collected. Leaves collected in Hiroshima on the Pacific Ocean coast yielded several known saponins based on 20(S)-protopanaxatriol (**36**) which were characterized as Re (0.1% yield), F1 (0.01%), F3 (0.1%) (see Table II) (*121*). New saponins derived from (**36**) and named chikusetsusaponins-L5 (C-L5, 0.7%), -L10 (C-L10, 0.1%), -L9bc (C-L9bc, 0.2%) and -L9a (C-L9a, 0.1%) were also obtained, the latter two of which have an oxygenated side chain. The structure of the sugar moiety of C-L5 was established by means of ^{13}C-NMR spectrometry using the partially relaxed Fourier transform technique (PRFT). Structures of these new saponins are shown in Table II and Chart 15; it should be noted that C-L10 is the first example of a dammarane saponin having a glycosyl linkage on the 12-hydroxyl group.

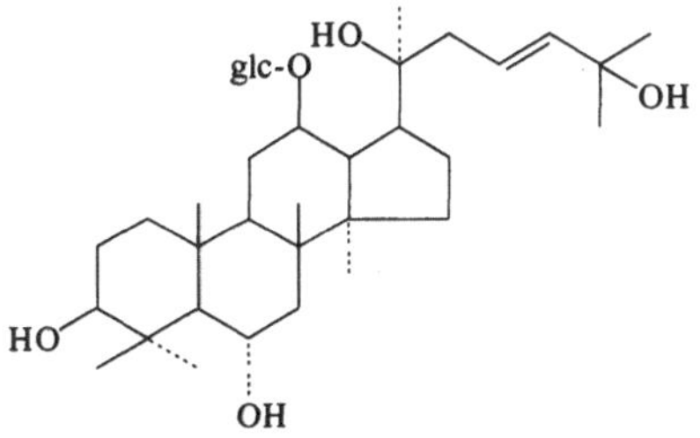

C-L9a   *P. japonicus*
         leaves (*121*)

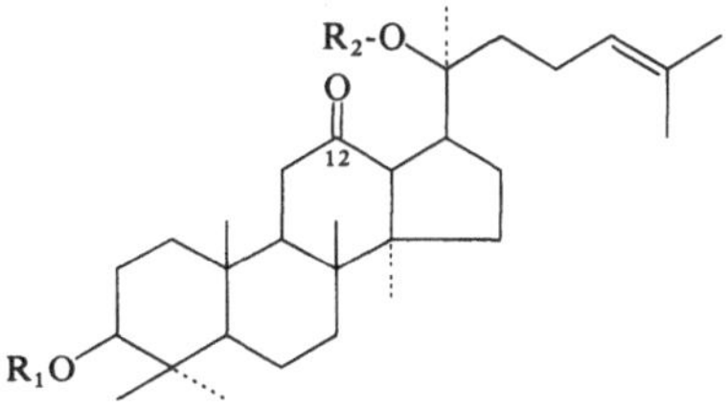

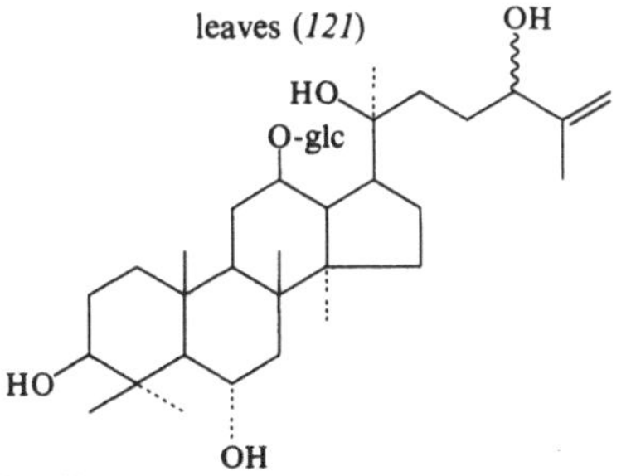

C-L9bc: *P. japonicus*
         leaves (*121*)

C-LT5: R_1 = -glc R_2 = -glc^6 — glc   *P. japonicus*
                                                 leaves (*122*)

C-LT8: R_1 = -glc R_2 = -glc   *P. japonicus*
                                                 leaves (*122*)

C-LN4: R_1 = -glc^6 — xyl   *P. japonicus*
         R_2 = -glc^6 — ara (pyr)   leaves (*122*)

C: Chikusetsusaponin

Chart 15

In contrast to leaves from Hiroshima which contain saponins of (36) only, the major saponins of leaves collected on the Japan Sea coast were all based on the 12-ketone (67) of 20(S)-protopanaxadiol. Leaves collected in Tottori and Kyoto gave the new saponins chikusetsusaponins-LT5 (C-LT5, 0.5 – 2.0% yield) and -LT8 (C-LT8, 0.1%) and leaves collected in Niigata gave a new saponin named chikusetsusaponin-LN4 (C-LN4, 1.4%yield) (*122*). Structures were established as shown in Chart 15. The rhizomes of these collections exhibited no significant differences in saponin composition.

X. Syntheses of Dammarane Sapogenins

After STORK *et al.* (*123*) had carried out the total synthesis of a α-onocerin (68), TSUDA and HATTORI (*124*) in 1967 succeeded in converting (68) to hydroxyhopanone (69) *via* gammaceran-3-on-21-ol (70) (see Chart 16).

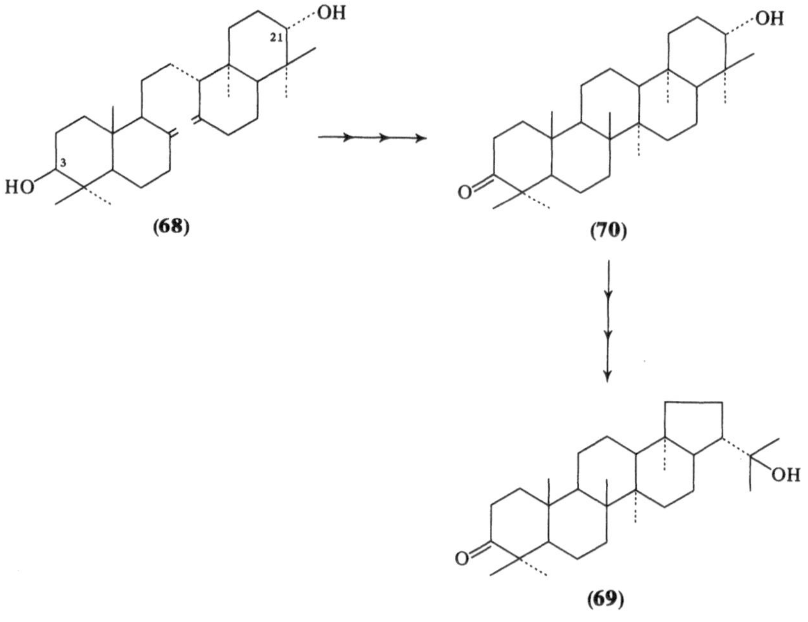

Chart 16

A synthesis of dammarenediol-II (15), a constituent of dammar resin (*125*), was carried out by FUJIMOTO and TANAKA (*126*) in 1970 as follows (see Chart 17). Acid treatment of (69) (*127*) afforded hopenone-I (71) which was reduced with LiAlH₄, to afford hop-17(21)-en-3β-ol (72). The acetate (73)

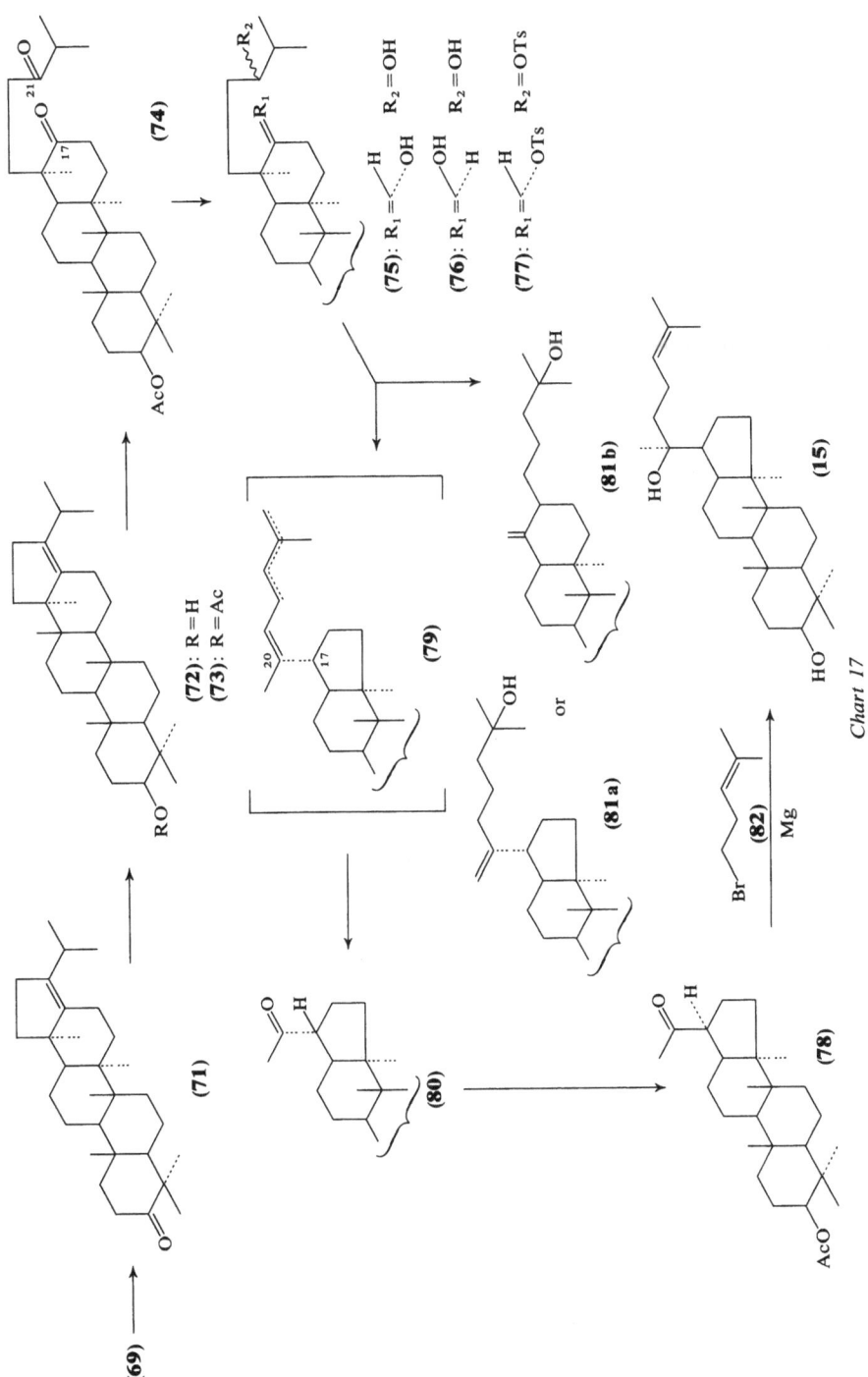

Chart 17

of (72) was subjected to ozonolysis to give the E-*seco*-diketone (74) which was reduced with NaBH$_4$, yielding a mixture of epimeric acetoxydiols. This mixture was separated by column chromatography to afford the acetoxydiols (75), (76) and their C-21-epimers. The α-equatorial configuration of the 17-hydroxyl group of (75) and its 21-epimer was substantiated by ^{1}H-NMR spectrometry.

The ditosylate (77) of (75) was subjected to Wagner-Meerwein type rearrangement; solvolysis of (77) by refluxing with CaCO$_3$ in aqueous dioxane followed by ozonolysis of the products afforded 3β-acetoxyhexakisnordammaran-20-one (78) which had already been prepared from (15) (*127*). Taking the stereochemistry of a rearrangement of this type into account, the intermediate of this reaction should be formulated as (79). Ozonolysis of (79) would yield primarily the 17β-H ketone (80) which would be readily isomerized to give the more stable 17α-H ketone (78). In this rearrangement, a compound which should be either (81a) or (81b) was also isolated (*126*).

Reaction of (78) with the Grignard reagent prepared from 4-methyl-3-pentenyl bromide (82) furnished synthetic (15) (*128*). Although formation to some extent of the 20-epimer dammarenediol-I (14) was also expected, it could not be detected.

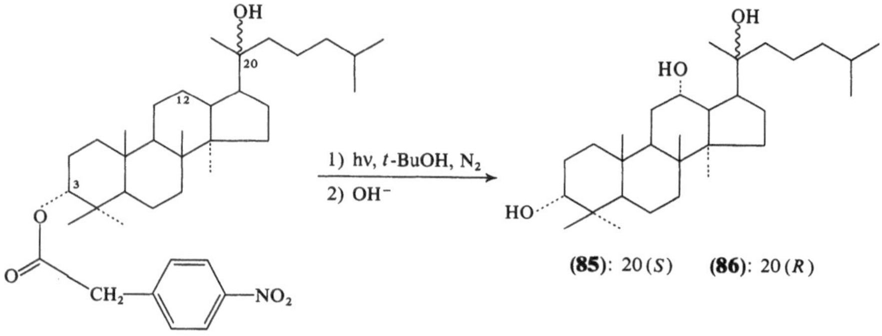

Chart 18

In an attempt toward syntheses of betulafolienetriol (13) and 20(*S*)-protopanaxadiol (26), the genuine sapogenin, introduction of an oxygen function at the unactivated methylene at C-12 of the dammarane skeleton was achieved (*128, 129*) by means of remote oxidation (*130*) using a photoexcited aromatic nitro group (*131, 132, 133*) (see Chart 18). Irradiation of the *p*-nitrophenyl-acetate (83) of 3-epidammaranediol-II (84)

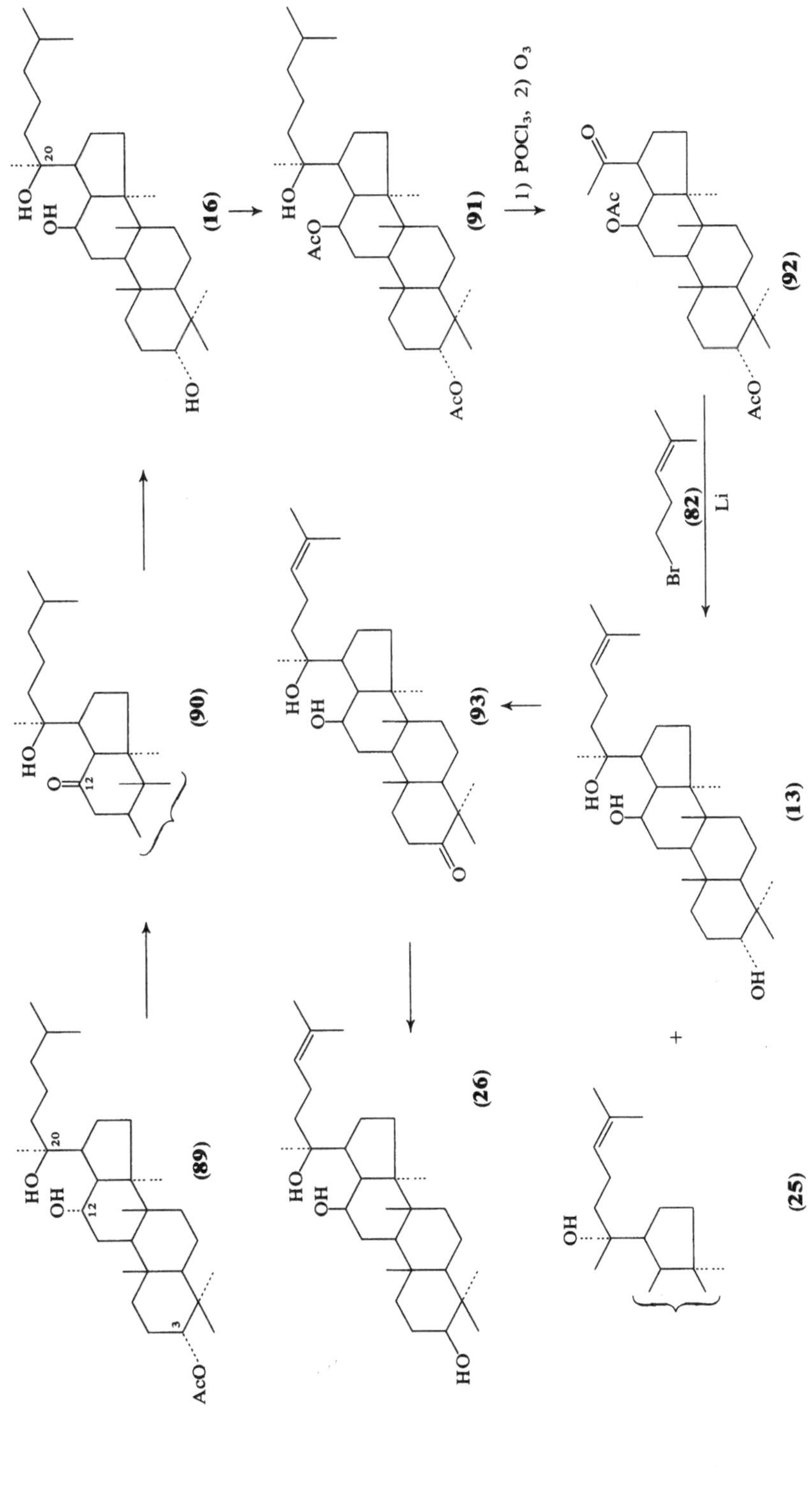

Chart 19

prepared from (15) and saponification of the reaction mixture gave the 12α-hydroxy compound (85) in 13% yield, (84) being recovered in 44% yield. The structure of (85) was substantiated by ¹H-NMR spectrometry and its preparation from (13) as follows (129): Selective acetylation of the 3-hydroxyl group of (13) was accomplished via the bromine derivative (60) (see Chapter IX-2, Chart 14) whose 12-hydroxyl group is sterically hindered. The resulting 3-acetate of (60) was reduced with zinc in acetic acid to give the 3-acetate of (13) in quantitative yield. Oxidation of the 12-hydroxyl group to the ketone followed by reduction with LiAlH₄ and subsequent catalytic hydrogenation afforded (85). The 12α-hydroxy compound (86) was obtained from the p-nitrophenylacetate (87) of 3-epidammaranediol-I (88) in the same manner as (85).

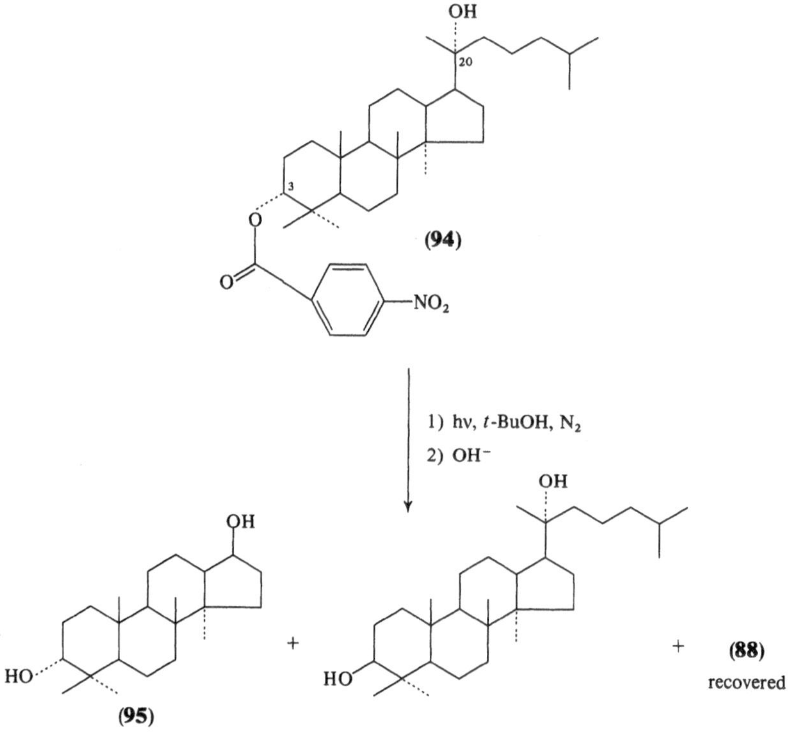

Chart 20

Because irradiation of the p-nitrophenylacetate of 3-epidammarenediol-II resulted in intermolecular addition of the photoexcited nitro group to the side chain double bond in preference to intramolecular hydroxylation of C-

12, betulafolienetriol (13) was prepared from (85) by the following sequence (*129*) (see Chart 19). The 3-monoacetate (89) of (85) was oxidized to 12-ketone (90), which after reduction with NaBH$_4$ followed by alkaline saponification yielded (16). The diacetate (91) of (16) was dehydrated with POCl$_3$ and successively degraded with ozone to give the hexakisnorketone (92). A modified Grignard reaction of (92) with (82) and Li afforded triol (13) and its 20-epimer (25). The 3-ketone (93) prepared from (13) was reduced with NaBH$_4$ to give 20(*S*)-protopanaxadiol (26), which accomplished the synthesis of the genuine sapogenin.

In connection with this synthesis, an unexpected photochemical cleavage of the 17,20-bond in the dammarane side chain was encountered (*128, 129*). Irradiation of the *p*-nitrobenzoate (94) of (88) (see Chart 20) under the usual conditions followed by saponification gave a diol (95) in 12% yield [70% of (88) being recovered] which was identified as 17-octakisnordammarane-3α, 17β-diol by comparison with an authentic sample. Epimerization of the 3α-hydroxyl group to the 3β-configuration also occurred in 5% yield. Since such a reaction was not observed when (88) was irradiated in the presence of methyl *p*-nitrobenzoate, the cleavage seems to be the result of an intramolecular reaction, although the mechanism is obscure.

XI. Other Constituents of Ginseng Roots

1. Ether-Soluble Compounds

From the ether-soluble fraction, TAKAHASHI *et al.* (*134, 135*) isolated sitosterol-β-D-glucoside, sitosterol, β-elemene (96), eremophilene (97) and a new polyacetylenic alcohol named panaxynol (98), the structure of which was elucidated as heptadeca-1,9(*Z*)-diene-4,6-diyn-3-ol by degradative and synthetic procedures. It is identical with carotatoxin from carrots (*136*) and falcarinol from Falcaria vulgaris (*137*). Recently two other polyacetylenic alcohols, 9,10-epoxy-3-hydroxyheptadeca-1-en-4,6-diyne [(99), panaxydol] (*138*) and heptadeca-1-en-4,6-diyn-3,9-diol (100) (*139*) were isolated from Ginseng roots by WROBEL *et al.*

Campesterol, stigmasterol and sitosterol have been identified in Ginseng roots by GLC (*140*) and an analysis of the fatty acids has been reported (*141*).

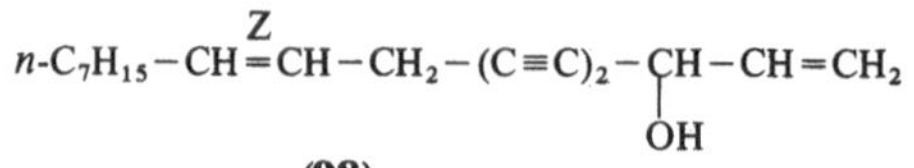

$$n\text{-}C_7H_{15}-CH\overset{Z}{=}CH-CH_2-(C\equiv C)_2-\underset{OH}{CH}-CH=CH_2$$

(98)

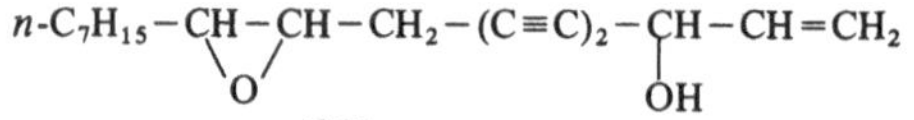

$$n\text{-}C_7H_{15}-\underset{\diagdown O \diagup}{CH-CH}-CH_2-(C\equiv C)_2-\underset{OH}{CH}-CH=CH_2$$

(99)

$$n\text{-}C_8H_{17}-\underset{OH}{CH}-CH_2-(C\equiv C)_2-\underset{OH}{CH}-CH=CH_2$$

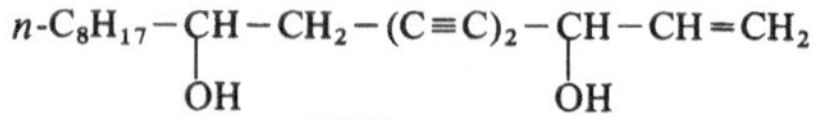

(100)

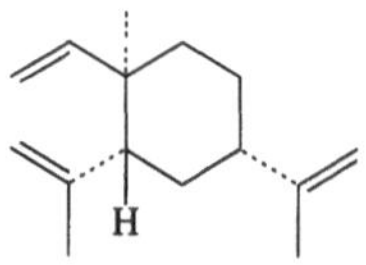

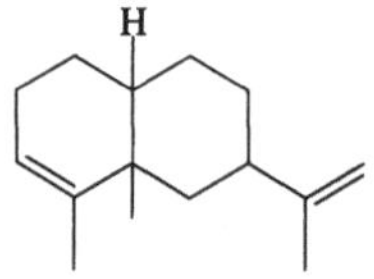

(96)   **(97)**

Chart 21

2. Carbohydrates

Sucrose, fructose and glucose have been isolated from Ginseng roots in yields of 8.5, 0.5 and 0.4% (*142*). As mentioned in Chapter VI-4, the sucrose content of roots collected in winter (20 – 25%) was much higher than in the summer (5 – 6%) (*76*). Three new trisaccharides from the roots were shown to be o-α-D-glucopyranosyl-(1→2)-o-β-D-fructofuranosyl-(1→2)-β-D-fructofuranoside,   o-α-D-glucopyranosyl-(1→6)-o-α-D-glucopyranosyl-(1→4)-o-α-D-glucopyranose and α-maltosyl-β-D-fructofuranoside (*143, 144*). The polysaccharides of Ginseng have also been studied (*145, 146*).

3. Nitrogen-Containing Compounds

In connection with the study on cold storage of Ginseng roots mentioned earlier, the amino acid composition was investigated (*77*). This showed that the arginine content was significant. It is noteworthy that arginine suppresses the unfavorable effect of saponins on rectal mucosa.

Uracil, uridine, adenine, adenosine and guanine have been identified in the roots (*147*). Choline was isolated from the roots as the reineckate in 0.55% yield (*148*) and α-pyrrolidone was identified by GLC analysis (*149*).

Gstirner and Vogt described isolation and amino acid composition of peptides of Ginseng roots (*150, 151*) and recently Okuda *et al.*, reported presence of an anti-lipolytic peptide like compound in Red Ginseng (*152*). Structures of these compounds are still unknown. Kosuge *et al.* recently isolated the neurotoxin β-N-oxalo-L-α,β-diaminopropionic acid as the hemostatic principle of Sanchi-Ginseng (*153*).

Table XIII. *Amino Acid Contents (ng/mg) in Methanolic Extract of Roots of Panax ginseng*

	Total[a]		Free amino acid	
	A[b]	B[c]	A[b]	B[c]
Lysine	876	876	380	512
Histidine	2,433	1,907	248	403
Arginine	37,410	28,014	22,890	27,593
Aspartic acid	5,320	4,150	1,025	998
Threonine	1,690	941	1,930	633
Serine	1,344	1,281	1,335	—
Glutamic acid	14,053	10,158	1,265	2,545
Proline	421	432	368	242
Glycine	1,320	855	128	150
Alanine	3,702	4,379	3,163	3,394
Cysteine	73	—	—	—
Valine	1,580	952	340	902
Methionine	359	359	—	—
Isoleucine	1,821	1,087	512	1,012
Leucine	1,755	1,611	603	840
Tyrosine	977	905	544	417
Phenylalanine	941	990	380	463
Total (mg/mg)	0.0761	0.0590	0.0351	0.0401

[a] Amino acids after hydrolyzed with 6N HCl at 110° for 15 hr (free form + combined form).
[b] Fresh roots harvested in August.
[c] Roots[b] treated at $2° \pm 1°$ for 30 days.

XII. Pharmaceutical Studies of Ginseng Saponins

As already mentioned in Chapters II-2 and III-1, the glycosyl linkage on the 20-hydroxyl group of dammarane type saponins is unstable and is readily hydrolyzed to yield a mixture of prosapogenin c-20-epimers. In an effort to mimic the way in which the saponins might be affected by gastric juice, the ginsenosides were exposed to 0.1 N HCl at 37° for 2 hr (*154*). As a result of this treatment, Rg_1 afforded c-20-epimeric mixtures of prosapogenins, (**101** = Rh_1) and (**102**) (see Chart 22) whose structures were determined by the ^{13}C-NMR spectrometry (see Chapter IV-2,3, Table IV). According to a private communication from Pro. J. SHOJI, Showa University, dammarane saponins undergo a similar degradation during the decoction of Red Ginseng with boiling water.

Synthesis of ^{14}C-labeled Rg_1 for metabolic studies has been reported (*155*) and several workers have examined the absorption, and distribution of Rg_1 in the rat (*156, 157*).

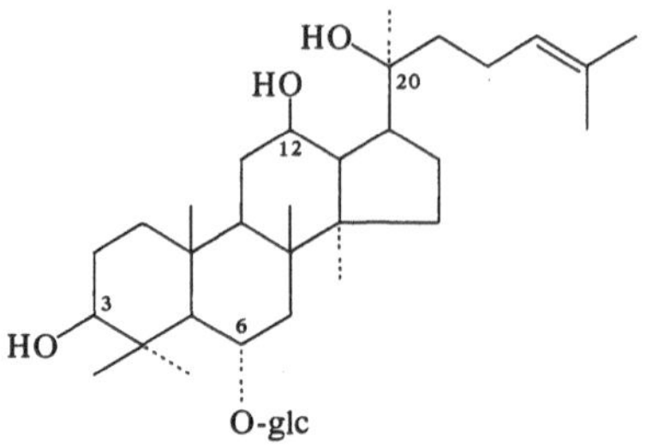

O-glc
Rh$_1$ (101)

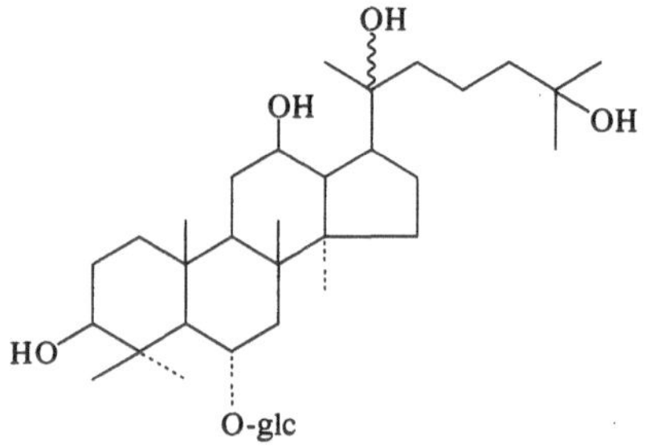

O-glc
(102)

Chart 22

XIII. Biological Activities of Ginseng

Ginseng is generally known as a tonic. However, the Chinese classics dealing with traditional medicine emphasize that the drug should be used for increasing mental efficiency, recovering physical balance and stimulating metabolic function in man. Since BRECKHMAN (*158*) and PETKOV (*159, 160*) earlier summarized the pharmacology of Ginseng, a multitude of papers on various biological activities of Ginseng and its congeners have been published by Japanese, Korean, Chinese, American, and European pharmacologists and biochemists. A listing of the observed activities of Ginseng or its crude extracts would include effects on the central nervous system, tranquillizing action, histamine-like action, blood pressure elevation, serotonine-like action, analgesic and antipyretic action, antihistamine-like action, papaverine-like action, ganglion stimulant action, protection against physical and chemical stress, anti-cancer activity and others. Because it is very difficult for the authors who are phytochemists to cover all of these papers and to summarize and evaluate their results within a few pages, the present chapter is confined to a description of some recent work mainly by Japanese scientists on the activities of partially or completely purified saponins. For a complete survey of the literature of this field, the introduction in the reference section on page 65 should be consulted.

Rb$_1$, a representative of the saponins derived from 20(*S*)-proto-panaxadiol (**26**) (Table I) reportedly exhibited central nervous system-depressant and antipsychotic activity, protection of stress ulcer, increase of gastrointestinal motility and weak anti-inflammatory action, while Rg$_1$, the major saponin of 20(*S*)-protopanaxatriol (**36**) (Table II) showed weak central nervous system-stimulant action, antifatigue action and aggravation of stress ulcer (*161, 162, 163, 164, 165, 166*). KAKU *et al.* (*167*), on investigating the pharmacology of the major saponins, reported the following results: diminution of acetylcholine-induced contraction of isolated guinea pig ileum, inducement in rats of a decrease in heart rate and biphasic action on the blood pressure especially for Rg_1, vasodilator action of Rg_1 and Re, facilitation of the conditioned avoidance response by repeated administration of Rf, Re and Rd, significant suppression in mice of fighting induced by foot shock by Rg_1, Rf, Re and Rd (Rb_1, Rb_2 and Rc exhibited little effect), antifatigue action for most of the saponins, etc.

Potentiation of the nerve growth factor (NGF) by Rb_1, Rd and some other saponins of (26) has been observed (*168*). Rg_1, the saponin of (36) showed no activity.

Effects of saponins on nucleic acid biosynthesis and serum protein metabolism have been extensively studied (*169 – 182*). The presence in Ginseng of substances with adenosine triphosphate activity has been reported (*183*). Effects by the saponins on cholesterol metabolism and that of other lipids have been described (*184 – 188*).

Stimulation of the pituitary-adrenocortical system by the saponins has been recently observed (*189, 190*) and the effect of the saponins on the action of adrenalin, ACTH and insulin in lipolysis and lipogenesis were investigated (*191*). Chemopharmacological studies of the hypoglycemic principles in Ginseng have been carried out (*192, 193*).

Haemolytic activities of Ginseng saponins have been reported (*194, 195*) and cytotoxic activities of dammarane type triterpenes have been investigated (*196*). Of interest also is a Ginseng abuse syndrome (*197*) which was commented upon at the third International Ginseng Symposium in Seoul, Korea.

Pharmacological activities of the callus (*198*) and leaves (*199*) of *Panax ginseng* and pharmacological properties of saponins of *Panax japonicus* (Chikusetsu-ninjin) have also been described (*200, 201*).

References

For a complete survey of the literature on cultivation, chemistry, pharmacology and related matters regarding to Ginseng and related crude drugs, see "Abstracts of Korean Ginseng Studies (1687 – 1975)" published by the Research Institute, Office of Monopoly, Republic of Korea; 112 Ini-Dong, Chongro-Ku, Seoul. S. FULDER "The Root of Being. Ginseng and the Pharmacology of Harmony" Hutchinson and Co. Publ. Ltd. London, Melbourne, Sydney, Auckland, Johannesburg: 1980, is also a valuable reference.

1. GARRIQUES, S.: On Panaquilon, a New Vegetable Substance. Ann. Chem. Pharm. **90**, 231 (1854).
2. ASAHINA, Y., and B. TAGUCHI: The Constituents of Ginseng. Yakugaku Zasshi (J. Pharm. Soc. Japan) **26**, 549 (1906).
3. KONDO, H., and O. TANAKA: Constituents of the Korean Ginseng III. Yakugaku Zasshi **40**, 1027 (1920).

4. KOTAKE, M.: Glucosides I. Glucoside of *Panax ginseng*. Nippon Kagaku Zasshi (J. Chem. Soc. Japan) **51**, 357 (1930).
5. WAGNER-JAUREGG, T., and M. ROTH: Panaxol, a New Constituent of Red Ginseng Root. Pharm. Acta Helv. **37**, 352 (1962).
6. HÖRHAMMER, L., H. WAGNER, and B. LAY: Contents of *Panax ginseng* Root. Preliminary Report. Pharm. Ztg., Ver. Apoth. Ztg. **106**, 1307 (1961).
7. SHIBATA, S., M. FUJITA, H. ITOKAWA, O. TANAKA, and T. ISHII: Studies on the Constituents of Japanese and Chinese Crude Drugs XI. Panaxadiol, a Sapogenin of Ginseng Roots (1). Chem. Pharm. Bull. (Japan) **11**, 759 (1963).
8. SHIBATA, S., O. TANAKA, M. NAGAI, and T. ISHII: The Stereochemistry of Panaxadiol. Tetrahedron Letters **1962**, 1239; *idem.*: Studies on the Constituents of Japanese and Chinese Crude Drugs XII. Panaxadiol, a Sapogenin of Ginseng Roots (2). Chem. Pharm. Bull. (Japan) **11**, 762 (1963).
9. SHIBATA, S., O. TANAKA, M. SADO, and S. TSUSHIMA: On the Genuine Sapogenin of Ginseng. Tetrahedron Letters **1963**, 795.
10. SHIBATA, S., O. TANAKA, T. ANDO, M. SADO, S. TSUSHIMA, and T. OHSAWA: Chemical Studies on the Oriental Plant Drugs XIV. Protopanaxadiol, a Genuine Sapogenin of Ginseng Saponins. Chem. Pharm. Bull. (Japan) **14**, 595 (1966).
11. TANAKA, O., M. NAGAI, and S. SHIBATA: The Stereochemistry of Protopanaxadiol, a Genuine Sapogenin of Ginseng. Tetrahedron Letters **1964**, 2291; *idem.*: Chemical Studies on the Oriental Plant Drugs XVI. The Stereochemistry of Protopanaxadiol, a Genuine Sapogenin of Ginseng. Chem. Pharm. Bull. (Japan) **14**, 1150 (1966).
12. NAGAI, M., O. TANAKA, and S. SHIBATA: The Stereochemistry of Protopanaxadiol. The Absolute Configuration of C(20) of Dammarenediols-I and -II. Tetrahedron Letters **1966**, 4797; *idem.*: Chemical Studies on the Oriental Plant Drugs XXVI. Saponins and Sapogenins of Ginseng. The Absolute Configurations of Cinenic Acid and Panaxadiol. Chem. Pharm. Bull. (Japan) **19**, 2349 (1971).
13. FISCHER, F. G., and N. SEILER: Die Triterpenealkohole der Birkenblätter. Liebigs Ann. Chem. **644**, 146 and 162 (1961).
14. SHIBATA, S., M. NAGAI, and O. TANAKA: On the Constituents of the Leaves of *Betula platyphylla* Sukatchev var. *japonica* Hara. Shoyakugaku Zasshi **18**, 27 (1964).
15. BIELMANN, J.-F.: Chiralite du Dipterocarpol en C-20. Tetrahedron Letters **1966**, 4803; *idem.*: Configuration en C-20 du Dipterocarpol. Bull. Soc. Chim. France **1967**, 3459.
16. TANAKA, O., M. NAGAI, T. OHSAWA, N. TANAKA, and S. SHIBATA: Stereochemistry of Protopanaxadiol. Acid Catalyzed Epimerization of C-20 Hydroxyl of Betulafolienetriol, Protopanaxadiol, and Their Derivatives. Tetrahedron Letters **1967**, 391.
17. TANAKA, O., M. NAGAI, T. OHSAWA, N. TANAKA, K. KAWAI, and S. SHIBATA: Chemical Studies on the Oriental Plant Drugs XXVII. The Acid Catalyzed Reaction and the Absolute Configuration at C(20) of Dammarane Type Triterpenes. Chem. Pharm. Bull. (Japan) **20**, 1204 (1972).
18. SAKAMOTO, I., K. MORIMOTO, and O. TANAKA: Quantitative Analysis of Dammarane Type Saponins of Ginseng and its Application to the Evaluation of the Commercial Ginseng Tea and Ginseng Extracts. Yakugaku Zasshi **95**, 1456 (1975).
19. NAGAI, M., T. ANDO, N. TANAKA, O. TANAKA, and S. SHIBATA: Chemical Studies on the Oriental Plant Drugs XXVIII. Saponins and Sapogenins of Ginseng: Stereochemistry of the Sapogenin of Ginsenosides-Rb$_1$, -Rb$_2$ and -Rc. Chem. Pharm. Bull. (Japan) **20**, 1212 (1972).
20. GOLDSTEIN, I. J., G. W. HAY, B. A. LEWIS, and F. SMITH: Controlled Degradation of Polysaccharides by Periodate Oxidation, Reduction, and Hydrolysis. Methods in Carbohydr. Chem. **5**, 361. New York, San Francisco, London: Academic Press. 1965.
21. NAGAI, M., T. ANDO, O. TANAKA, and S. SHIBATA: 20-Epiprotopanaxadiol, a Genuine Sapogenin of Ginsenosides-Rb$_1$, -Rb$_2$ and -Rc. Tetrahedron Letters **1967**, 3579.

22. TANAKA, O., N. TANAKA, T. OHSAWA, Y. IITAKA, and S. SHIBATA: Stereochemistry of the Side Chain of Dammarane Type Triterpenes. Tetrahedron Letters **1968**, 4235.

23. OHSAWA, T., N. TANAKA, O. TANAKA, and S. SHIBATA: Chemical Studies on the Oriental Plant Drugs XXIX. Saponins and Sapogenins of Ginseng: Further Study on the Chemical Properties of the Side Chain of Dammarane Type Triterpenes. Chem. Pharm. Bull. (Japan) **20**, 1890 (1972).

24. SHIBATA, S., T. ANDO, and O. TANAKA: Chemical Studies on the Oriental Plant Drugs XVII. The Prosapogenin of the Ginseng Saponins (Ginsenosides-Rb$_1$, -Rb$_2$, and -Rc). Chem. Pharm. Bull. (Japan) **14**, 1157 (1966).

25. HAKOMORI, S.: A Rapid Permethylation of Glycolipid and Polysaccharide Catalyzed by Methylsulfinyl Carbanion in Dimethyl Sulfoxide. J. Biochemistry (Tokyo) **55**, 205 (1964).

26. KAKU, T., and Y.KAWASHIMA: Isolation and Characterization of Ginsenoside-Rg$_3$, 20(R)-Prosapogenin, 20(S)-Prosapogenin and Δ^{20}-Prosapogenin. Arzneim.-Forsch. (Drug Res.) **30**, 936 (1980).

27. ARMOUR, C., C. A. BUNTON, S. PATAI, L. H. SELMAN, and C. A. VERNON: Mechanism of Reaction in the Sugar Series. Part III. The Acid-Catalyzed Hydrolysis of t-Butyl-β-D-Glucopyranoside and Other Glycosides. J. Chem. Soc. (London) **1961**, 412.

28. COCKER, D., and M. L. SINNOTT: Steric Acceleration in the Acid-catalyzed Hydrolysis of 1-Adamantyl-β-D-Glucopyranoside. The Origin of the High Rates of Hydrolysis of Tertiary Glycosides. J. C. S. Chem. Comm. **1972**, 414.

29. IIDA, Y., O. TANAKA, and S. SHIBATA: Studies on Saponins of Ginseng: The Structure of Ginsenoside-Rg$_1$. Tetrahedron Letters **1968**, 5449; NAGAI (n'ee IIDA), Y., O. TANAKA, and S. SHIBATA: Chemical Studies on the Oriental Plant Drugs XXIV. Structure of Ginsenoside-Rg$_1$, a Neutral Saponin of Ginseng Roots. Tetrahedron **27**, 881 (1971).

30. SANADA, S., N. KONDO, J. SHOJI, O. TANAKA, and S. SHIBATA: Studies on the Saponins of Ginseng. I. Structures of Ginsenosides-Ro, -Rb$_1$, -Rb$_2$, -Rc and -Rd. Chem. Pharm. Bull. (Japan) **22**, 421 (1974).

31. SANADA, S., and J. SHOJI: Studies on the Saponins of Ginseng III. Structures of Ginsenoside-Rb$_3$ and 20-Gluco-ginsenoside-Rf. Chem. Pharm. Bull. (Japan) **26**, 1694 (1976).

32. SANADA, S., K. KONDO, J. SHOJI, O. TANAKA, and S. SHIBATA: Studies on the Saponins of Ginseng II. Structures of Ginsenosides-Re, -Rf and Rg$_2$. Chem. Pharm. Bull. (Japan) **22**, 2407 (1974).

33. SHIBATA, S., O. TANAKA, K. SOMA, Y. IIDA, T. ANDO, and H. NAKAMURA: Studies on Saponins and Sapogenins of Ginseng. The Structure of Panaxatriol. Tetrahedron Letters **1965**, 207.

34. KOMORI, T., O. TANAKA, and Y. NAGAI: Studien über die Saponine aus der Arznei-Ginseng-Wurzel: Massenspektren von Ginsenoside-Rg$_1$-decaacetat und verwandten Verbindungen. Org. Mass Spectrom. **9**, 744 (1974).

35. ELYAKOV, G. B., L. I. STRINGINA, E. V. SHAPKINA, N. T. ALADYINA, S. A. KORNILOVA, and A. K. DZIZENKO: The Probable Structure of the True Aglycones of Ginseng Glycosides. Tetrahedron **24**, 5483 (1968).

36. HAN, B., and Y. HAN: Partial Structure of *Panax* Saponin C. Kor. J. Pharmacog. **3**, 211 (1972).

37. YOSIOKA, I., M. FUJIO, M. OSAMURA, and I. KITAGAWA: A Novel Cleavage Method of Saponin with Soil Bacteria, Intending to the Genuine Sapogenin: On *Senega* and *Panax* Saponins. Tetrahedron Letters **1966**, 6303.

38. YOSIOKA, I., T. SUGAWARA, K. IMAI, and I. KITAGAWA: Soil Bacterial Hydrolysis Leading to Genuine Aglycone V. On Ginsenosides-Rb$_1$, -Rb$_2$ and -Rc of the Ginseng Root Saponins. Chem. Pharm. Bull. (Japan) **20**, 2418 (1972).

39. KOHDA, H., and O. TANAKA: Enzymic Hydrolysis of Ginseng Saponins and Their Related Glycosides. Yakugaku Zasshi **95**, 246 (1975).

40. Okada, S., K. Kishi, M. Higashihara, and J. Fukumoto: Studies on Flavanoid-hydrolyzing Enzymes Part I. Crystallization of Naringinase I and Hesperidinase I and Their Actions. Nippon Nogeikagaku Kaishi (J. Agric. Chem. Soc. Japan) **37**, 84 (1963).

41. — — — — Part II. Substrate Specificities of Naringinase I and Hesperidinase I. Nippon Nogeikagaku Kaishi (J. Agric. Chem. Soc. Japan) **37**, 142 (1963).

42. Sakamoto, I., H. Kohda, K. Murakami, and O. Tanaka: Quantitative Analysis of Stevioside. Yakugaku Zasshi **95**, 1507 (1975).

43. Asakawa, J., R. Kasai, K. Yamasaki, and O. Tanaka: ^{13}C-NMR Study of Ginseng Sapogenins and Their Related Dammarane Type Triterpenes. Tetrahedron **33**, 1935 (1977).

44. Beierbeck, H., and J. K. Saunders: A Reinterpretation of beta, gamma, and delta Substituent Effects on ^{13}C-Chemical Shifts. Canad. J. Chem. **54**, 2985 (1976).

45. Kasai, R., M. Suzuo, J. Asakawa, and O. Tanaka: Carbon-13 Chemical Shifts of Isoprenoid-β-D-glucopyranosides and -β-D-mannopyranosides. Stereochemical Influences of Aglycone Alcohols. Tetrahedron Letters **1977**, 175.

46. Kasai, R., M. Okihara, J. Asakawa, K. Mizutani, and O. Tanaka: ^{13}C-NMR Study of α- and β-Anomeric Pairs of D-Mannopyranosides and L-Rhamnopyranosides. Tetrahedron **35**, 1427 (1979).

47. Mizutani, K., R. Kasai, and O. Tanaka: ^{13}C-NMR Spectroscopy of α- and β-Anomeric Series of Alkyl L-Arabinopyranosides. Carbohydr. Res. **87**, 19 (1980).

48. Zhou, J., M. Wu, S. Taniyasu, H. Besso, O. Tanaka, Y. Saruwatari, and T. Fuwa: Dammarane-saponins of Sanchi-Ginseng, Roots of *Panax notoginseng* (Burk.) F. H. Chen (Araliaceae): Structures of New Saponins, Notoginsenosides-R1 and -R2, and Identification of Ginsenosides-Rg$_2$ and -Rh$_1$. Chem. Pharm. Bull. (Japan) **29**, 2844 (1981).

49. Kasai, R., K. Matsuura, O. Tanaka, S. Sanada, and J. Shoji: Mass Spectra of Trimethylsilyl Ethers of Dammarane-type Ginseng-sapogenins and Their Related Compounds.Chem. Pharm. Bull. (Japan) **25**, 3277 (1977).

50. Komori, T., M. Kawamura, K. Miyahara, T. Kawasaki, O. Tanaka, and S. Yahara: Field Desorption Mass Spectrometry of Physiologically Active Steroid- and dammarane-saponins. Z. Naturforsch. **34c**, 1094 (1979).

51. Schulten, H. R., and F. Soldati: Identification of Ginsenosides from *Panax ginseng* in Fractions Obtained by High-performance Liquid Chromatography by Field Desorption Mass Spectrometry, Multiple Internal Reflection Infrared Spectroscopy and Thin-layer Chromatography. J. Chromatogr. **212**, 37 (1981).

52. Besso, H., R. Kasai, Y. Saruwatari, T. Fuwa, and O. Tanaka: Ginsenoside-Ra$_1$ and Ginsenoside-Ra$_2$, New Dammarane-Saponins of Ginseng Roots. Chem. Pharm. Bull. (Japan) **30**, 2380 (1982).

53. Koizumi, H., S. Sanada, Y. Ida, and J. Shoji: Studies on the Saponins of Ginseng IV. On the Structure and Enzymic Hydrolysis of Ginsenoside-Ra$_1$. Chem. Pharm. Bull. (Japan) **30**, 2393 (1982).

54. Matsuura, H., R. Kasai, Y. Saruwatari, T. Fuwa, and O. Tanaka: Ginsenoside-Ra$_3$, a Minor Saponins of Ginseng Roots. Chem. Pharm. Bull. (Japan) **32**, 1188 (1984).

55. Yahara, S., K. Kaji, and O. Tanaka: Further Study on Dammarane Type Saponins of Roots, Leaves, Flower buds, and Fruits of *Panax ginseng* C. A. Meyer. Chem. Pharm. Bull. (Japan) **27**, 88 (1979).

56. Kasai, R., H. Besso, O. Tanaka, Y. Saruwatari, and T. Fuwa: Saponins of Red Ginseng. Chem. Pharm. Bull. (Japan) **31**, 2120 (1983).

57. Yoshimoto, K., Y. Itatani, and Y. Tsuda: ^{13}C-Nuclear Magnetic Resonance (NMR) Spectra of O-Acylglucoses. Additivity of Shift Parameters and its Application to Structure Elucidation. Chem. Pharm. Bull. (Japan) **28**, 2065 (1980).

58. Woo, L., B. Han, D. Baik, and D. Park: Characterization of Ginseng Extracts. Yakhak Hoeji 17, 129 (1973).

59. Hiai, S., H. Oura, H. Hamanaka, and Y. Odaka: A Color Reaction of Panaxadiol with Vanillin and Sulfuric Acid. Planta Medica 28, 131 (1975).

60. Hiai, S., H. Oura, Y. Odaka, and T. Nakajima: A Colorimetric Estimation of Ginseng Saponins. Planta Medica 28, 363 (1975).

61. Hiai, S., H. Oura, and T. Nakajima: Color Reaction of Some Sapogenins and Saponins with Vanillin and Sulfuric Acid. Planta Medica 29, 116 (1976).

62. Saruwatari, Y., H. Besso, K. Futamura, T. Fuwa, and O. Tanaka: Thin Layer Chromatographic Determination of Panaxadiol and Panaxatriol by Ultraviolet Derivatization. Chem. Pharm. Bull. (Japan) 27, 147 (1979).

63. Sanada, S., J. Shoji, and S. Shibata: Quantitative Analysis of Ginseng Saponins. Yakugaku Zhassi 98, 1048 (1978).

64. Tanaka, O.: Steviol-glycosides: New Natural Sweeteners. Trends Anal. Chem. 1, 246 (1982).

65. Lee, T. M., and A. D. Marderosian: Two-dimensional TLC Analysis of Ginsenosides from Root of Dwarf Ginseng (Panax trifolius L.) Araliaceae. J. Pharm. Sci. 70, 89 (1981).

66. Nagasawa, T., T. Yokosawa, Y. Nishino, and H. Oura: Application of High Performance Liquid Chromatography to the Isolation of Ginsenosides-Rb_1, -Rb_2, -Rc, - Rd, -Re and -Rg_1 from Ginseng Saponins. Chem. Pharm. Bull. (Japan) 28, 2059 (1980).

67. Nagasawa, T., J. H. Choi, Y. Nishino, and H. Oura: Application of High Performance Liquid Chromatography to the Isolation of Ginsenosides-Rf, -Rg_2 and -Rh_1 from a Crude Saponin Mixture of Ginseng. Chem. Pharm. Bull. (Japan) 28, 3701 (1980).

68. Sticher, O., and F. Soldati: HPLC Trennung und quantitative Bestimmung der Ginsenoside von Panax ginseng, Panax quinquefolium und von Ginseng-Spezialitäten. 1. Mitteilung. Planta Medica 36, 30 (1979).

69. Soldati, F., and O. Sticher: HPLC Separation and Quantitative Determination of Ginsenosides from Panax ginseng and from Ginseng Drug Preparations, 2nd Communication. Planta Medica 39, 348 (1980).

70. Besso, H., Y. Saruwatari, K. Futamura, K. Kunihiro, T. Fuwa, and O. Tanaka: High Performance Liquid Chromatographic Determination of Ginseng Saponin by Ultraviolet Derivatisation. Planta Medica 37, 226 (1979).

71. Kunihiro, K., H. Yamaguchi, and S. Hara: Retention Behavior of Panax ginseng Sapogenins in Silica Gel Liquid-solid Chromatography. Bunseki Kagaku 31, 83 (1982).

72. Kaizuka, H., and K. Takahashi: A New Separation Method for Wide Varieties of Naturally Occurring Glycosides. J. Chromatogr. 258, 135 (1983).

73. Sankawa, U., C. Sung, B. Han, T. Akiyama, and K. Kawashima: Radioimmunoassay for the Determination of Ginseng Saponin, Ginsenoside-Rg_1. Chem. Pharm. Bull. (Japan) 30, 1907 (1982).

74. Kubo, M., T. Tani, T. Katsuki, K. Ishizaki, and S. Arichi: Histochemistry. I. Ginsenosides in Ginseng (Panax ginseng C. A. Meyer Root). J. Natural Products 43, 278 (1980).

75. Otsuka, H., T. Morita, Y. Ogihara, and S. Shibata: The Evaluation of Ginseng and its Congeners by Droplet Counter-current Chromatography (DCC). Planta Medica 32, 9 (1977).

76. Kim, S., I. Sakamoto, K. Morimoto, M. Sakata, K. Yamasaki, and O. Tanaka: Seasonal Variation of Saponins, Sucrose and Monosaccharides in Cultivated Ginseng Roots. Planta Medica 42, 181 (1981).

77. Kim, S., C. Hiyama, K. Yamasaki, R. Hiraoka, O. Tanaka, J. Kim, and I. Kim: Content Variation of Constituents by Cold Treatment of Fresh Root of Panax ginseng C. A. Meyer. Shoyakugaku Zasshi 33, 245 (1979).

78. Furuya, T., H. Kojima, K. Syono, T. Ishii, K. Uotani, and M. Nishino: Isolation of

Saponins and Sapogenins from Callus Tissue of *Panax ginseng*. Chem. Pharm. Bull. (Japan) **21**, 98 (1973).

79. HARA, H.: On the Asiatic Species of the Genus *Panax*. J. Jap. Bot. **45**, 197 (1970).

80. ZHOU, J., W. HUANG, M. WU, T. YANG, G. FENG, and C. WU: Triterpenoids from *Panax* Linn. and Their Relationship with Taxonomy and Geographical Distribution. Acta Phytotaxonomica Sinica **13**, 29 (1975).

81. SHIBATA, S., T. ANDO, O. TANAKA, Y. MEGURO, K. SOMA, and Y. IIDA: Saponins and Sapogenins of *Panax ginseng* C. A. Meyer and Some *Panax* spp. Yakugaku Zasshi **85**, 7753 (1965).

82. ANDO, T., O. TANAKA, and S. SHIBATA: Chemical Studies on the Oriental Plant Drugs XXV. Comparative Studies on the Saponins and Sapogenins of Ginseng and Related Crude Drugs. Syoyakugaku Zasshi **25**, 28 (1971).

83. KIM, J., and E. J. STABA: Studies on the Ginseng Plants (1). Saponins and Sapogenins from American Ginseng. Korean J. Pharmacog. **4**, 193 (1973).

84. LUI, J. H., and E. J. STABA: The Ginsenosides of Various Ginseng Plants and Selected Products. J. Natural Products **43**, 340 (1980).

85. SANADA, S., and J. SHOJI: Comparative Studies on the Saponins of Ginseng and Related Crude Drugs (I). Shoyakugaku Zasshi **32**, 96 (1978).

86. BESSO, H., R. KASAI, J. WEI, J. WANG, Y. SARUWATARI, T. FUWA, and O. TANAKA: Further Studies on Dammarane Saponins of American Ginseng, Roots of *Panax quinquefolium* L. Chem. Pharm. Bull. (Japan) **30**, 4534 (1982).

87. TAKEMOTO, T., S. ARIHARA, T. NAKAJIMA, and M. OKUHIRA: Studies on the Constituents of *Gymnostemma Pentaphyllum* Makino I. Structures of Gypenoside I – XIV. Yakugaku Zasshi **103**, 173 (1983).

88. WU, M.: Studies on the Saponins Components of Plants in Yunnan IV. Two Saponins of *Panax notoginseng*. Acta Botanica Yunnanica **1**, 119 (1979).

89. WEI, J., J. WANG, L. CHANG, and Y. DU: Chemical Studies of San-chi, *Panax notoginseng* (Burk.) F. H. Chen I. Studies on the Constituents of San-chi Root Hairs. Acta Pharm. Sinica **15**, 359 (1980).

90. KOCH, H. J., and R. S. STUART: The Synthesis of Per-C-Deuterated D-Glucose. Carbohydr. Res. **64**, 127 (1978).

91. MIZUTANI, K., H. KAJITA, T. TASHIMA, and O. TANAKA: Studies on ^{13}C-NMR Spectroscopy of Carbohydrates, Application of Selective Deuteration. Nippon Kagaku Kaishi (J. Chem. Soc. Japan) **1982**, 1595.

92. MATSUURA, H., R. KASAI, O. TANAKA, Y. SARUWATARI, T. FUWA, and J. ZHOU: Further Studies on Dammarane Saponins of Sanchi-Ginseng. Chem. Pharm. Bull. (Japan) **31**, 2281 (1983).

93. AOYAMA, S.: Saponin of *Panax repens* Maxim. IV. Panaxsapogenin. Yakugaku Zasshi **50**, 1065.

94. — Saponin of *Panax repens* Maxim. V. Panaxsapogenin. Yakugaku Zasshi **50**, 1163 (1930).

95. KITASATO, Z., and G. SONE: Constitution of Hederagenin and Oleanolic Acid I. Acta Phytochim. (Tokyo) **6**, 179 (1932).

96. MURAYAMA, Y., and T. ITAGAKI: The Saponin of *Panax repens* Maxim. I. Yakugaku Zasshi **43**, 783 (1923).

97. MURAMAYA, Y., and S. TANAKA: Saponin of *Panax repens* Maxim. II. Yakugaku Zasshi **47**, 526 (1927).

98. KOTAKE, M., and Y. KIMOTO: Saponin Series III. Sapogenins of *Panax repens* Maxim. and *Aralia chinensis* L. Sci. Papers Inst. Phys. Chem. Res. (Tokyo) **18**, 83 (1932).

99. KONDO, N., and J. SHOJI: Studies on the Constituents of Panacis japonici Rhizoma I. Isolation and Purification of the Saponins (1). Yakugaku Zasshi **88**, 325 (1968).

100. KONDO, N., J. SHOJI, N. NAGUMO, and N. KOMATSU: Studies on the Constituents of

Panacis japonici rhizoma II. The Structure of Chikusetsusaponin IV and Some Observations on the Structural Relationship with Araloside A. Yakugaku Zasshi **89**, 846 (1969).

101. KONDO, N., K. AOKI, H. OGAWA, R. KASAI, and J. SHOJI: Studies on the Constituents of Panacis japonici Rhizoma III. The Structure of Chikusetsusaponin III. Chem. Pharm. Bull. (Japan) **18**, 1558 (1970).

102. KONDO, N., Y. MARUMOTO, and J. SHOJI: Studies on the Constituents of Panacis japonici Rhizoma IV. The Structure of Chikusetsusaponin V. Chem. Pharm. Bull. (Japan) **19**, 1103 (1971).

103. LIN, T. D., N. KONDO, and J. SHOJI: Studies on the Constituents of Panacis japonici Rhizoma V. The Structures of Chikusetsusaponin I, Ia, Ib, IVa and Glycoside P_1. Chem. Pharm. Bull. (Japan) **24**, 253 (1976).

104. MORITA, T., R. KASAI, H. KOHDA, O. TANAKA, J. ZHOU, and T. YANG: Chemical and Morphological Study on Chinese *Panax japonicus* C. A. Meyer (Zhujie-shen). Chem. Pharm. Bull. (Japan) **31**, 3205 (1983).

105. MORITA, T., R. KASAI, O. TANAKA, J. ZHOU, T. YANG, and J. SHOJI: Saponins of Zu-tiseng, Rhizomes of *Panax japonicus* C. A. Meyer var. *major* (Burk.) C. Y. Wu et K. M. Feng, Collected in Yunnan, China. Chem. Pharm. Bull. (Japan) **30**, 4341 (1982).

106. YANG, T., Z. JIANG, M. WU, J. ZHOU, and O. TANAKA: Studies on Saponins of Rhizomes of *Panax zingiberensis* Wu et Feng. Acta Pharm. Sinica **19**, 232 (1984).

107. KONDO, N., J. SHOJI, and O. TANAKA: Studies on the Constituents of Himalayan Ginseng, *Panax pseudo-ginseng* I. The Structures of the Saponins (1). Chem. Pharm. Bull. (Japan) **21**, 2702 (1973).

108. KONDO, N., and J. SHOJI: Studies on the Constituents of Himalayan Ginseng, *Panax pseudo-ginseng* II. The Structure of the Saponins (2). Chem. Pharm. Bull. (Japan) **23**, 3282 (1975).

109. YAHARA, S., O. TANAKA, and T. KOMORI: Saponins of the Leaves of *Panax ginseng* C. A. Meyer. Chem. Pharm. Bull. (Japan) **24**, 2204 (1976).

110. YAHARA, S., K. MATSUURA, R. KASAI, and O. TANAKA: Saponins of Buds and Flowers of *Panax ginseng* C. A. Meyer (1). Isolation of Ginsenosides-Rd, -Re, and -Rg$_1$. Chem. Pharm. Bull. (Japan) **24**, 3212 (1976).

111. TANAKA, O., and S. YAHARA: Dammarane Saponins of Leaves of *Panax pseudo-ginseng* subsp. *himalaicus*. Phytochem. **17**, 1353 (1978).

112. KOCHETKOV, N. K., O. S. CHIZHOV, and N. V. MOLODTSOV: Mass Spectrometry of Oligosaccharides. Tetrahedron **24**, 5587 (1968).

113. NAGAI, M., N. TANAKA, O. TANAKA, and S. ICHIKAWA: Triterpenes of *Betula platyphylla* var. *japonica* and the Configuration at C-24 of Ocotillol-II and its Related Compounds. Tetrahedron Letters **1968**, 4239; *idem.:* Triterpenes of Japanese white birch and the configuration at C-24 of ocotillol II and its related compounds. Chem. Pharm. Bull. (Japan) **21**, 2061 (1973).

114. RAO, M. M., H. MESHULAM, R. ZELNIK, and D. LAVIE: *Cabrarea eichleriana* DC. (Meliaceae)-I, Structure and Stereochemistry of Wood Extractives. Tetrahedron **31**, 333 (1975).

115. YOSIOKA, I., H. YAMAUCHI, and I. KITAGAWA: Lichen Triterpenoids V. On the Neutral Triterpenoids of *Pyxine endochrysina* Nyl. Chem. Pharm. Bull. (Japan) **20**, 502 (1972).

116. CHEN, S. E., and E. J. STABA: American Ginseng I. Large Scale Isolation of Ginsenosides from Leaves and Stems. Lloydia **41**, 361 (1978).

117. CHEN, S. E., E. J. STABA, S. TANIYASU, R. KASAI, and O. TANAKA: Further Study on Dammarane Saponins of Leaves and Stems of American Ginseng, *Panax quinquefolium*. Planta Medica **42**, 406 (1981).

118. YANG, T., R. KASAI, J. ZHOU, and O. TANAKA: Dammarane Saponins of Leaves and Seeds of *Panax notoginseng*. Phytochem. **22**, 1473 (1983).

119. Taniyasu, S., O. Tanaka, T. Yang, and J. Zhou: Dammarane Saponins of Flower Buds of *Panax notoginseng* (Sanchi-Ginseng). Planta Medica **44**, 124 (1982).

120. Yang, T., R. Kasai, J. Zhou, and O. Tanaka: Acta Botanica Yunnanica **6**, 118 (1984).

121. Yahara, S., R. Kasai, and O. Tanaka: New Dammarane Type Saponins of Leaves of *Panax japonicus* C. A. Meyer. (1). Chikusetsusaponin-L5, -L9a and -L10. Chem. Pharm. Bull. (Japan) **25**, 2041 (1977).

122. Yahara, S., O. Tanaka, and I. Nishioka: Dammarane Type Saponins of Leaves of *Panax japonicus* C. A. Meyer. (2). Saponins of the Specimens Collected in Tottori-ken Kyoto-shi, and Niigata-ken. Chem. Pharm. Bull. (Japan) **26**, 3010 (1978).

123. Stork, G., J. E. Davies, and A. Meisels: Total Syntheses of Polycyclic Triterpenes: The Total Synthesis of (+)-α-Onocerin. J. Amer. Chem. Soc. **85**, 3419 (1963).

124. Tsuda, Y., and M. Hattori: Total Synthesis of Hydroxyhopanone. Chem. Pharm. Bull. (Japan) **15**, 1073 (1967).

125. Mills, J. S., and A. E. A. Werner: The Chemistry of Dammar Resin. J. Chem. Soc. (London) **1955**, 3132.

126. Fujimoto, H., and O. Tanaka: Synthesis of the Skeleton of Dammarane Type Triterpene. Chem. Pharm. Bull. (Japan) **18**, 1440 (1970).

127. Mills, J. S.: The Constitution of the Neutral Tetracyclic Triterpenes of Dammar Resin. J. Chem. Soc. (London) **1956**, 2196.

128. Kasai, R., K. Shinzo, O. Tanaka, and K. Kawai: Synthetic Study of the Ginseng Sapogenins: Preparation of Dammarenediol-II and Remote Oxidation of Dammaranediols with Photoexcited Nitrobenzene Derivatives. Chem. Pharm. Bull. (Japan) **22**, 1213 (1974).

129. Kasai, R., K. Shinzo, and O. Tanaka: Syntheses of Betulafolienetriol and the Ginseng Sapogenin, 20(*S*)-Protopanaxadiol. Chem. Pharm. Bull. (Japan) **24**, 400 (1976).

130. Breslow, R., S. Baldwin, T. Flechtner, P. Kalicky, S. Liu, and W. Washburn: Remote Oxidation of Steroids by Photolysis of Attached Benzophenone groups. J. Amer. Chem. Soc. **95**, 3251 [see also Breslow, R.: Biomimetic Chemistry. Chem. Soc. Review **1**, 553 (1972)].

131. Döpp, D., and E. Brugger: Photolyse von β-deuterierten 1-*t*-Butyl-2-Nitrobenzolen. Chem. Ber. **106**, 2166 (1973).

132. Weller, J. W., and G. A. Hamilton: The Photo-oxidation of Alkanes by Nitrobenzene. Chem. Comm. **1970**, 1390.

133. Scholl, P. C., and M. R. Van de Mark: Remote Oxidation with Photoexcited Nitrobenzene Derivatives. J. Organ. Chem. (USA) **38**, 2376 (1973).

134. Takahashi, M., and M. Yoshikura: Extraction of Constituents of *Panax ginseng*. Yakugaku Zasshi **84**, 752, 757 (1964).

135. — — Studies on the Components of *Panax ginseng* C. A. Meyer. On the Structure of a New Acetylenic Derivative "Panaxynol". Yakugaku Zasshi **86**, 1051, 1053 (1966).

136. Bentley, R. K., D. Bhattacharjee, Sir R. H. Jones, and V. Thaller: Natural Acetylenes. Part XXVIII. C-17-Polyacetylenic Alcohols from the Umbellifer *Daucus carota* L. (Carrot): Alkylation of Benzene by Acetylene (Vinyl) Carbinols in the Presence of Toluene-*p*-sulphonic Acid. J. Chem. Soc. Section C (London) **1969**, 685.

137. Bohlmann, F., U. Niedballa, and J. Schneider: Polyacetylenverbindungen LXXXIV. Synthese natürlich vorkommender C-17-Polyine. Chem. Ber. **98**, 3010 (1965).

138. Poplawski, J., J. T. Wrobel, and T. Glinka: Panaxydol, a new Polyacetylenic Epoxide from *Panax ginseng* Roots. Phytochem. **19**, 1539 (1980).

139. Dabrowski, Z., J. T. Wrobel, and K. Wojtasiewicz: Structure of an Acetylenic Compound from *Panax ginseng*. Phytochem. **19**, 2464 (1980).

140. Chung, B. S.: Studies on the Constituents of Korean ginseng (1). On the Constituents of Korean Ginseng Sterols. Korean J. Pharmacog. **5**, 175 (1974).

141. COOK, C. H., and S. H. AN: Studies on the Components in the Etheral Extracts of *Panax ginseng* (1). Studies on the Free Fatty Acids. Korean J. Pharmacog. **6,** 15 (1975).

142. LEE, T. N., and T. W. KWON: Studies on the Carbohydrates of Ginseng (1). Free Sugars in Ginseng. Insam Munhun Teukjip (Seoul) **1,** 161 (1962).

143. TAKIURA, K., and I. NAKAGAWA: Oligosaccharides. IV. Separation of Oligosaccharides and Identification of Disaccharides in Ginseng Root. Yakugaku Zasshi **83,** 298 (1963).

144. TAKIURA, K., and I. NAKAGAWA: Ginseng Trisaccharides. Yakugaku Zasshi **83,** 301 and 305 (1963).

145. OVODOV, YU. S., and T. E. SOLOV'EVA: Polysaccharides of *Panax ginseng.* Khim. Prir. Soedin. **2,** 299 (1966).

146. SOLOV'EVA, T. E., L. V. ARSENYUK, and YU. S. OVODOV: Structural features of *Panax ginseng* Pectin. Carbohydr. Res. **10,** 13 (1969).

147. HIYAMA, C., S. MIYAI, H. YOSHIDA, K. YAMASAKI, and O. TANAKA: Application of High Speed Liquid Chromatography and Dual Wave Length Thin Layer Chromatograph Densitometry to Analysis of Crude Drug: Nucleosides and Free Bases of Nucleic Acids in Ginseng Roots. Yakugaku Zasshi **98,** 1132 (1978).

148. TAKATORI, K., T. KATO, S. ASANO, M. OZAKI, and T. NAKASHIMA: Choline in *Panax ginseng* C. A. Meyer. Chem. Pharm. Bull. (Japan) **11,** 1342 (1963).

149. SUH, C. S.: Determination of α-Pyrrolidone in *Panax ginseng* Extracts. Yakhak Hoeji (Korea) **13,** 111 (1969).

150. GSTIRNER, F., and H. J. VOGT: Peptides in White Korean Ginseng. Arch. Pharmaz. **299,** 936 (1966).

151. — — Comprehensive Chemical Investigation of Ginseng Drugs. Arch. Pharmaz. **300,** 371 (1967).

152. ANDO, T., T. MURAOKA, N. YAMASAKI, and H. OKUDA: Preparation of Antilipolytic Substance from *Panax ginseng.* Planta Medica **38,** 18 (1980).

153. KOSUGE, T., M. YOKOTA, and A. OCHIAI: Studies on Antihemorrhagic Principles in Crude Drugs for Hemostatis. II. On the Antihemorrhagic Principle of Sanchi Ginseng Radix. Yakugaku Zasshi **101,** 629 (1981).

154. HAN, B. H., M. H. PARK, Y. N. HAN, L. K. WOO, U. SANKAWA, S. YAHARA, and O. TANAKA: Degradation of Ginseng Saponins under Mild Acidic Conditions. Planta Medica **44,** 146 (1982).

155. WOO, L. K.: The Synthesis of ^{14}C-Labeled Dammarane Glycoside of Ginseng. Yakhak Hoeji **17,** 123 (1973).

156. TAKINO, Y., T. ODANI, H. TANIZAWA, and T. HAYASHI: Studies on the Absorption, Distribution, Excretion and Metabolism of Ginseng Saponins. I. Quantitative Analysis of Ginsenoside Rg_1 in Rats. Chem. Pharm. Bull. (Japan) **30,** 2196 (1982).

157. ODANI, T., H. TANIZAWA, and Y. TAKINO: Studies on the Absorption, Distribution, Excretion and Metabolism of Ginseng saponins. II. The Absorption, Distribution and Excretion of Ginsenoside Rg_1 in the Rat. Chem . Pharm. Bull. (Japan) **31,** 292 (1983).

158. BREKMAN, I. I.: "Zen-Shen", State Publ. House for Med. Lit., Leningrad (1957) and references cited therein.

159. PETKOV, W.: Pharmacological Influence of Reactivity (Experiments with Ginseng) Bulgarska Akad. Nauk. **5,** 57 (1962).

160. PETIKOV, W.: Über den Wirkungsmechanismus des *Panax ginseng* C. A. Meyer. Zur Frage einer Pharmakologie der Reaktivität. 2. Mitteilung. Arzneim. Forsch. (Drug Res.) **11,** 418 (1961) and references cited therein.

161. TAKAGI, K., H. SAITO, and H. NABATA: Pharmagological Studies of *Panax ginseng* Root: Estimation of Pharmacological Actions of *Panax ginseng* Root. Japan. J. Pharmacol. **22,** 245 (1972).

162. TAKAGI, K., H. SAITO, and M. TSUCHIYA: Pharmacological Studies of *Panax ginseng* Root: Pharmacological Properties of a Crude Saponin Fraction. Japan. J. Pharmacol. **22**, 339 (1972).

163. NABATA, H., H. SAITO, and K. TAKAGI: Pharmacological Studies of Neutral Saponins (GNS) of *Panax ginseng* Root. Japan. J. Pharmacol. **23**, 29 (1973).

164. SAITO, H., Y. YOSHIDA, and K. TAKAGI: Effect of *Panax ginseng* Root on Exhaustive Exercise in Mice. Japan J. Pharmacol. **24**, 119 (1974).

165. TAKAGI, K., H. SAITO, and M. TSUCHIYA: Effect of *Panax ginseng* Root on Spontaneous Movement and Exercise in Mice. Japan. J. Pharmacol. **24**, 41 (1974).

166. SAITO, H. M. TSUCHIYA, S. NAKA, and K. TAKAGI: Effects of *Panax ginseng* Root on Conditioned Avoidance Response in Rats. Japan. J. Pharmacol. **27**, 509 (1977).

167. KAKU, T., T. MIYATA, T. URUNO, I. SAKO, and A. KINOSHITA: Chemico-Pharmacological Studies on Saponins of *Panax ginseng* C. A. Meyer. Arzneim. Forsch. (Drug Res.) **25**, 343 and 539 (1975).

168. SAITO, H., K. SUDA, M. SCHWAB, and H. THOENEN: Potentiation of the NGF-Mediated Nerve Fiber Outgrowth by Ginsenoside-Rb$_1$ in Organ Cultures of Chicken Dorsal Root Ganglia. Japan. J. Pharmacol. **27**, 445 (1977).

169. OURA, H., S. HIAI, S. NAKASHIMA, and K. TSUKADA: Stimulation Effect of the Roots of *Panax ginseng* C. A. Meyer on the Incorporation of Labeled Precursors into Rat Liver RNA. Chem. Pharm. Bull. (Japan) **19**, 453 (1971).

170. OURA, H., S. HIAI, and H. SENO: Synthesis and Characterization of Nuclear RNA Induced by Radix Ginseng Extract in Rat Liver. Chem. Pharm. Bull. (Japan) **19**, 1598 (1971).

171. HIAI, S., H. OURA, K. TSUKADA, and Y. HIRAI: Stimulating Effect of *Panax ginseng* Extract on RNA Polymerase Activity in Rat Liver Nuclei. Chem. Pharm. Bull. (Japan) **19**, 1656 (1971).

172. OURA, H., K. TSUKADA, and H. NAKAGAWA: Effect of Radix Ginseng Extract on Cytoplasmic Polysome in Rat Liver. Chem. Pharm. Bull. (Japan) **20**, 219 (1972).

173. OURA, H., S. NAKASHIMA, K. TSUKADA, and Y. OHTA: Effect of Radix Ginseng Extract on Serum Protein Synthesis. Chem. Pharm. Bull. (Japan) **20**, 980 (1972).

174. OURA, H., S. HIAI, S. NABETANI, H. NAKAGAWA, Y. KURATA, and N. SASAKI: Effect of Ginseng Extract on Endoplasmic Reticulum and Ribosome. Planta Medica **28**, 76 (1975).

175. OURA, H., S. HIAI, Y. ODAKA, and T. YOKOZAWA: Studies on the Biochemical Action of Ginseng Saponin I. Purification from Ginseng Extract of the Active Component Stimulating Serum Protein Biosynthesis. J. Biochemistry (Tokyo) **77**, 1057 (1975).

176. NAGASAWA, T., H. OURA, S. HIAI, and K. NISHINAGA: Effect of Ginseng Extract on Ribonucleic Acid and Protein Synthesis in Rat Kidney. Chem. Pharm. Bull. (Japan) **25**, 1665 (1977).

177. HIAI, S., S. SASAKI, and H. OURA: Effect of Ginseng Saponin on Rat Adrenal Cyclic AMP. Planta Medica **37**, 15 (1979).

178. YAMAMOTO, M., A. KUMAGAI, and Y. YAMAMURA: Stimulatory Effect of *Panax ginseng* Principles on DNA and Protein Synthesis in Rat Testis. Arzneim. Forsch. (Drug Res.) **27**, 1404 (1977).

179. YAMAMOTO, M., N. TAKEUCHI, A. KUMAGAI, and Y. YAMAMURA: Stimulatory Effect of *Panax ginseng* Principles on DNA, RNA, Protein and Lipid Synthesis in Rat Bone Marrow. Arzneim. Forsch. (Drug Res.) **27**, 1169 (1977).

180. YAMAMOTO, M., M. MASAKA, K. YAMADA, Y. HAYASHI, H. HIRAI, and A. KUMAGAI: Stimulatory Effect of Ginsenosides on DNA, Protein and Lipid Synthesis in Rat Bone Marrow and Participation of Cyclic Nucleosides. Arzneim. Forsch. (Drug Res.) **28**, 2238 (1978), and reference cited therein.

181. IIJIMA, M. T. HIGASHI, S. SANADA, and J. SHOJI: Effect of Ginseng Saponins on Nucleic

Acid (RNA) Metabolism. I. RNA Synthesis in Rats Treated with Ginsenosides. Chem. Pharm. Bull. (Japan) **24**, 2400 (1976).

182. SHIBATA, Y., T. NOZAKI, T. HIGASHI, S. SANADA, and J. SHOJI: Stimulation of Serum Protein Synthesis in Ginsenosides Treated Rat. Chem. Pharm. Bull. (Japan) **24**, 2818 (1976).

183. TAKAHASHI, M., and J. CYONG: Studies on Adenosine Triphosphate Activity in Ginseng Radix. Shoyakugaku Zasshi **36**, 177 (1982).

184. SAKAKIBARA, K., Y. SHIBATA, T. HIGASHI, S. SANADA, and J. SHOJI: Effect of Ginseng Saponins on Cholesterol Metabolism. I. The Level and the Synthesis of Serum and Liver Cholesterol in Rats Treated with Ginsenosides. Chem. Pharm. Bull. (Japan) **23**, 1009 (1975).

185. GOMMORI, K., F. MIYAMOTO, Y. SHIBATA, T. HIGASHI, S. SANADA, and J. SHOJI: Effect of Ginseng Saponins on Cholesterol Metabolism. II. Effect of Ginsenosides on Cholesterol Synthesis by Liver Slice. Chem. Pharm. Bull. (Japan) **24**, 2985 (1976).

186. IKEHARA, M., Y. SHIBATA, T. HIGASHI, S. SANADA, and J. SHOJI: Effect of Ginseng Saponins on Cholesterol Metabolism. III. Effect of Ginsenoside-Rb_1 on Cholesterol Synthesis in Rats Fed on High-fat Diet. Chem. Pharm. Bull. (Japan) **26**, 2844 (1978).

187. YOKOZAWA, T., H. SENO, and H. OURA: Effect of Ginseng Extract on Lipid and Sugar Metabolism. I. Metabolic Correlation between Liver and Adipose Tissue. Chem. Pharm. Bull. (Japan) **23**, 3095 (1975).

188. YOKOZAWA, T., and H. OURA: Effect of Ginseng Extract on Lipid and Sugar Metabolism. II. Nutritional States in Rats. Chem. Pharm. Bull. (Japan) **24**, 987 (1976).

189. HIAI, S., H. YOKOZAWA, H. OURA, and S. YANO: Stimulation of Pituitary-adrenocortical System by Ginseng Saponin: Endocrinol. Japan. **26**, 661 (1979).

190. HIAI, S., H. YOKOZAWA, and H. OURA: Features of Ginseng Saponin-induced Corticosterone Secretion. Endocrinol. Japan. **26**, 737 (1979).

191. OHMINAMI, H., Y. KIMURA, H. OKUDA, T. TANI, S. ARICHI, and T. HAYASHI: Effects of Ginseng Saponins on the Actions of Adrenalin, ACTH and Insulin on Lipolysis and Lipogenesis in Adipose Tissue. Planta Medica **41**, 351 (1981).

192. KIMURA, M., I. WAKI, O. TANAKA, Y. NAGAI, and S. SHIBATA: Pharmacological Sequential Trial for the Fractionation of Components with Hypoglycemic Activity in Alloxan Diabetic Mice from Ginseng Radix. J. Pharm. Dyn. (Japan) **4**, 402 (1981).

193. KIMURA, M., I. WAKI, T. CHUJO, T. KIKUCHI, C. HIYAMA, K. YAMASAKI, and O. TANAKA: Effects of Hypoglycemic Components in Ginseng Radix on Blood Insulin Level in Alloxan Diabetic Mice and on Insulin Release from Perfused Rat Pancreas. J. Pharm. Dyn. (Japan) **4**, 410 (1981).

194. NAMBA, T., M. YOSHIZAKI, T. TOMIMORI, K. KOBASHI, K. MITSUI, and J. HASE: Hemolytic and its Protective Activity of Ginseng Saponins. Chem. Pharm. Bull. (Japan) **21**, 459 (1973).

195. — — — — — Fundamental Studies on the Evaluation of the Crude Drugs. (1). Hemolytic and its Protective Activity of Ginseng Saponins. Planta Medica **25**, 28 (1974).

196. ANISIMOV, M. M., E. B. SHENTSOVA, V. V. SHCHEGLOV, L. I. STRIGINA, N. I. UVAROVA, E. V. LEVINA, G. I. OSHITOK, and G. B. ELYAKOV: A Comparative Study of the Cytotoxic Effect of Dammarane Row Triterpenoids and Betulin on Early Embryogenesis of the Sea Urchin. Toxicon **16**, 31 (1978).

197. SIEGEL, R. K.: Ginseng Abuse Syndrome, Problems with the Panacea. JAMA **241**, 1614 (1979).

198. SAITO, H., T. KOHNO, K. TOZUKA, and K. TAKAGI: Biological Evaluation of *Panax ginseng* Callus (I). Shoyakugaku Zasshi **34**, 177 (1980) (in English).

199. SAITO, H., M. MORITA, and K. TAKAGI: Pharmacological Studies of *Panax ginseng* Leaves. Japan. J. Pharmacol. **23**, 43 (1973).

200. SAITO, H., Y. LEE, K. TAKAGI, S. SHIBATA, J. SHOJI, and N. KONDO: Pharmacological Studies of Panacis Japonici Rhizoma. I. Chem. Pharm. Bull. (Japan) **25,** 1017 (1977).
201. LEE, Y., H. SAITO, K. TAKAGI, S. SHIBATA, J. SHOJI, and N. KONDO: Pharmacological Studies of Panacis Japonici Rhizoma. II. Chem. Pharm. Bull. (Japan) **25,** 1391 (1977).

(Received April 5, 1983)

Addendum

The following important papers appeared in 1983 after submission of our original manuscript.

V.   Structure of Ginseng Saponins (2)
     2. Saponins of Red Ginseng

KITAGAWA, I., M. YOSHIKAWA, M. YOSHIHARA, T. HAYASHI, and T. TANIYAMA: Chemical Studies on Crude Drug Processing. I. On the Constituents of Ginseng Radix Rubra (1). Yakugaku Zasshi **103,** 612 (1983).

VII.   Production of Saponins by Tissue Culture

FURUYA, T., T. YOSHIKAWA, T. ISHII, and K. KAJII: Effects of Auxins on Growth and Saponin Production in Callus Cultures of *Panax Ginseng.* Planta Medica **47,** 183 (1983).

FURUYA, T., T. YOSHIKAWA, T. ISHII, and K. KAJII: Regulation of Saponin Production in Callus Cultures of *Panax Ginseng* [1]. Planta Medica **47,** 200 (1983).

FURUYA, T., T. YOSHIKAWA, Y. ORIHARA, and H. ODA: Saponin Production in Cell Suspension Cultures of *Panax Ginseng.* Planta Medica **48,** 83 (1983).

XI.   Other Constituents of Ginseng Roots
      1. Ether Soluble Compounds

SHIM, S., H. KOH, and B. HAN: Polyacetylenes from *Panax Ginseng* Roots. Phytochemistry **22,** 1817 (1983).

XII.   Pharmaceutical Studies of Ginseng Saponins

ODANI, T., H. TANIZAWA, and Y. TAKINO: Studies on the Absorption, Distribution, Excretion and Metabolism of Ginseng Saponins. III. The Absorption, Distribution and Excretion of Ginsenoside Rb_1 in the Rat. Chem. Pharm. Bull. (Japan) **31,** 1059 (1983).

Diterpenoids of *Rabdosia* Species

By E. FUJITA and M. NODE

Institute for Chemical Research, Kyoto University, Uji, Kyoto-fu, Japan

With 6 Figures

Contents

I. Introduction

Members of the genus *Rabdosia* (Labiatae) grow naturally in eastern Asia, but only species found in Japan and China heve been investigated chemically. In Japan, the leaves of *Rabdosia japonica* and *R. trichocarpa* are used as a common household medicine for gastrointestinal disorders. The drug is called "enmei-so", which means "grass effective for prolongation of human life". In China, some *Rabdosia* species are used as antitumor and antiphlogistic agents. Species now included in the genus were formerly referred to *Plectranthus* and *Isodon,* but since 1972 the genus name *Rabdosia* has been adopted in accordance with a recommendation by HARA (*1*).

The species under consideration were originally included in *Plectranthus,* but were separated in 1929 as an independent genus *Isodon* by KUDO. Although in 1934 a new genus *Amethystanthus* was established by NAKAI, HARA reached the conclusion that *Amethystanthus* should be absorbed in *Isodon.* Subsequently, in 1971, BLAKE pointed out that *Rabdosia* which had been established in 1842 as a Javanese genus and *Isodon* were identical. As a result, in 1972 HARA transferred all species of *Isodon* which are found mainly in the Himalayas, India, Thailand, China and Japan to *Rabdosia.*

The first investigation of the bitter principles in "enmei-so" was carried out in 1910 and isolation of a crystalline bitter substance was reported by YAGI (*2*). In 1954, antibacterial activity was reported for the extract (*3*). In 1958, isolation of enmein, one of the major diterpenoid constituents, initiated structure determination and investigation of other constituents. Since then, our knowledge of the diterpenoids of *Rabdosia* species has developed to a remarkable degree. Particular interest has centered on their antitumor activity. For previous reviews of the chemistry of *Rabdosia* diterpenoids, the reader is referred to Ref. (*4—7*).

II. Isolation and Structure Determination

The chemical investigation of *Rabdosia* plants began with isolation of the bitter principles of "enmei-so" *(R. japonica* and *R. trichocarpa).* During the first period (1958—1966), the main effort was devoted to the structure elucidation of enmein, one of the major constituents. After this was

References, pp. 149—157

accomplished in 1966, the minor diterpenoids of "enmei-so" were attacked; subsequently, the investigation was extended to the components of other *Rabdosia* species.

1. Structure Determination of Enmein in Outline

Investigations on the structure of enmein were carried out by many Japanese researchers. Enmein isolated from *R. trichocarpa* is usually contaminated by dihydroenmein and complete separation of the two compounds is difficult. On the other hand, *R. japonica* does not contain dihydroenmein and thus pure enmein can be isolated more simply from its leaves (*8*). Enmein, $C_{20}H_{26}O_6$, has mp 308—312° (decomp.) and $[\alpha]_D$ −136.3°. UV and IR spectra showed λ_{max} 233 nm (8200) and v_{max} 3460, 1755, 1710, and 1640 cm^{-1}.

Research on the structure determination of enmein consisted of three stages. In 1958, isolation, characterization of five of the six oxygen atoms, and proposals for a partial structure were reported by three groups at Kanazawa University (*9, 10*), Osaka City University (*11*) and Kyoto University (*12—14*). The functional groups assumed by the three groups are summarized below (**I**).

(**I**)

In 1961, two experiments which provided the basis for deducing the skeleton of enmein were carried out by KANATOMO (*15, 16*). In the first he obtained 1-ethyl-4(3,3-dimethylcyclohexyl)-benzene (**II**) whose structure was proved by synthesis (*15*). In the second he isolated retene (**III**) by selenium dehydrogenation of the material obtained by LiAlH$_4$ reduction of enmein (*16*). Thus enmein was proved to be a diterpene and a phyllocladene (**IV**) skeleton was proposed for it (*16*). KUBOTA and coworkers (*17*) deduced the presence of a hemiacetal ring (**V**) and partial structure (**VI**), with the lactone function part of a six-membered ring, on the basis of chemical evidence and spectral data of various derivatives, and advanced four formulas including (**VII**) as possible structures for enmein.

(II)   (III)   (IV)

(V)   (VI)   (VII)

Finally in 1964, cooperative studies of the three research groups resulted in a successful conclusion to this part of the work. On the basis of chemical reactions and the spectroscopic properties of the transformation products, the planar structural formula of enmein was shown to be (VII) with stereochemistry (VIIIa) or the C-9 epimer (VIIIb) (*18, 19*). At the same time, an X-ray analysis of dihydroenmein 3-acetate 6-bromoacetate permitted depiction of the structure and absolute configuration of enmein as (VIIIa) (*20*). A review of this work by Uyeo has been published (*21*).

(VIIIa)   (VIIIb)

2. Listing of *Rabdosia* Diterpenoids and Their Sources

Up to the present, twelve Japanese *Rabdosia* species and ten species in China and Taiwan have been investigated and 108 different diterpenoids have been isolated and characterized from them. These *Rabdosia* diterpenoids can be classified into four groups according to their structures: (1) *ent*-kauranes (**IX**), (2) *ent*-6,7-secokauranes (**X**), (3) *ent*-8,9-secokauranes (**XI**), and (4) others. Group (1) is divided into C-20-non-oxygenated kauranes and C-20-oxygenated kauranes; group (2) is divided into compounds of the enmein type and compounds of the spirolactone type and group (4) is divided into C-7, C-20-bonded *ent*-kauranes (**XII**) and *ent*-gibberellanes (**XIII**).

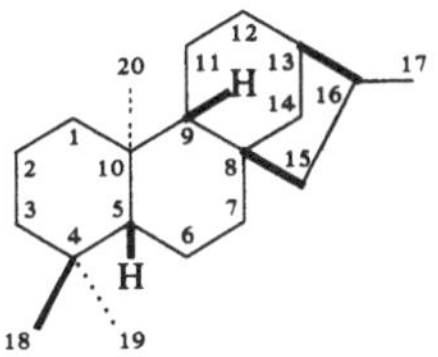

(IX) 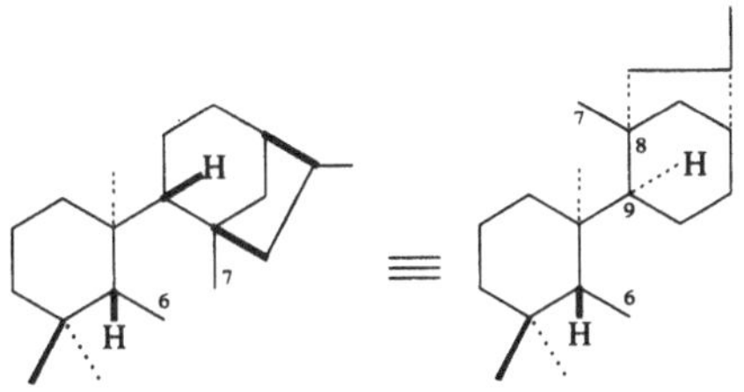 (X)

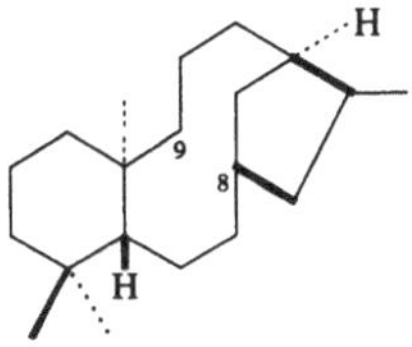

(XI) 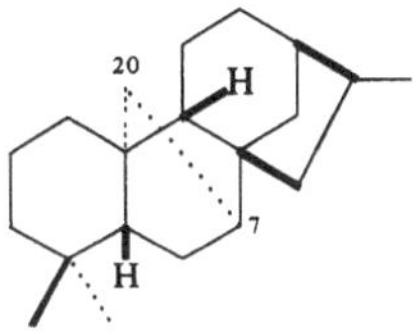 (XII)

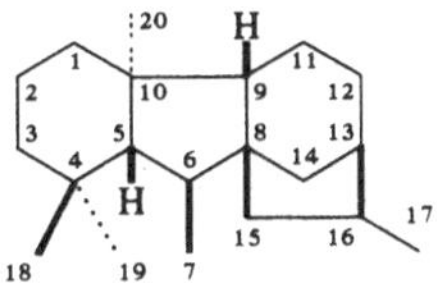

(XIII)

Tables I—IV list structures of all *Rabdosia* diterpenoids known at present on the basis of this classification.

Table I. ent-*Kaurane Type Diterpenoids*

a) C-20-Non-oxygenated *ent*-Kauranes

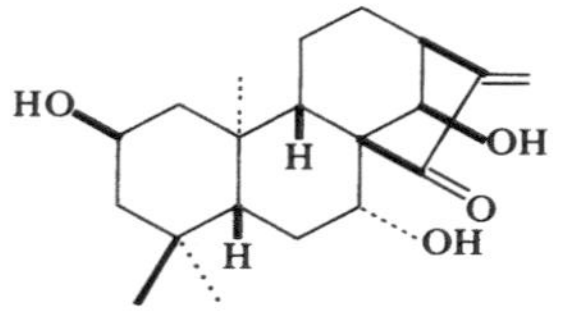

glaucocalyxin A
— (48)
—
—

(4)

excisanin A
$C_{20}H_{30}O_5$ (29)
m.p. 262—264°
$[a]_D$ —27.7° (Pyr.)

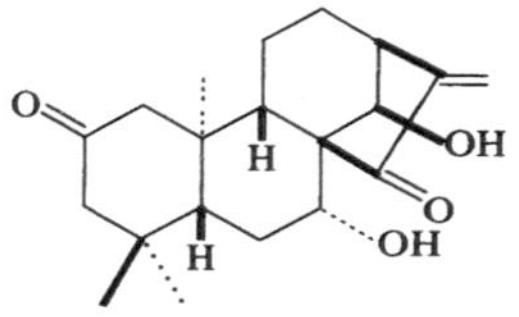

(8)

14-acetylumbrosin B
— (22)
—
—

(3)

kamebanin
$C_{20}H_{30}O_4$ (91)
m.p. 266—267°
$[a]_D$ —108° (dioxane)

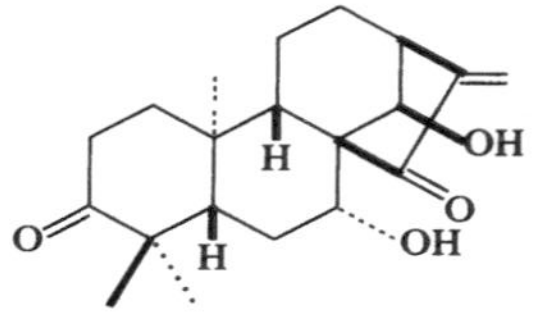

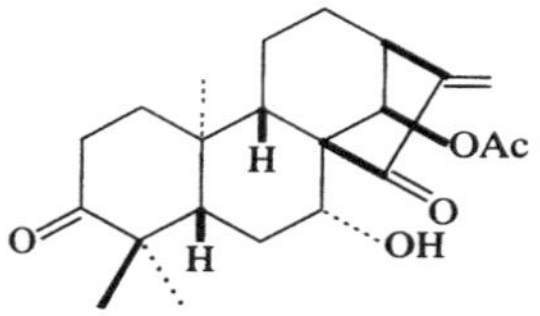

(7)

umbrosin B
$C_{20}H_{28}O_4$ (88)
m.p. 262—265° (dec.)
$[a]_D$ —150° (Pyr.)

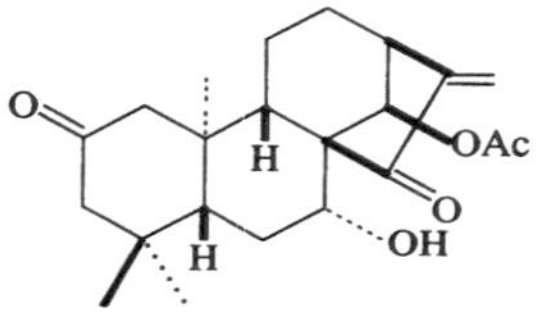

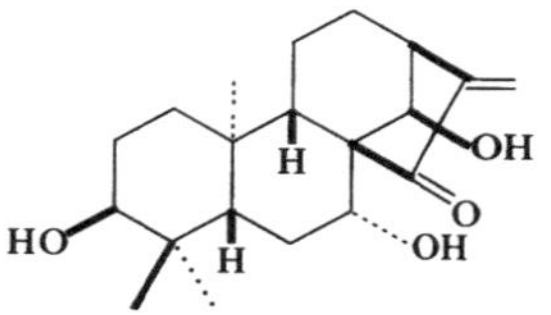

(2)

wangzaozin A
$C_{20}H_{30}O_4$ (23)
m.p. 225—227°
$[a]_D$ —77° (CHCl$_3$)

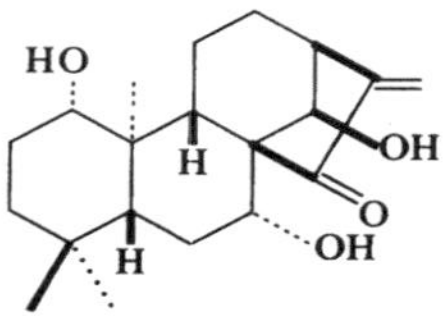

(6)

umbrosin A
$C_{20}H_{30}O_4$ (88)
m.p. 225—228°
$[a]_D$ —126° (Pyr.)

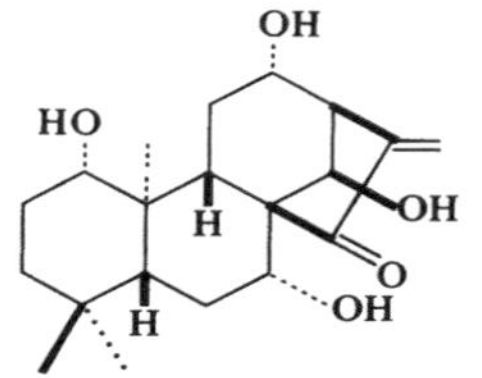

(1)

glaucocalyxin B
— (48)
—
—

(5)

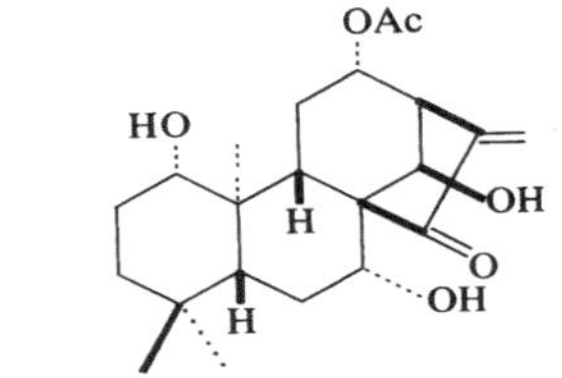

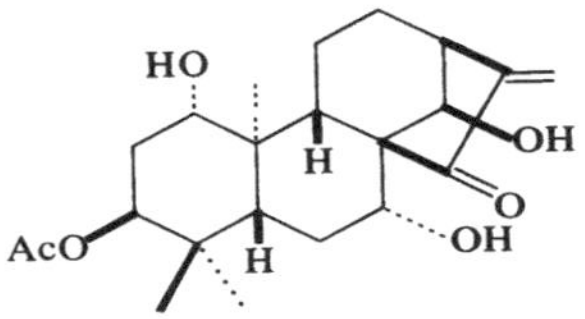

(12)

isodomedin
$C_{22}H_{32}O_6$ *(69)*
m.p. 217–218°
$[\alpha]_D$ −59° (EtOH)

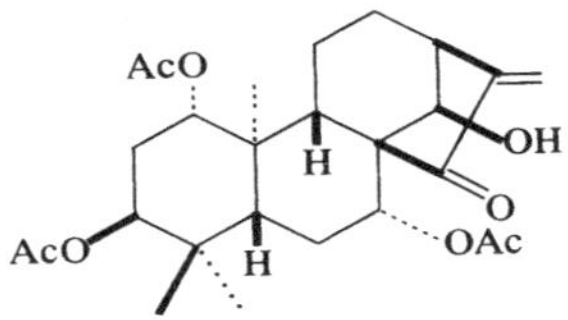

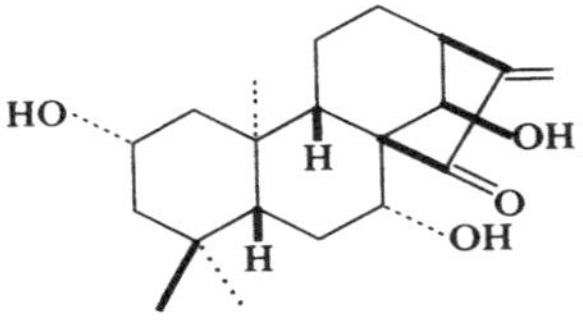

(16)

mebadonin
$C_{20}H_{30}O_4$ *(90)*
m.p. 271–273° (dec.)
$[\alpha]_D$ −158° (dioxane)

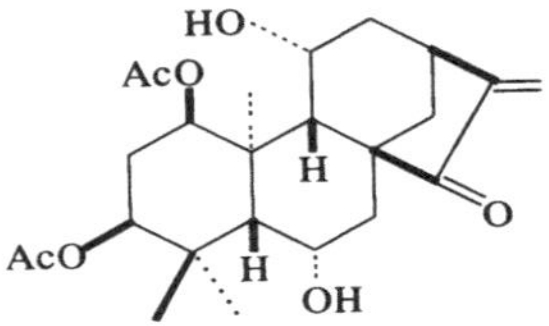

(11)

inflexinol
$C_{24}H_{34}O_7$ *(31)*
amorphous
$[\alpha]_D$ −43° (MeOH)

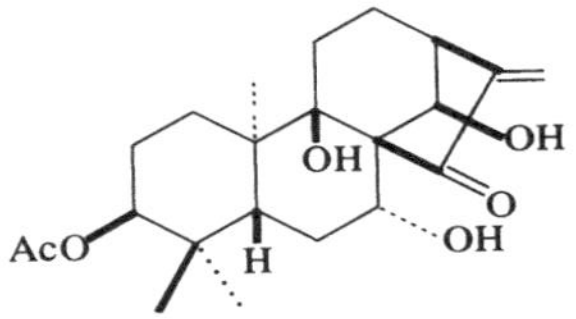

(15)

shikoccidin
$C_{22}H_{32}O_6$ *(73)*
m.p. 181–182°
$[\alpha]_D$ −4.2° (EtOAc)

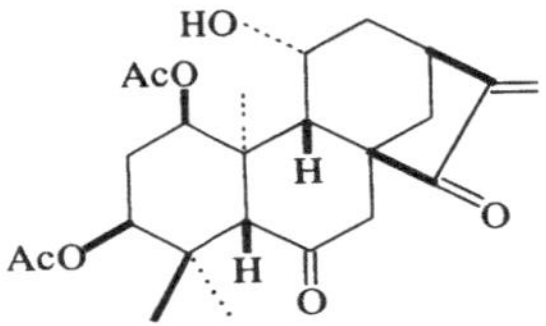

(10)

inflexin
$C_{24}H_{32}O_7$ *(30)*
m.p. 203–205°
$[\alpha]_D$ −47° (EtOH)

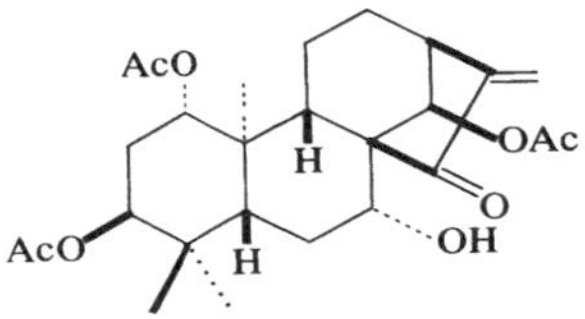

(14)

shikodokaurin B
$C_{26}H_{36}O_8$ *(70)*
amorphous
—

(9)

excisanin B
$C_{22}H_{32}O_6$ *(29)*
m.p. 240–243°
$[\alpha]_D$ −13.9° (Pyr.)

(13)

shikodokaurin A
$C_{26}H_{36}O_8$ *(70)*
m.p. 232–234°
—

Table I (continued)

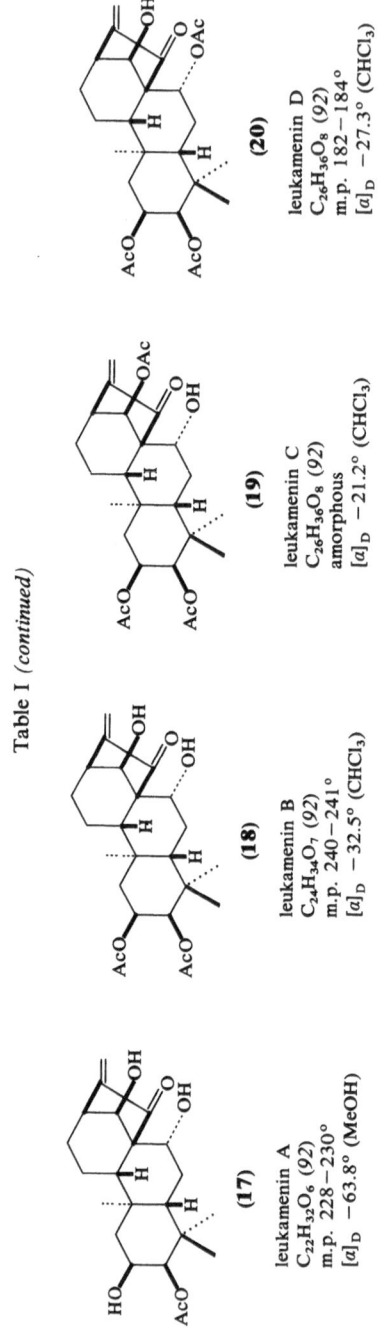

(20)

leukamenin D
$C_{26}H_{36}O_8$ (92)
m.p. 182–184°
$[\alpha]_D$ −27.3° (CHCl₃)

(19)

leukamenin C
$C_{26}H_{36}O_8$ (92)
amorphous
$[\alpha]_D$ −21.2° (CHCl₃)

(18)

leukamenin B
$C_{24}H_{34}O_7$ (92)
m.p. 240–241°
$[\alpha]_D$ −32.5° (CHCl₃)

(17)

leukamenin A
$C_{22}H_{32}O_6$ (92)
m.p. 228–230°
$[\alpha]_D$ −63.8° (MeOH)

(22)

leukamenin F
$C_{20}H_{28}O_4$ (92)
m.p. 223–224°
$[\alpha]_D$ −170° (MeOH)

(21)

leukamenin E
$C_{22}H_{32}O_5$ (92)
m.p. 148–149°
$[\alpha]_D$ −55.8° (MeOH)

b) C-20-Oxygenated *ent*-Kauranes

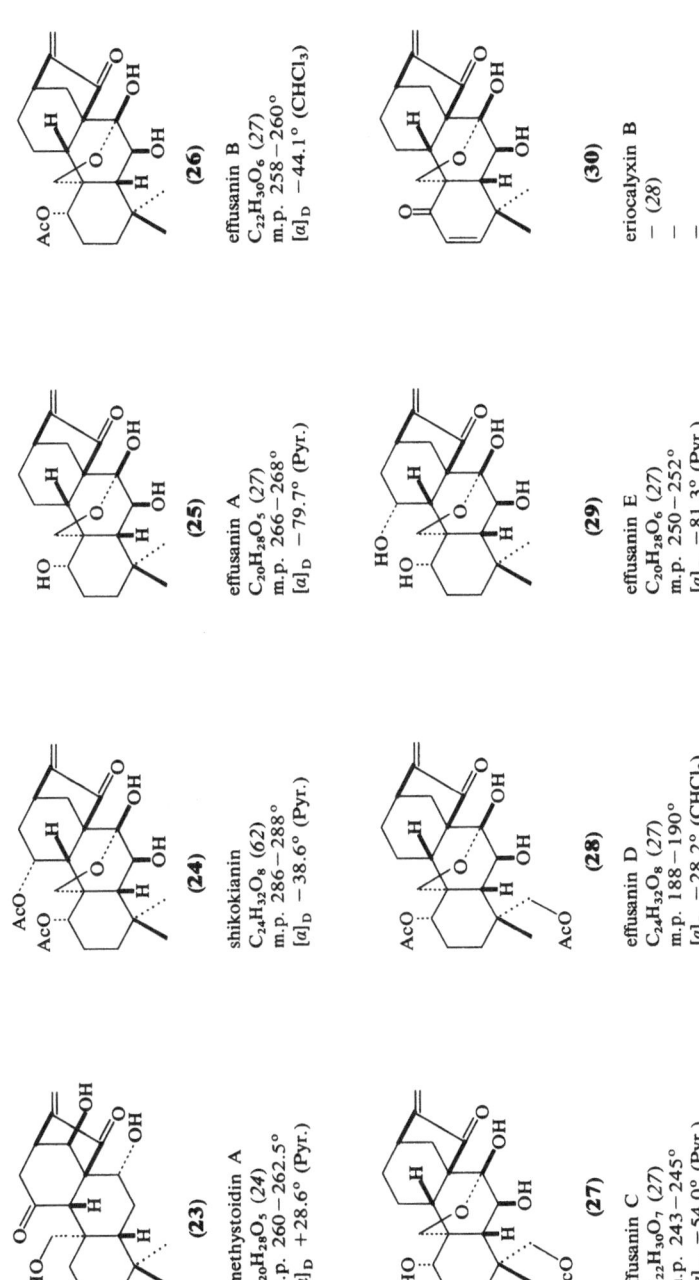

(26) effusanin B
$C_{22}H_{30}O_6$ (27)
m.p. 258–260°
$[a]_D$ −44.1° (CHCl₃)

(30) eriocalyxin B
— (28)
—
—

(25) effusanin A
$C_{20}H_{28}O_5$ (27)
m.p. 266–268°
$[a]_D$ −79.7° (Pyr.)

(29) effusanin E
$C_{20}H_{28}O_6$ (27)
m.p. 250–252°
$[a]_D$ −81.3° (Pyr.)

(24) shikokianin
$C_{24}H_{32}O_8$ (62)
m.p. 286–288°
$[a]_D$ −38.6° (Pyr.)

(28) effusanin D
$C_{24}H_{32}O_8$ (27)
m.p. 188–190°
$[a]_D$ −28.2° (CHCl₃)

(23) amethystoidin A
$C_{20}H_{28}O_5$ (24)
m.p. 260–262.5°
$[a]_D$ +28.6° (Pyr.)

(27) effusanin C
$C_{22}H_{30}O_7$ (27)
m.p. 243–245°
$[a]_D$ −54.0° (Pyr.)

Table I (continued)

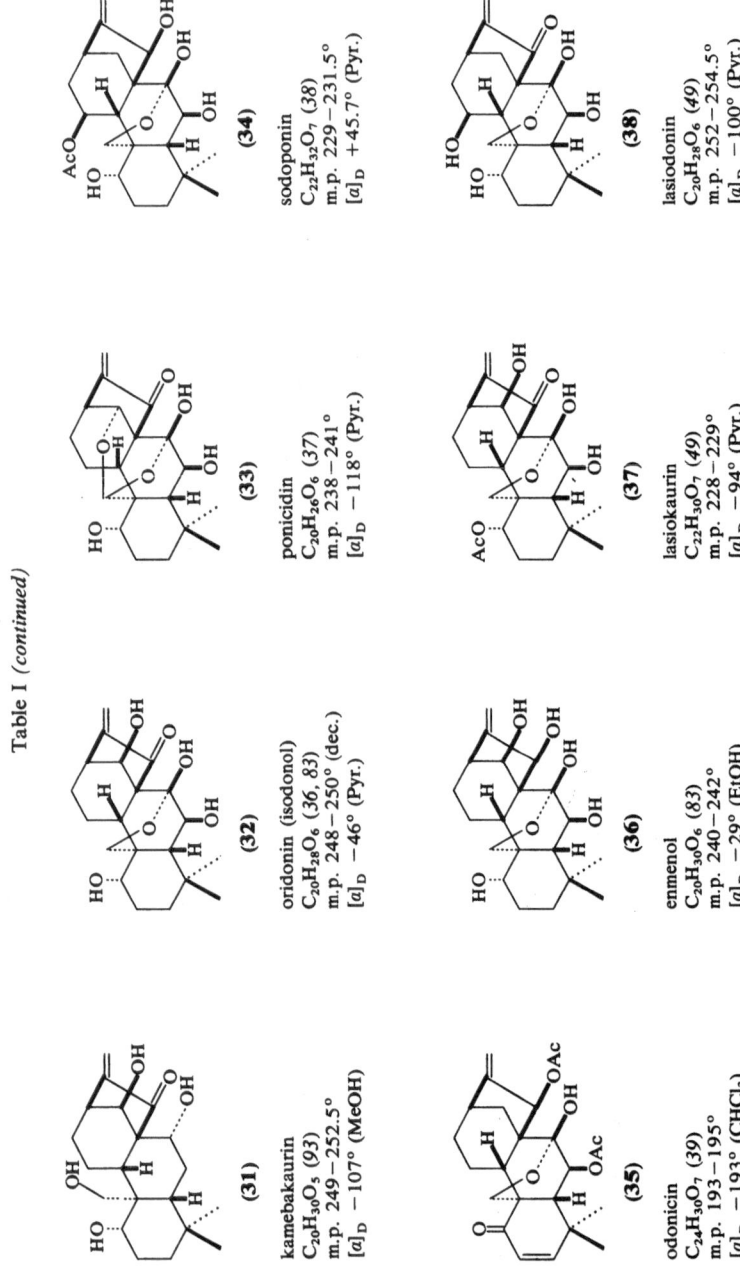

sodoponin (38)
$C_{22}H_{32}O_7$ (38)
m.p. 229–231.5°
$[\alpha]_D$ +45.7° (Pyr.)

(34)

lasiodonin (49)
$C_{20}H_{28}O_6$ (49)
m.p. 252–254.5°
$[\alpha]_D$ –100° (Pyr.)

(38)

ponicidin (37)
$C_{20}H_{26}O_6$ (37)
m.p. 238–241°
$[\alpha]_D$ –118° (Pyr.)

(33)

lasiokaurin (49)
$C_{22}H_{30}O_7$ (49)
m.p. 228–229°
$[\alpha]_D$ –94° (Pyr.)

(37)

oridonin (isodonol) (36, 83)
$C_{20}H_{28}O_6$ (36, 83)
m.p. 248–250° (dec.)
$[\alpha]_D$ –46° (Pyr.)

(32)

enmenol (83)
$C_{20}H_{30}O_6$ (83)
m.p. 240–242°
$[\alpha]_D$ –29° (EtOH)

(36)

kamebakaurin (93)
$C_{20}H_{30}O_5$ (93)
m.p. 249–252.5°
$[\alpha]_D$ –107° (MeOH)

(31)

odonicin (39)
$C_{24}H_{30}O_7$ (39)
m.p. 193–195°
$[\alpha]_D$ –193° (CHCl$_3$)

(35)

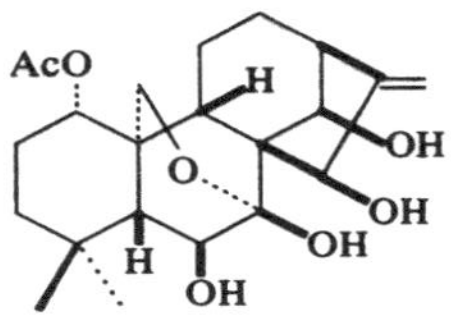

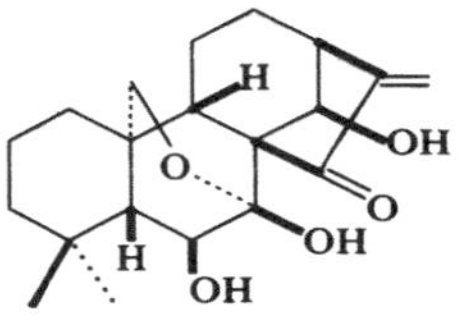

(42)

longikaurin A
$C_{20}H_{28}O_5$ (52)
m.p. 223–225°
$[a]_D$ −91.1° (Pyr.)

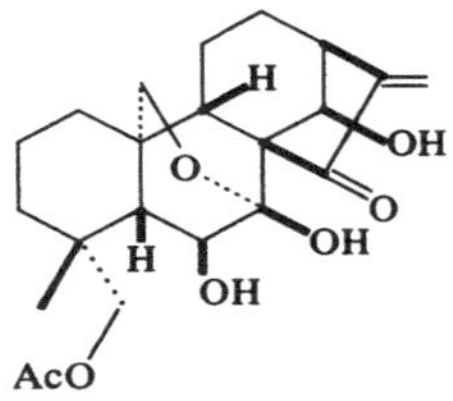

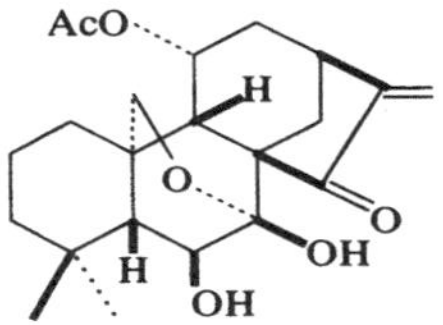

(46)

longikaurin E
$C_{22}H_{30}O_6$ (53)
m.p. 252–254°
$[a]_D$ −78.6° (Pyr.)

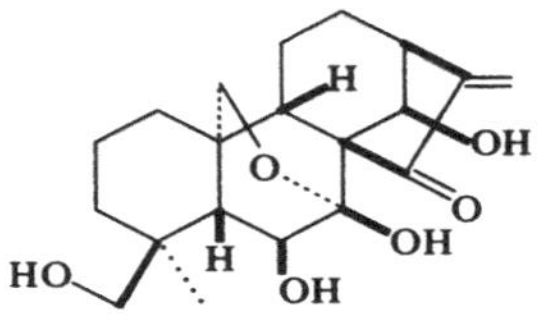

(41)

lasiocarpanin
$C_{20}H_{28}O_6$ (51)
amorphous
$[a]_D$ −59.4° (MeOH)

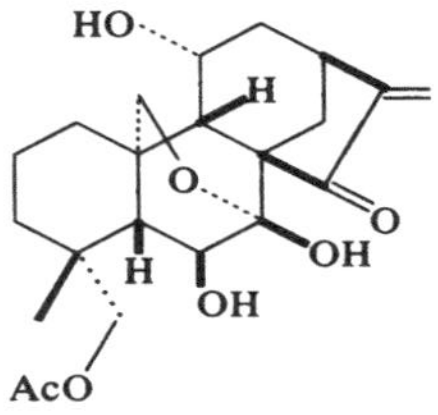

(45)

longikaurin D
$C_{22}H_{30}O_7$ (53)
m.p. 262–264°
$[a]_D$ −109.0° (Pyr.)

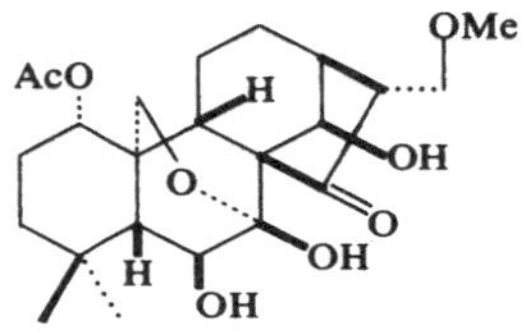

(40)

lasiokaurinin
$C_{23}H_{34}O_8$ (50)
m.p. 219–222° (dec.)
$[a]_D$ −14° (MeOH)

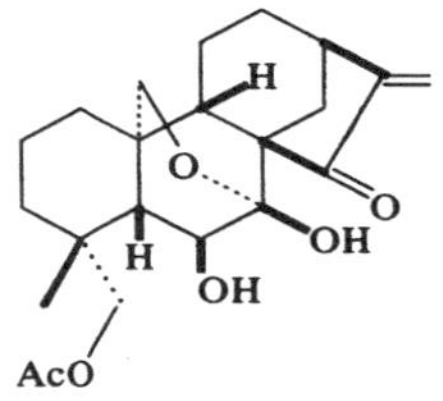

(44)

longikaurin C
$C_{22}H_{30}O_6$ (53)
m.p. 248–250°
$[a]_D$ −137.5° (Pyr.)

(39)

lasiokaurinol
$C_{22}H_{32}O_7$ (50)
m.p. 143–147° and 218–221°
$[a]_D$ −12° (MeOH)

(43)

longikaurin B
$C_{22}H_{30}O_7$ (52)
m.p. 238–239.5°
$[a]_D$ −115.9° (Pyr.)

Table I (continued)

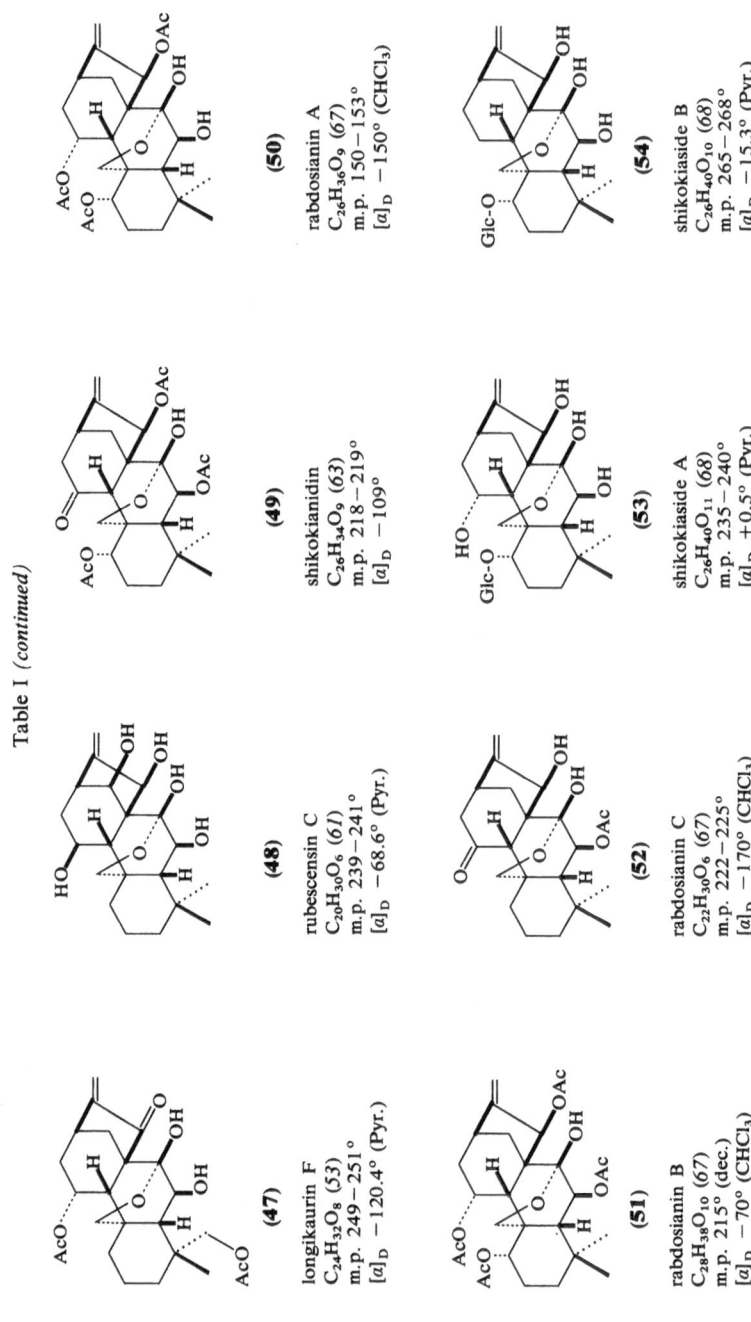

(50) rabdosianin A
$C_{26}H_{36}O_9$ (67)
m.p. 150—153°
$[\alpha]_D$ −150° (CHCl₃)

(54) shikokiaside B
$C_{26}H_{40}O_{10}$ (68)
m.p. 265—268°
$[\alpha]_D$ −15.3° (Pyr.)

(49) shikokianidin
$C_{26}H_{34}O_9$ (63)
m.p. 218—219°
$[\alpha]_D$ −109°

(53) shikokiaside A
$C_{26}H_{40}O_{11}$ (68)
m.p. 235—240°
$[\alpha]_D$ +0.5° (Pyr.)

(48) rubescensin C
$C_{20}H_{30}O_6$ (61)
m.p. 239—241°
$[\alpha]_D$ −68.6° (Pyr.)

(52) rabdosianin C
$C_{22}H_{30}O_6$ (67)
m.p. 222—225°
$[\alpha]_D$ −170° (CHCl₃)

(47) longikaurin F
$C_{24}H_{32}O_8$ (53)
m.p. 249—251°
$[\alpha]_D$ −120.4° (Pyr.)

(51) rabdosianin B
$C_{28}H_{38}O_{10}$ (67)
m.p. 215° (dec.)
$[\alpha]_D$ −70° (CHCl₃)

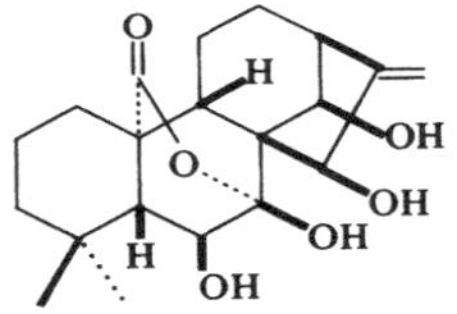

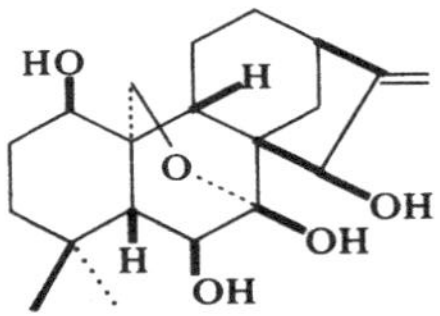

(58)

enmelol
$C_{20}H_{30}O_5$ (84)
m.p. 263–265°
$[\alpha]_D$ −48° (EtOH)

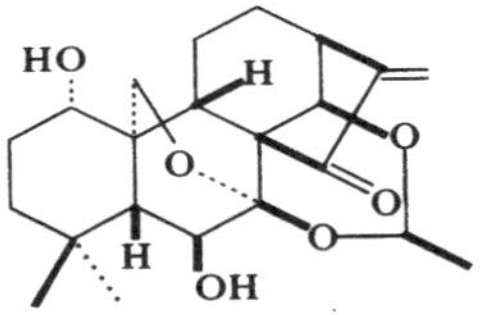

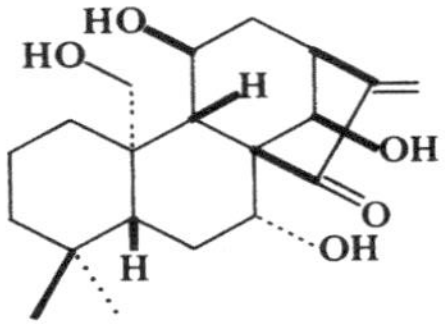

(57)

enmedol
$C_{22}H_{30}O_6$ (83)
m.p. 297–299°
$[\alpha]_D$ −45° (MeOH)

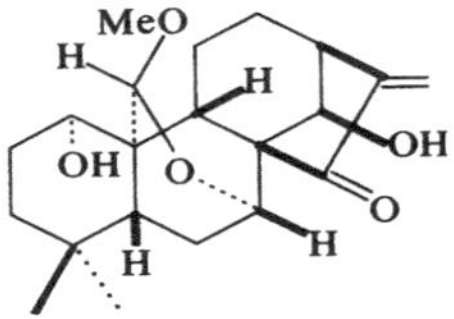

(61)

kamebacetal B
$C_{21}H_{30}O_5$ (95)
m.p. 230–232°
$[\alpha]_D$ −58° (MeOH)

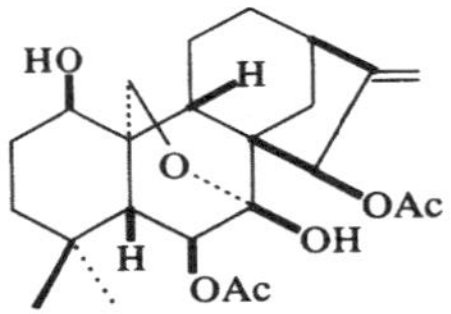

(56)

trichokaurin (enmenin)
$C_{24}H_{34}O_7$ (78, 84)
m.p. 184–185° (dec.)
$[\alpha]_D$ −93° (CHCl₃)

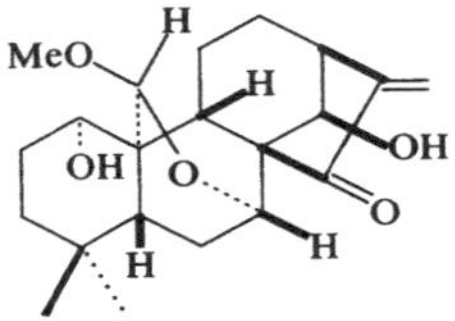

(60)

kamebacetal A
$C_{21}H_{30}O_5$ (95)
m.p. 253–256°
$[\alpha]_D$ −40° (MeOH)

(55)

rabdoternin A
$C_{20}H_{28}O_6$ (76)
m.p. 250–252°
$[\alpha]_D$ −45.6° (MeOH)

(59)

kamebakaurinin
$C_{20}H_{30}O_5$ (94)
m.p. 236–239°
$[\alpha]_D$ −101° (Pyr.)

Table II. *ent-6,7-Secokaurane Type Diterpenoids*

a) Enmein Type

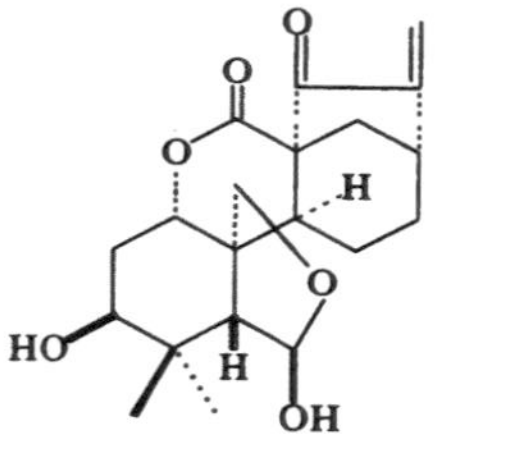

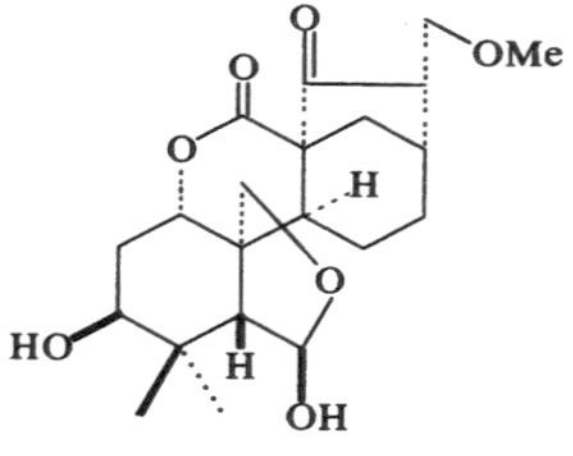

(65)
isodotricin
$C_{21}H_{30}O_7$ *(32—34)*
m.p. 240—245° (dec.)
$[\alpha]_D$ −114° (Pyr.)

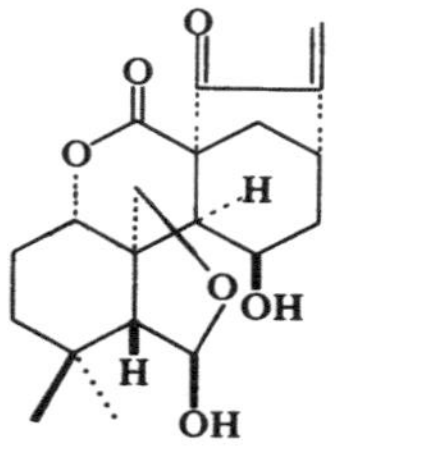

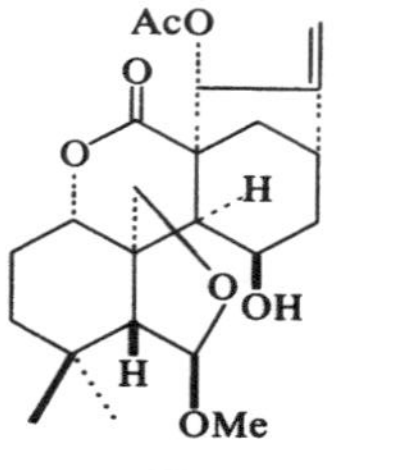

(69)
nodosinin
$C_{23}H_{32}O_7$ *(39)*
m.p. 281—284°
$[\alpha]_D$ −211° (CHCl$_3$)

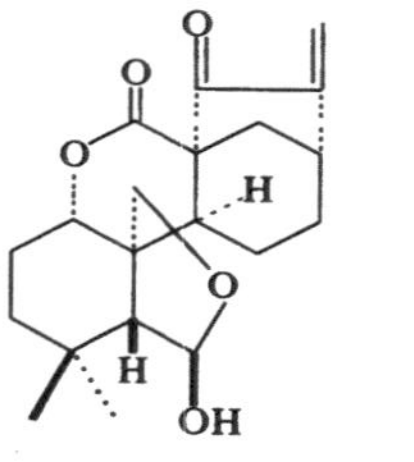

(64)
isodocarpin
$C_{20}H_{26}O_5$ *(33)*
m.p. 270—273° (dec.)
$[\alpha]_D$ −172° (CHCl$_3$)

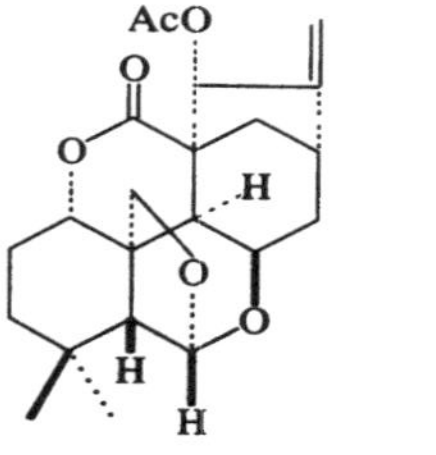

(68)
isodoacetal
$C_{22}H_{28}O_6$ *(39)*
m.p. >300°
$[\alpha]_D$ −134° (CHCl$_3$)

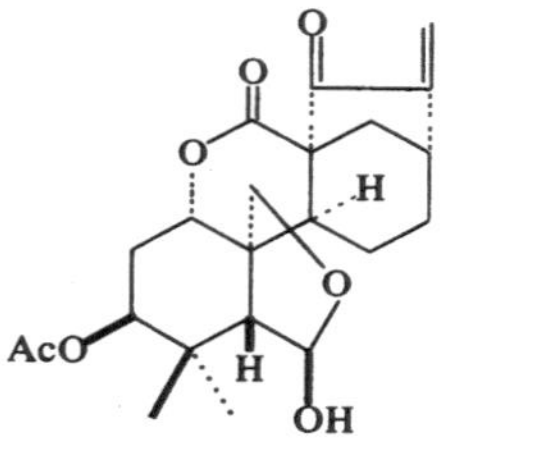

(63)
enmein-3-acetate
$C_{22}H_{28}O_7$ *(8)*
m.p. 267—271° (dec.)
$[\alpha]_D$ −112° (Pyr.)

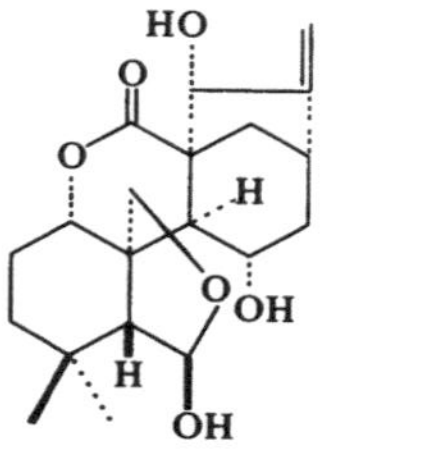

(67)
epinodosinol
$C_{20}H_{28}O_6$ *(38)*
m.p. 244—247°
$[\alpha]_D$ −87.5° (Pyr.)

(62)
enmein
$C_{20}H_{26}O_6$ *(8)*
m.p. 308—312° (dec.)
$[\alpha]_D$ −136° (Pyr.)

(66)
nodosin
$C_{20}H_{26}O_6$ *(32, 35)*
m.p. 275—280° (dec.)
$[\alpha]_D$ −203° (Pyr.)

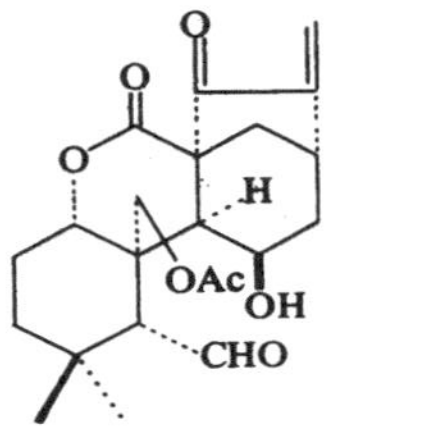

(73)
rabdosin A
− (44)
−
−

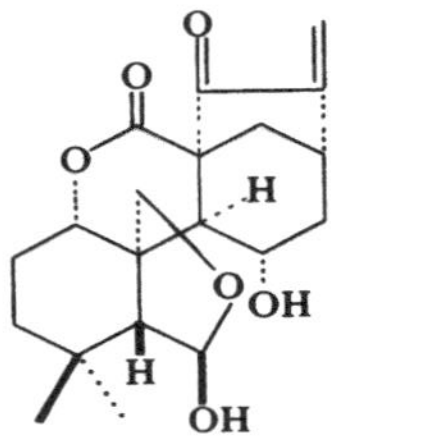

(72)
epinodosin
$C_{20}H_{26}O_6$ (41, 42, 38b)
m.p. 245−248° (dec.)
$[a]_D$ −173.7° (Pyr.)

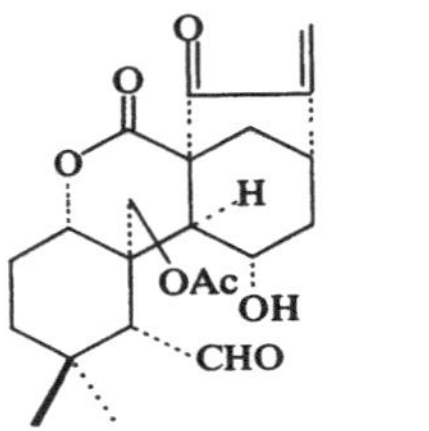

(71)
isodonal
$C_{22}H_{28}O_7$ (42, 43)
m.p. 245−247° (dec.)
$[a]_D$ +91.8° (Pyr.)

(70)
trichodonin
$C_{22}H_{28}O_7$ (41, 42)
m.p. 245−247° (dec.)
$[a]_D$ +10.0° (Pyr.)

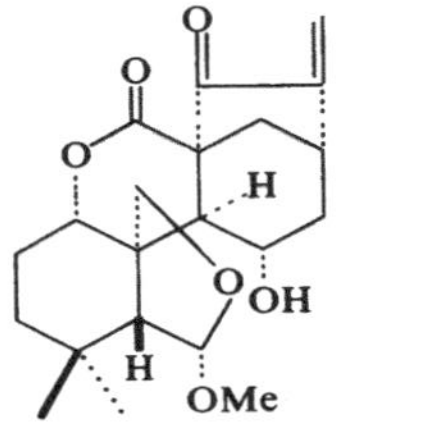

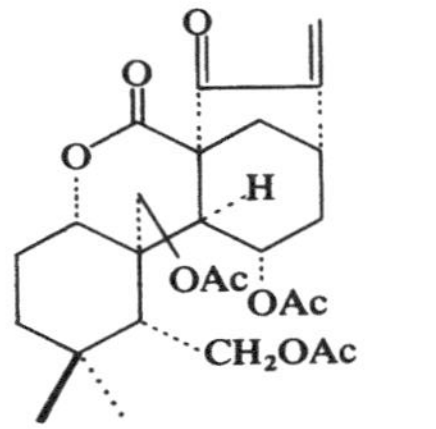

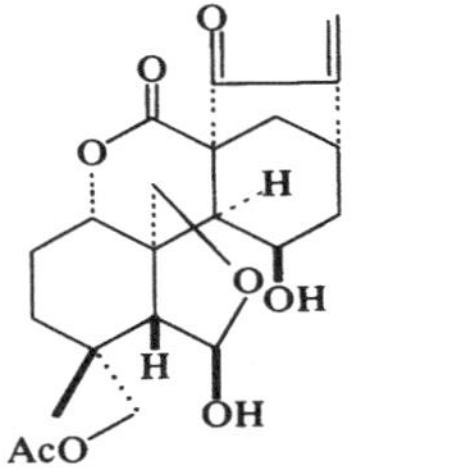

(77)
carpalasionin
$C_{22}H_{28}O_8$ (51)
m.p. 287−288°
$[a]_D$ −122.3° (MeOH)

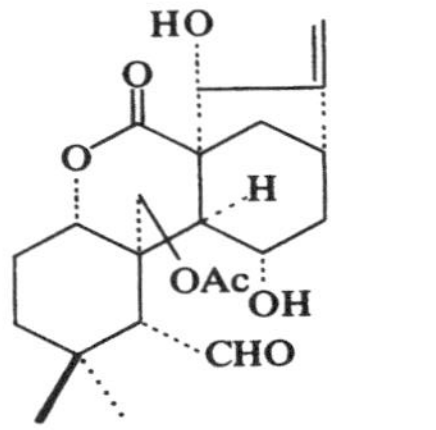

(76)
rabdolasional
$C_{22}H_{30}O_7$ (51)
amorphous
$[a]_D$ +7.5° (MeOH)

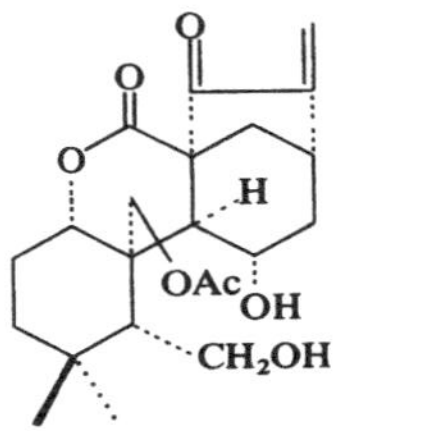

(75)
rabdophyllin G (rabdosin C)
− (45, 56, 57)
−
−

(74)
rabdosin B
− (44)
−
−

Table II *(continued)*

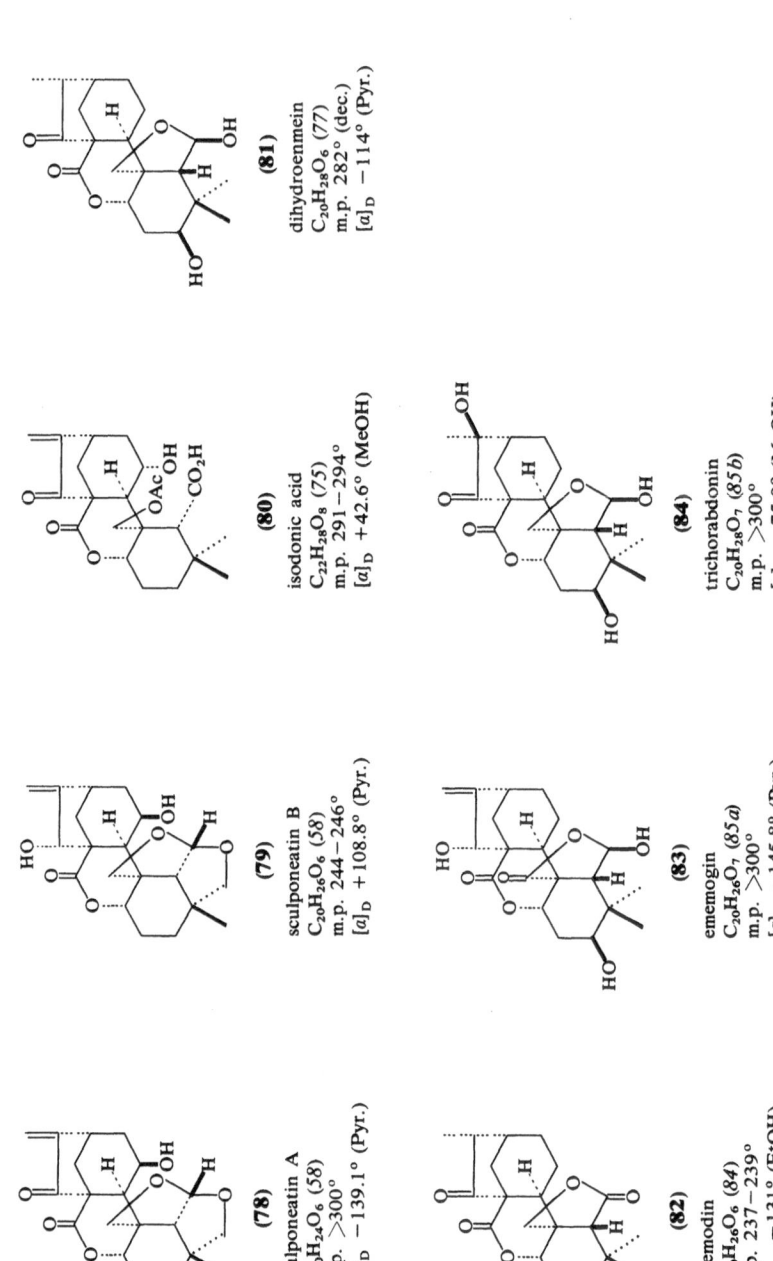

(81) dihydroenmein
$C_{20}H_{28}O_6$ (77)
m.p. 282° (dec.)
$[\alpha]_D$ −114° (Pyr.)

(80) isodonic acid
$C_{22}H_{28}O_8$ (75)
m.p. 291−294°
$[\alpha]_D$ +42.6° (MeOH)

(84) trichorabdonin
$C_{20}H_{28}O_7$ (85b)
m.p. >300°
$[\alpha]_D$ −75.0° (MeOH)

(79) sculponeatin B
$C_{20}H_{26}O_6$ (58)
m.p. 244−246°
$[\alpha]_D$ +108.8° (Pyr.)

(83) ememogin
$C_{20}H_{26}O_7$ (85a)
m.p. >300°
$[\alpha]_D$ −145.8° (Pyr.)

(78) sculponeatin A
$C_{20}H_{24}O_6$ (58)
m.p. >300°
$[\alpha]_D$ −139.1° (Pyr.)

(82) ememodin
$C_{20}H_{26}O_6$ (84)
m.p. 237−239°
$[\alpha]_D$ −131° (EtOH)

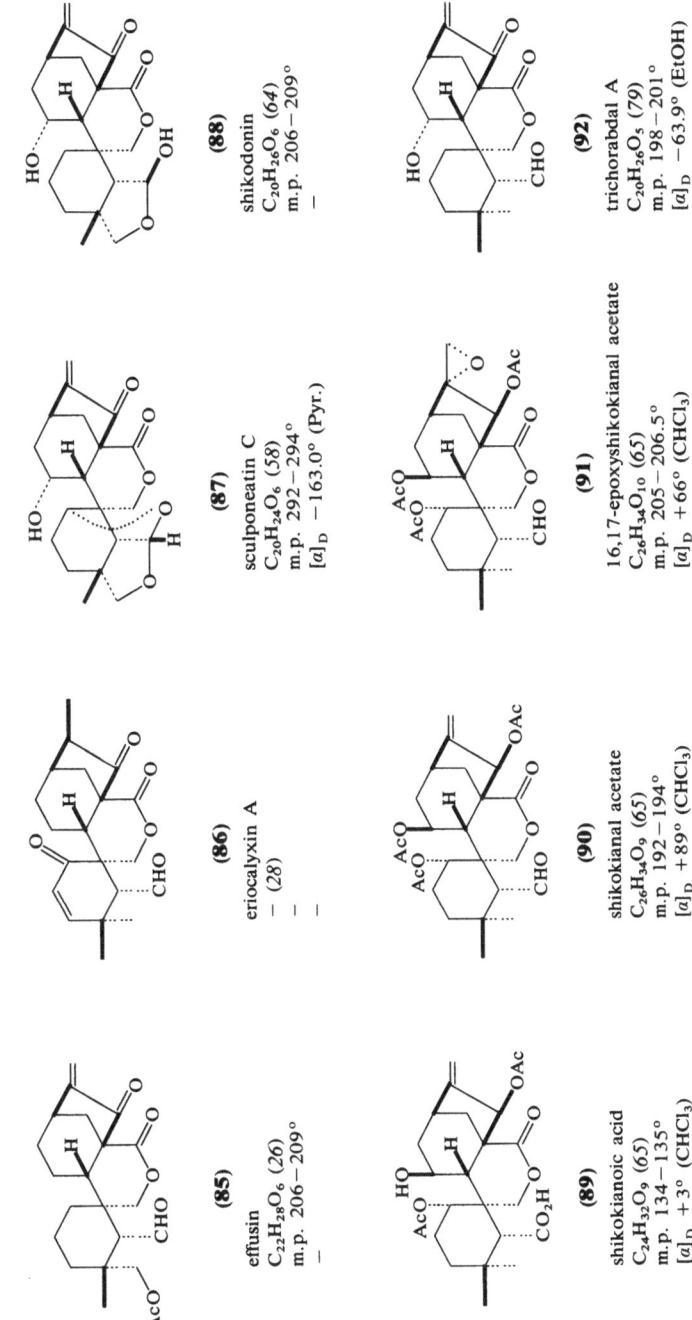

b) Spiro-lactone Type

(88)

shikodonin
$C_{20}H_{26}O_6$ *(64)*
m.p. 206–209°
—

(87)

sculponeatin C
$C_{20}H_{24}O_6$ *(58)*
m.p. 292–294°
$[\alpha]_D$ −163.0° (Pyr.)

(86)

eriocalyxin A
— *(28)*
—
—

(85)

effusin
$C_{22}H_{28}O_6$ *(26)*
m.p. 206–209°
—

(92)

trichorabdal A
$C_{20}H_{26}O_5$ *(79)*
m.p. 198–201°
$[\alpha]_D$ −63.9° (EtOH)

(91)

16,17-epoxyshikokianal acetate
$C_{26}H_{34}O_{10}$ *(65)*
m.p. 205–206.5°
$[\alpha]_D$ +66° (CHCl₃)

(90)

shikokianal acetate
$C_{26}H_{34}O_9$ *(65)*
m.p. 192–194°
$[\alpha]_D$ +89° (CHCl₃)

(89)

shikokianoic acid
$C_{24}H_{32}O_9$ *(65)*
m.p. 134–135°
$[\alpha]_D$ +3° (CHCl₃)

Table II *(continued)*

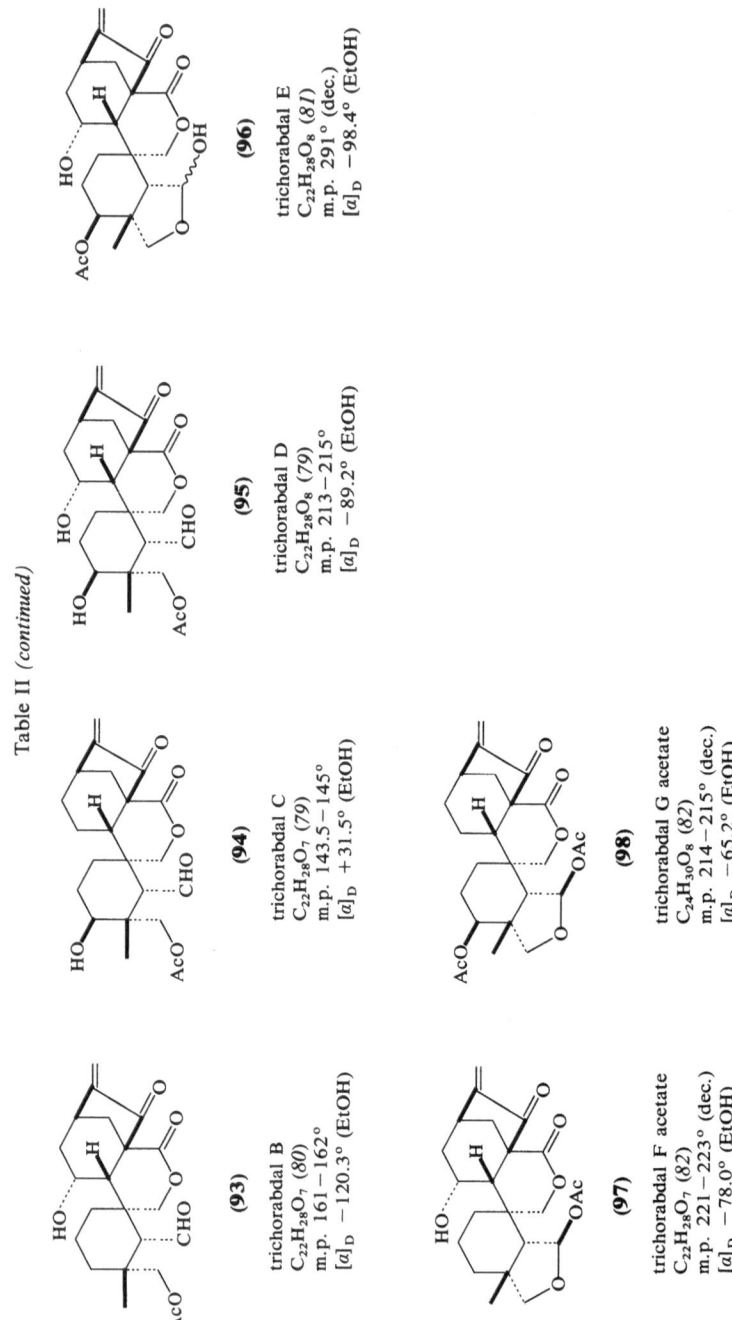

(96)

trichorabdal E
$C_{22}H_{28}O_8$ *(81)*
m.p. 291° (dec.)
$[\alpha]_D$ −98.4° (EtOH)

(95)

trichorabdal D
$C_{22}H_{28}O_8$ *(79)*
m.p. 213−215°
$[\alpha]_D$ −89.2° (EtOH)

(94)

trichorabdal C
$C_{22}H_{28}O_7$ *(79)*
m.p. 143.5−145°
$[\alpha]_D$ +31.5° (EtOH)

(98)

trichorabdal G acetate
$C_{24}H_{30}O_8$ *(82)*
m.p. 214−215° (dec.)
$[\alpha]_D$ −65.2° (EtOH)

(93)

trichorabdal B
$C_{22}H_{28}O_7$ *(80)*
m.p. 161−162°
$[\alpha]_D$ −120.3° (EtOH)

(97)

trichorabdal F acetate
$C_{22}H_{28}O_7$ *(82)*
m.p. 221−223° (dec.)
$[\alpha]_D$ −78.0° (EtOH)

Table III. ent-8,9-Secokaurane Type

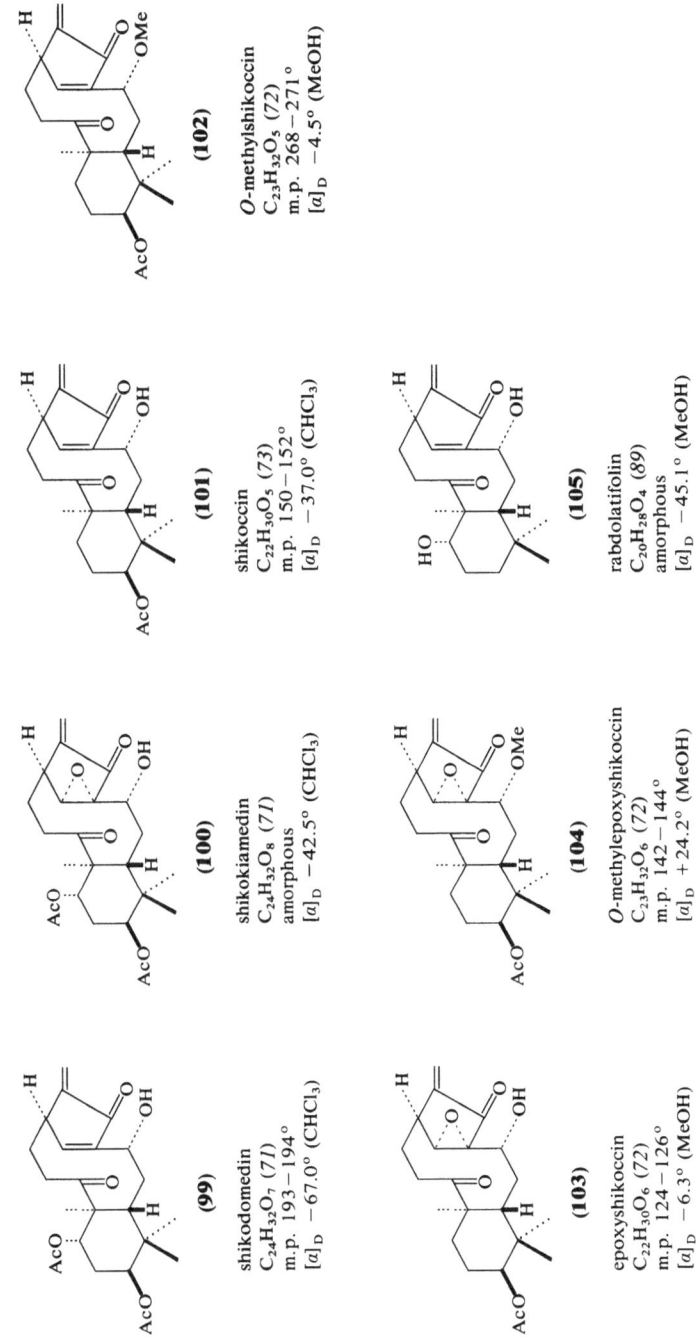

O-methylshikoccin
$C_{23}H_{32}O_5$ (72)
m.p. 268–271°
$[a]_D$ −4.5° (MeOH)

(102)

shikoccin
$C_{22}H_{30}O_5$ (73)
m.p. 150–152°
$[a]_D$ −37.0° (CHCl$_3$)

(101)

rabdolatifolin
$C_{20}H_{28}O_4$ (89)
amorphous
$[a]_D$ −45.1° (MeOH)

(105)

shikokiamedin
$C_{24}H_{32}O_8$ (71)
amorphous
$[a]_D$ −42.5° (CHCl$_3$)

(100)

O-methylepoxyshikoccin
$C_{23}H_{32}O_6$ (72)
m.p. 142–144°
$[a]_D$ +24.2° (MeOH)

(104)

shikodomedin
$C_{24}H_{32}O_7$ (71)
m.p. 193–194°
$[a]_D$ −67.0° (CHCl$_3$)

(99)

epoxyshikoccin
$C_{22}H_{30}O_6$ (72)
m.p. 124–126°
$[a]_D$ −6.3° (MeOH)

(103)

Table IV. *Others*

a) 7,20-Bonded *ent*-Kaurane Type   b) *ent*-Gibberellane Type

(106)	(107)	(108)
phyllostachysin A	rubescensin D	rabdoepigibberellolide
$C_{20}H_{26}O_6$ (58)	$C_{20}H_{26}O_6$ (58)	$C_{26}H_{34}O_9$ (66)
m.p. 264–265°	m.p. 266°	m.p. 255.5–265.5°
$[a]_D$ −30.5° (Pyr.)	$[a]_D$ −57.2° (Pyr.)	$[a]_D$ −89.0° (CHCl$_3$)

Table V lists those *Rabdosia* species which have been investigated in alphabetical order and the diterpenoids isolated from each species.

Table V

Species name	Origin	Isolated diterpenoids	Reference
Rabdosia amethystoides	China	umbrosin A (1)	(22)
(Benth.) Hara		umbrosin B (2)	(22)
		14-acetylumbrosin B (3)	(22)
		glaucocalyxin A (4)	(23)
		glaucocalyxin B (5)	(23)
		wangzaozin A (6)	(23)
		amethystoidin A (23)	(24)
R. effusa (Maxim.) Hara	Japan	shikokianin (24)	(27)
		effusanin A (25)	(27)
		effusanin B (26)	(27)
		effusanin C (27)	(27)
		effusanin D (28)	(27)
		effusanin E (29)	(27)
		longikaurin E (46)	(27)
		longikaurin F (47)	(27)
		effusin (85)	(25, 26)
R. eriocalyx (Dunn) Hara	China	eriocalyxin A (86)	(28)
		eriocalyxin B (30)	(28)
R. excisa (Maxim.) Hara	China	kamebanin (7)	(29)
		excisanin A (8)	(29)
		excisanin B (9)	(29)
		kamebakaurin (31)	(29)
		kamebacetal B (61)	(29)
R. inflexa (Thumb.) Hara	Japan	inflexin (10)	(25, 30, 31)
		inflexinol (11)	(31)

Table V *(continued)*

Species name	Origin	Isolated diterpenoids	Reference
R. japonica (Burman f.) Hara	Japan and China	oridonin (**32**)	(*32, 36, 45*)
		ponicidin (**33**)	(*32, 37*)
		sodoponin (**34**)	(*38*)
		odonicin (**35**)	(*39*)
		enmenol (**36**)	(*44*)
		lasiokaurin (**37**)	(*45*)
		enmein (**62**)	(*8, 9, 32*)
		enmein-3-acetate (**63**)	(*8, 32*)
		isodocarpin (**64**)	(*33*)
		isodotricin (**65**)	(*34*)
		nodosin (**66**)	(*32, 35*)
		epinodosinol (**67**)	(*38, 44*)
		isodoacetal (**68**)	(*39*)
		nodosinin (**69**)	(*39*)
		trichodonin (**70**)	(*41, 42*)
		isodonal (**71**)	(*38, 42, 43*)
		epinodosin (**72**)	(*38, 41, 42, 45*)
		rabdosin A (**73**)	(*44*)
		rabdosin B (**74**)	(*44*)
		rabdosin C (**75**)	(*45*)
R. japonica (Burman f. Hara var. glaucocalyx* (Maxim.) Hara	China	glaucocalyxin A (**4**)	(*48*)
		glaucocalyxin B (**5**)	(*48*)
R. lasiocarpa (Hayata) Hara	Taiwan	oridonin (**32**)	(*49, 51*)
		lasiokaurin (**37**)	(*49, 51*)
		lasiodonin (**38**)	(*49, 51*)
		lasiokaurinol (**39**)	(*50*)
		lasiokaurinin (**40**)	(*50*)
		lasiocarpanin (**41**)	(*51*)
		enmein (**62**)	(*51*)
		nodosin (**66**)	(*51*)
		epinodosin (**72**)	(*51*)
		rabdolasional (**76**)	(*51*)
		carpalasionin (**77**)	(*51*)
R. longituba (Miquel) Hara	Japan	kamebakaurin (**31**)	(*52*)
		oridonin (**32**)	(*25*)
		lasiokaurin (**37**)	(*25*)
		longikaurin A (**42**)	(*52*)
		longikaurin B (**43**)	(*52*)
		longikaurin C (**44**)	(*53*)
		longikaurin D (**45**)	(*53*)
		longikaurin E (**46**)	(*53*)
		longikaurin F (**47**)	(*53*)
		isodocarpin (**64**)	(*25*)
		nodosin (**66**)	(*25*)

Table V *(continued)*

Species name	Origin	Isolated diterpenoids	Reference
R. macrophylla (Migo)	China	amethystoidin A (23)	*(56, 57)*
		oridonin (32)	*(54, 55)*
		enmenol (36)	*(56, 57)*
		isodonal (71)	*(54, 55)*
		rabdophyllin G (75)	*(56, 57)*
R. phyllostachys (Diels) Hara	China	phyllostachysin A (106)	*(58)*
R. rubescens (Hemsl.) Hara	China	rubescensin A (32)[1]	*(58, 59)*
		rubescensin B (33)[2]	*(58, 59, 60, 61)*
		rubescensin C (48)	*(58, 61)*
		rubescensin D (107)	*(58)*
R. sculponeata (Vaniot) Hara	China	enmein (62)	*(58)*
		sculponeatin A (78)	*(58)*
		sculponeatin B (79)	*(58)*
		sculponeatin C (87)	*(58)*
R. shikokiana (Makino) Hara	Japan	shikokianin (24)	*(62)*
		effusanin B (26)	*(67)*
		oridonin (32)	*(62)*
		longikaurin E (46)	*(67)*
		shikokianidin (49)	*(63)*
		rabdosianin A (50)	*(67)*
		rabdosianin B (51)	*(67)*
		rabdosianin C (52)	*(67)*
		shikokiaside A (53)[3]	*(68)*
		shikokiaside B (54)	*(68)*
		shikodonin (88)	*(64)*
		shikokianoic acid (89)	*(65)*
		shikokianal acetate (90)[4]	*(65)*
		16,17-epoxyshikokianal acetate (91)[4]	*(65)*
		rabdoepigibberellolide (108)[5]	*(66)*
R. shikokiana (Makino) Hara *var. intermedia* (Kudo) Hara	Japan	isodomedin (12)	*(25, 69)*
		shikodokaurin A (13)	*(70)*
		shikodokaurin B (14)	*(70)*
		shikodomedin (99)	*(71)*
		shikokiamedin (100)	*(71)*
R. shikokiana (Makino) Hara *var. occidentalis* (Murata) Hara	Japan	isodomedin (12)	*(72)*
		shikoccidin (15)	*(73)*
		leukamenin E (21)	*(72)*
		shikoccin (101)	*(73)*
		O-methylshikoccin (102)	*(72)*
		epoxyshikoccin (103)	*(72)*
		O-methylepoxyshikoccin (104)	*(72)*

Table V *(continued)*

Species name	Origin	Isolated diterpenoids	Reference
R. ternifolia (W. W. Smith) Hara	China	oridonin (**32**)	*(76)*
		ponicidin (**33**)	*(76)*
		longikaurin A (**42**)	*(75)*
		longikaurin E (**46**)	*(75)*
		rabdoternin A (**55**)	*(76)*
		isodonal (**71**)	*(75)*
		isodonic acid (**80**)	*(75)*
R. trichocarpa (Maxim.) Hara	Japan	oridonin (isodonol) (**32**)	*(32, 36, 83, 87)*
		enmenol (**36**)	*(83)*
		longikaurin D (**45**)	*(80)*
		trichokaurin (enmenin) (**56**)	*(78, 84)*
		enmedol (**57**)	*(83)*
		enmelol (**58**)	*(84)*
		enmein (**62**)	*(9, 11, 12, 32, 87)*
		isodocarpin (**64**)	*(32, 33)*
		isodotricin (**65**)	*(32, 34)*
		nodosin (**66**)	*(32, 35)*
		trichodonin (**70**)	*(32)*
		dihydroenmein (**81**)	*(77)*
		ememodin (**82**)	*(84)*
		ememogin (**83**)	*(83, 85)*
		trichorabdonin (**84**)	*(85)*
		trichorabdal A (**92**)	*(79)*
		trichorabdal B (**93**)	*(80)*
		trichorabdal C (**94**)	*(79)*
		trichorabdal D (**95**)	*(79)*
		trichorabdal E (**96**)	*(81)*
		trichorabdal F acetate (**97**)[4]	*(81, 82)*
		trichorabdal G acetate (**98**)[4]	*(81, 82)*
R. umbrosa (Maxim.) Hara	Japan	umbrosin A (**1**)	*(88)*
		umbrosin B (**2**)	*(88)*
R. umbrosa (Maxim.) Hara var. *latifolia* (Okuyama) Hara	Japan	umbrosin A (**1**)	*(96)*
		kamebanin (**7**)	*(89, 96)*
		isodomedin (**12**)	*(89, 96)*
		shikoccidin (**15**)	*(89)*
		mebadonin (**16**)	*(96)*
		leukamenin E (**21**)	*(89)*
		kamebakaurinin (**59**)	*(96)*
		shikoccin (**101**)	*(89)*
		rabdoratifolin (**105**)	*(89)*
R. umbrosa (Maxim.) Hara var. *leucantha* (Murai) Hara f. *kameba* (Okuyama ex Ohwi) Hara	Japan	kamebanin (**7**)	*(25, 91)*
		isodomedin (**12**)	*(25)*
		mebadonin (**16**)	*(25, 90)*
		leukamenin A (**17**)	*(92)*
		leukamenin B (**18**)	*(92)*

Table V *(continued)*

Species name	Origin	Isolated diterpenoids	Reference
		leukamenin C (**19**)	*(92)*
		leukamenin D (**20**)	*(92)*
		leukamenin E (**21**)	*(92)*
		leukamenin F (**22**)	*(92)*
		kamebakaurin (**31**)	*(93)*
		kamebakaurinin (**59**)	*(94)*
		kamebacetal A (**60**)	*(95)*
		kamebacetal B (**61**)	*(95)*

[1] Originally named oridonin.
[2] Originally named ponicidin.
[3] First glycoside from a *Rabdosia* species.
[4] Acetate of the natural product.
[5] Only gibberellane found in a *Rabdosia* species.

A brief historical outline of work on three species, *R. japonica*, *R. trichocarpa* and *R. umbrosa* var. *leucantha* f. *kameba* is presented in the following paragraphs.

1. *R. japonica:* As mentioned earlier, plectranthin (*2*) in 1910 was the first substance isolated from this species. After about 40 years, four other diterpenoids were isolated, but their structures were not determined (*3, 12*). Enmein (**62**) was first isolated from *R. trichocarpa*, but in this species it is usually contaminated by dihydroenmein (**81**). The first isolation of pure enmein (**62**) from *R. japonica* was reported in 1966 (*8*). Enmein 3-acetate (**63**) (*8*), isodocarpin (**64**) (*33*), isodotricin (**65**) (*32, 34*), oridonin (**32**) (*32, 36*), ponicidin (**33**) (*32, 37*), epinodosinol (**67**) (*38*), sodoponin (**34**) (*38*), odonicin (**35**) (*39*), isodoacetal (**68**) (*39*), and nodosinin (**69**) (*39*) have been isolated and their structures elucidated. Trichodonin was first isolated in 1958 (*12*); structure (**70**) was proposed for it on the basis of spectroscopic evidence (*40*) and established chemically (*41, 42*). Isodonal (**71**) (*42, 43*) and epinodosin (**72**) (*41, 42, 38b*) were also isolated and their structures were clarified. Recently the known diterpenoids, enmenol (**36**) and lasiokaurin (**37**), and several new diterpenoids, rabdosin A (**73**) (*44*), rabdosin B (**74**) (*44*), and rabdosin C (**75**) (*45*) have been isolated by Chinese chemists. Constituents of *R. japonica* other than diterpenoids such as steroids, triterpenoids, and flavonoids have also been isolated (*46*). Isolation of steroids and triterpenoids from tissue cultures of *R. japonica* was also reported (*47*).

2. *R. trichocarpa:* Work on the constituents was reported in 1958 by three research groups (*9, 11, 12*) independently. The major component was named enmein by the first discoverers, Ikeda and Kanatomo (*9*). Isodonin

named thus by NAYA (*11*) was proved to be identical with enmein. The structure determination of enmein (**62**) is described in II - 1. Dihydroenmein (**81**) was found in the same plant (*77*).

Since 1966, the minor constituents have been investigated. E. FUJITA and coworkers isolated isodocarpin (**64**) (*33*), nodosin (**66**) (*32, 35*), trichodonin (**70**) (*32, 40*), and oridonin (**32**) (*32, 36*), which had been isolated previously from *R. japonica*, and a new diterpenoid, trichokaurin; they elucidated the structure (**56**) of the latter (*78*). Recently, they also isolated the following new diterpenoids, in addition to longikaurin D (**45**) which had been isolated from *R. longituba*, and determined their structures: trichorabdals A (**92**) (*79*), B (**93**) (*80*), C (**94**) (*79*), and D (**95**) (*79*). Furthermore, they obtained a crystalline mixture from a fraction of higher polarity and isolated trichorabdals E (**96**) (major constituent) and F and G as acetates (**97**) and (**98**), respectively (*81, 82*).

OKAMOTO and coworkers isolated isodonol (**32**) (*83*), enmedol (**57**) (*83*), enmenol (**36**) (*83*), enmein (**62**) (*84*), enmelol (**58**) (*84*), and ememodin (**82**) (*84*), and determined their structures. Isodonol and enmenin were proved to be identical with oridonin (**32**) and trichokaurin (**56**), respectively. They also isolated six other diterpenoids (*83*), among which "compound C" proved to be identical with trichodonin (**70**) and ememogin was shown to have structure (**83**) by T. FUJITA and coworkers (*85*). The latter authors also isolated a new diterpenoid, trichorabdonin, and determined its structure (**84**) (*85*).

Gas chromatographic analysis showed that the two main components, enmein and oridonin, were most abundant in June (*86*). All of the foregoing diterpenoids were isolated from the leaves. Investigation of the diterpenoids in the stems resulted in isolation of smaller quantities of enmein and oridonin (*87*).

3. *R. umbrosa var. leucantha f. kameba:* KUBOTA and coworkers isolated mebadonin (**16**) (*25, 90*) and kamebanin (**7**) (*25, 91*) in addition to the known isodomedin (**12**), and determined their structures. Afterwards, T. FUJITA and coworkers (*92*) isolated the following new diterpenoids and elucidated their structures: leukamenins A (**17**), B (**18**), C (**19**), D (**20**), E (**21**), and F (**22**). They also isolated and characterized kamebakaurin (**31**) (*93*), kamebakaurinin (**59**) (*94*), kamebacetals A (**60**) and B (**61**) (*95*).

The distribution of diterpenoids in five *Rabdosia umbrosus* varieties has been investigated from the chemotaxonomic point of view (*96*); *R. umbrosa var. umbrosa, R. umbrosus var. hakusanensis, R. umbrosus var. latifolia, R. umbrosus var. excisinflexa,* and *R. umbrosus f. kameba* were studied by reverse phase HPLC (*97*) and six of the known *ent*-kaurene type diterpenoids, kamebakaurinin, isodomedin, umbrosin A, mebadonin, kamebakaurin, and kamebanin were recognized. Kamebanin was found in all varieties, and umbrosin A (**1**) and its epimer mebadonin (**16**) were found to be present in pairs (*96*). The isolation and separation of *Rabdosia* diterpenoids by HPLC has been reported (*98*).

3. Properties and Reactions Important for Structure Determination

For structure determination of the *Rabdosia* diterpenoids, spectroscopic investigations and chemical reactions are useful in general and X-ray analysis is used by necessity.

a) Spectral Aspects and Important Reactions for Ring D Structure Determination

Oxidation patterns of ring D in *Rabdosia* diterpenoids derived from *ent*-16-kaurene are classified into 5 types shown in Table VI. Each class exhibits characteristic spectral data, from which the structural features of ring D are easily deduced. Table VI shows typical UV, IR, and NMR data for a typical example belonging to each class.

Chemical transformations used for ring D structure determination are summarized in Scheme 1. These reactions proceed stereoselectively by attack of the reagents from the less hindered side of the molecule. $NaBH_4$ reduction of compounds containing an α-methylenecyclopentanone generally yields the tetrahydroderivative, but an allylic alcohol was obtained as a minor product on reduction of lasiokaurin (*49*). Such an allylic alcohol

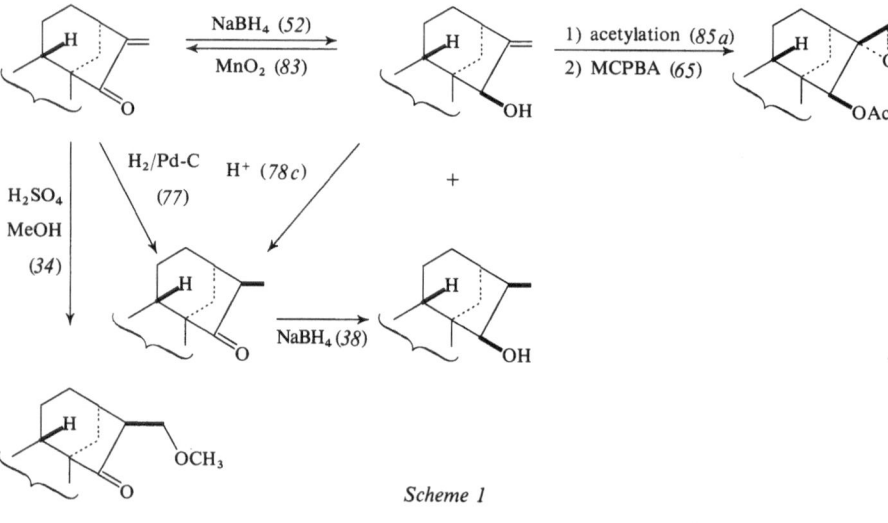

Scheme 1

Table VI. Oxidation Patterns of Ring D and Their Spectral Data

oxidation patterns of ring D				
	R = H or Ac	R = H or OMe		
the number of diterpenoids				
78	23	5	1	1
representative				
mebadonin (**16**)	epinodosinol (**67**) (R = H)	isodotricin (**65**) (R = OMe)	16,17-epoxyshiko-kianal acetate (**91**)	trichorabdonin (**84**)
ν max (cm⁻¹)				
1730 ($>$=O)	1663 ($>$=)	1755 ($>$=O)	—	1750 ($>$=O)
1640 ($>$=)				
λ max (nm)				
232 (ε 8230)	—	—	—	—
δ (ppm)				
5.29 (t*, Ha)	5.25 (br.s, Ha)	3.15 (s, OMe)	2.77 (d*, Ha)	1.49 (s, 16-Me)
5.87 (t*, Hb)	5.58 (br.s, Hb)	3.58 (AB part of ABX type)	2.97 (d*, Hb)	
*J = 1 Hz	5.54 (br.s, Hc)		5.28 (s, Hc)	
			*J = 4.6 Hz	

on exposure to acid undergoes a garryfoline-cuauchichicine type rearrangement to the methyl ketone in good yield (*78c*). The rearrangement is characteristic of β-orientation of the allylic 15β-hydroxyl group and presumably involves the non-classical carbonium ion shown in Fig. 1. For further reactions shedding light on the stereochemistry of ring D, the reader is referred to III-6-(a).

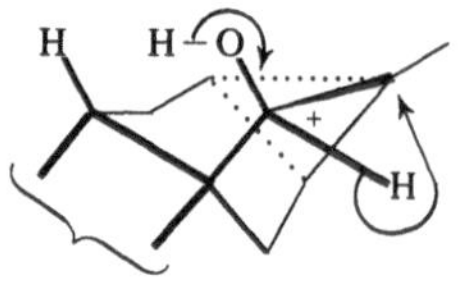

Fig. 1

b) Structure Determination of ent-*Kaurane-type Diterpenoids*

This type of diterpenoids contains C-20-non-oxygenated and C-20-oxygenated *ent*-kauranes. The latter are classified into three groups on the basis of their oxidation patterns. Table VII exemplifies representative diterpenoids of the four groups and their ^{1}H NMR data which are characteristic and therefore useful for structure determination.

The NMR spectra of umbrosin A type diterpenoids are characterized by three methyl singlets, the signal of an axial proton due to 7β-H, and the singlet-like resonance of 14α-H whose appearance is due to the dihedral angle of ca. 90° between 14α-H and 13H. Only two diterpenes in this group, inflexin (**10**) and inflexinol (**11**), lack the 7α-hydroxy and 14β-hydroxy or acetoxy groups. This structural feature indicating the presence of 7α- and 14β-hydroxy groups is also found in amethystoidin A (**23**), kamebakaurin (**31**), and kamebakaurinin (**59**) which belong to the C-20-oxygenated kaurane type and was useful in the structure determination of these diterpenes.

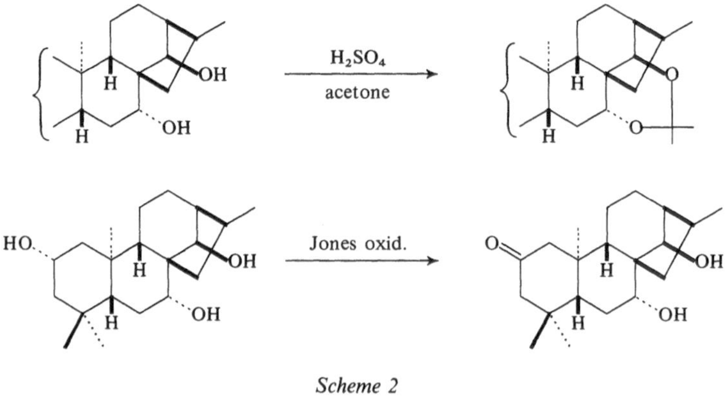

Scheme 2

Table VII. *Representative ent-Kaurane Diterpenoids and Their NMR Data*

	umbrosin A (1)	oridonin (32)	longikaurin C (44)	kamebacetal A (60)
δ (ppm) C-6-H	—	4.29 (1H, d, J=7 Hz)*	4.30 (1H, d, J=6.5 Hz)	—
C-7-H	4.45 (1H, dd, J=13 and 5 Hz)	—	—	4.70 (1H, dd, J=4 and 2 Hz)
C-14-H	4.90 (1H, br, s)	5.35 (1H, s)		5.10 (1H, d, J=1.5 Hz)
C-18-H₃	0.85, 0.90	1.13, 1.20	1.33 (3H, s)	0.76, 0.93
C-19-H₃ (or -H₂)	1.04 (each 3H, s)	(each 3H, s)	4.40, 4.68 (each 1H, AB type, J=11 Hz)	(each 3H, s)
C-20-H₃ (-H₂ or -H)		4.42, 4.78 (each 1H, AB type, J=10 Hz)	4.34, 4.62 (each 1H, AB type, J=11 Hz)	5.46 (1H, d, J=1 Hz)

* After addition of D₂O.

Useful chemical reactions of compounds in this group involve acetonide formation between 7α- and 14β-hydroxy groups (*88–94*) and selective oxidation by Jones reagent (*88*) of hydroxyl groups in positions other than C-7 and C-14. This resistance to oxidation shown in Scheme 2 is due to hydrogen bonding between the 7α- and 14β-hydroxyl groups.

Oridonin type diterpenoids (32 representatives) have two methyl groups on a tertiary carbon atom, 6β-hydroxyl and a hemiacetal function involving a hydroxyl on C-20 and a 7-ketone. Characteristic for this class is spin-spin coupling between 6α-H and 6β-OH with a large coupling constant ($J = 10$ Hz); addition of D_2O produces a doublet due to spin-spin coupling ($J = 6–7$ Hz) between 6α-H and 5-H. An AB-type signal ($J = 10–11$ Hz) due to the two protons on C-20 is also observed; one of these is long-range coupled ($J \sim 3$ Hz) to 9-H.

Longikaurin C-type diterpenoids (6 representatives) contain an oxygen function at C-18 or C-19 in addition to the oridonin-type oxidation pattern. Whether C-18 or C-19 is oxidized can be ascertained by establishing the presence or absence of an NOE between C-18 and C-20 or between C-19 and C-20 (*51, 52*).

The 6-hydroxy-7-hemiacetal structure which is common to oridonin-type and longikaurin C-type diterpenoids is oxidatively cleaved at the C-6, C-7 bond by periodate oxidation. If a hydroxyl is present at C-1, then relactonization occurs to give enmein-type diterpenes (*36, 78*). If no hydroxy group is present at C-1 (R = H or OAc in Scheme 3), then the oxidative cleavage gives spirolactone-type compounds [*26, 36, 79, 80*, see also III-5-(a)].

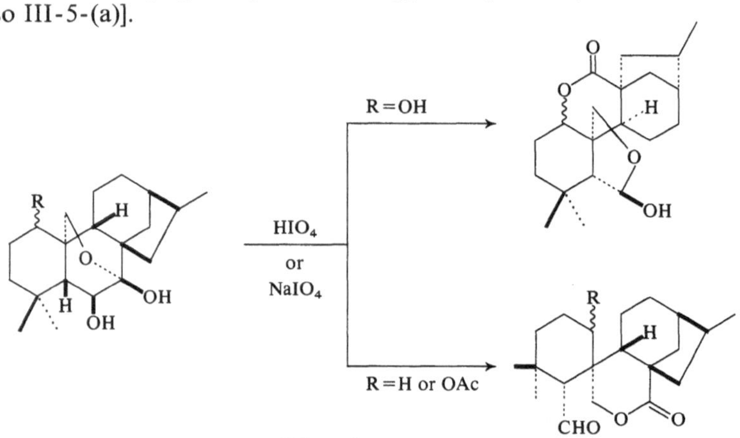

Scheme 3

In 7-hemiacetal compounds, a retro-Dieckmann or retro-Aldol reaction occurs under alkaline conditions and gives *ent*-abietane derivatives by opening of ring D as shown in Scheme 4 (*36b, 78c*).

References, pp. 149—157

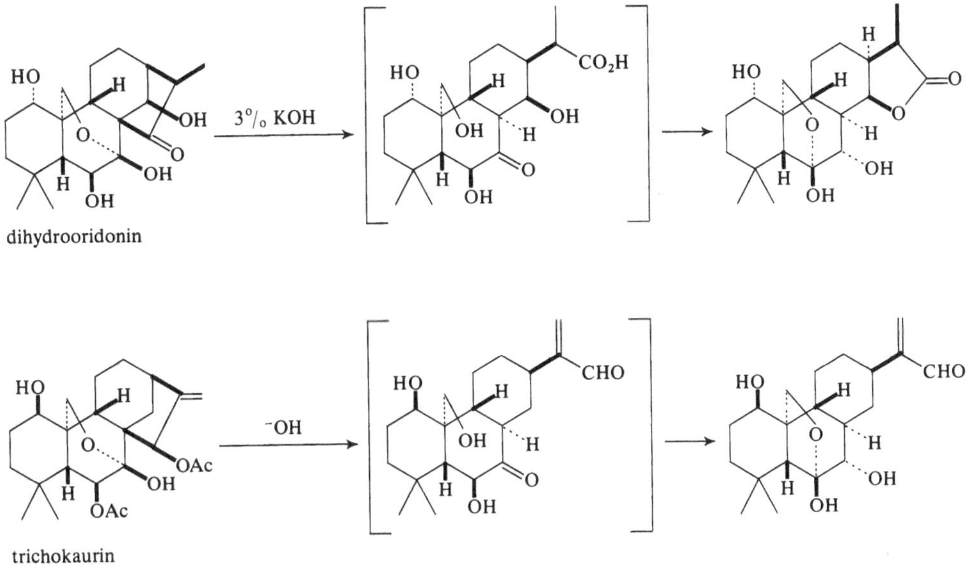

HO ... dihydrooridonin   3% KOH →   [HO ... CO₂H]   →   HO ... O

trichokaurin   ⁻OH →   [HO ... CHO]   →   HO ... CHO

Scheme 4

The last group in which C-20 is oxidized to an acetal includes ponicidin (**33**), kamebacetal A (**60**), and kamebacetal B (**61**). The structure of ponicidin (**33**) is similar to that of oridonin (**32**), except for the extra presence of an ether linkage between C-14 and C-20, and was determined by decoupling and use of the INDOR technique (*37*). The NMR spectra of this group of diterpenes show a singlet or a narrowly-split doublet due to 20-H. The small coupling is attributed to a W-shape long range coupling with 9-H. Selective oxidation at C-20 of the hemiacetal function is possible (*95*) (see Scheme 5).

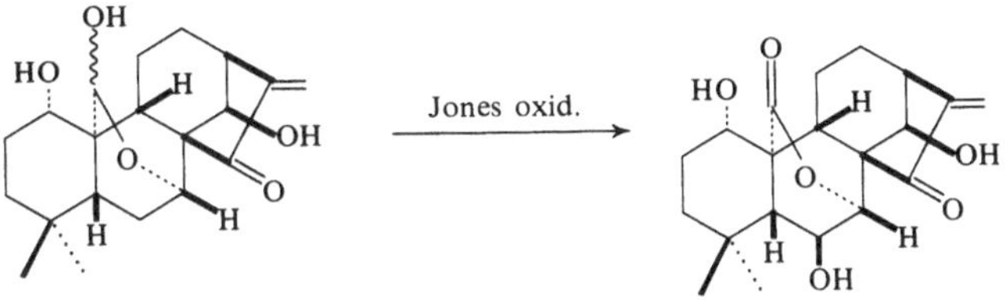

HO ... OH   Jones oxid. →   HO ... O

Scheme 5

The conformations of rings A, B, and C in these *ent*-kaurane-type diterpenoids were elucidated by X-ray analysis and use of NMR data. Thus, rings A, B, and C of shikoccidin (**15**) were shown to have chair conformations (*73*), while ring A of mebadonin (**16**) was shown to have a skew boat conformation because of the steric interaction between the 20-

methyl, 2-hydroxy, and 19-methyl groups (*90*). On the other hand, ring B of sodoponin (**34**) was shown to have a boat conformation due to the ether linkage between C-7 and C-20; ring C has also a boat-like conformation because of induced strain (*38*).

c) Structure Determination of ent-6,7-Secokaurane-type Diterpenoids

Before 1979, the only 6,7-secokauranes were of what we have called the enmein-type, but since then a number of spirolactone-type diterpenoids have been reported, the first being shikodonin (**88**). As shown in Table VIII, each group is classified into four types. NMR data for a representative of all eight types are also listed.

The NMR spectra of enmein-type diterpenoids in which 6α-H is part of a hemiacetal contain only a singlet assignable to this proton because of its ca. 90° dihedral angle with 5-H [*e.g.* isodocarpin (**64**), carpalasionin (**77**) in Table VIII, and enmein (**62**), isodotricin (**65**) *etc.* in Table II]. In spirolactone-type diterpenoids having a hemiacetal function 6-H is a doublet [*e.g.* trichorabdal E (**96**) in Table VIII and trichorabdals F (**97**) and G (**98**) in Table II]. In the enmein-type diterpenoids of this group, C-6 is *R* except for rabdosin A (**73**), while the spirolactone-type diterpenoids occur as a mixture of 6*R*- and 6*S*-isomers (*81*). Shikodonin has been reported to have structure (**88**) on the basis of X-ray analyses of its 6-O-methyl and 6-O-ethyl derivatives, but our experimental results are inconsistent with this proposal, and thus the stereochemistries of C-5 and C-6 in (**88**) seem questionable.

In both enmein- and spirolactone-types, the 20-methylene protons are observed as AB-type signals. The coupling constants vary *e.g.*, they are $8 \sim 10$ Hz in five-membered hemiacetal rings, $10 \sim 11$ Hz in δ-lactones, and ~ 13 Hz in non-cyclized acetates. These values allow one to deduce likely structures. One proton of the C-20 methylene involved in a hemiacetal ether or lactone ring exhibits W-type long-range coupling to 9-H; this results in a somewhat broadened singlet or a narrowly split doublet with a coupling constant below 3 Hz.

Chemical reactions which were useful for structure determination of enmein-type diterpenoids involve removal or chemical transformations of the functional groups which are illustrated in Scheme 6.

On the other hand, the periodate cleavage of known kauranes illustrated in Scheme 3 is crucial for the structure determination of spirolactone-type diterpenoids (*26, 79, 80*). Some other transformations employed in this series are shown in Schemes 7 and 8. Trichorabdals C (**94**) and D (**95**) having 3β-hydroxy and 19-O-acetate groups on heating with acetic acid undergo migration of the acetyl group from 19-O to 3-OH, hemiacetal ring formation between C-6 and C-19 and subsequent acetylation. Thus

Table VIII. ¹H NMR Data of Typical 6,7-Secokaurane Type Diterpenoids

enmein-type

δ (ppm)	isodocarpin (**64**)	isodonal (**71**)	carpalasionin (**77**)	sculponeatin A (**78**)
C-1-H	4.68 (1H, t, J = 8 Hz)	4.07 (1H, dd, J = 10 and 12 Hz)	—	5.62 (1H, dd, J = 6 and 10 Hz)
C-6-H	5.73 (1H, s)	9.90 (1H, d, J = 3 Hz)	5.98 (1H, s)	6.13 (1H, d, J = 5 Hz)
C-18-H₃	0.97, 1.02 (each 3H, s)	1.01, 1.10 (each 3H, s)	1.14 (3H, s)	1.06 (3H, s)
C-19-H₃ (C-19-H₂)			4.26, 4.58 (each 1H, ABd, J = 12 Hz)	3.47, 4.08 (each 1H, ABd, J = 8 Hz)
C-20-H₂	4.26, 4.41 (each 1H, ABd, J = 9 Hz)	5.08* (2H, AB type, J = 13 Hz)	4.38, 4.57 (each 1H, ABd, J = 9 Hz)	4.18, 4.34 (each 1H, ABd, J = 10 Hz)

spirolactone-type

δ (ppm)	trichorabdal A (**92**)	effusin (**85**)	trichorabdal E** (**96**)	sculponeatin C (**87**)
C-1-H	—	—	—	4.64 (1H, ddd, J = 1, 3, and 3 Hz)
C-6-H	10.03 (1H, d, J = 3 Hz)	9.75 (1H, d, J = 5 Hz)	5.60 (1H, d, J = 3 Hz)	5.84 (1H, d, J = 4 Hz)
C-18-H₃	0.97, 1.00 (each 3H, s)	1.02 (3H, s)	1.06 (3H, s)	1.11 (3H, s)
C-19-H₃ (C-19-H₂)		3.55, 3.77 (each 1H, ABd, J = 9 Hz)	3.53, 4.01 (each 1H, ABd, J = 9 Hz)	3.70, 3.89 (each 1H, ABd, J = 8 Hz)
C-20-H₂	4.71, 5.10 (each 1H, ABd, J = 11 Hz)	4.16, 4.22 (each 1H, ABd, J = 10 Hz)	4.41, 4.79 (each 1H, ABd, J = 11 Hz)	4.36, 5.33 (each 1H, ABd, J = 12 Hz)

*δ Value at center of the AB-type.   ** Major C-6-isomer.

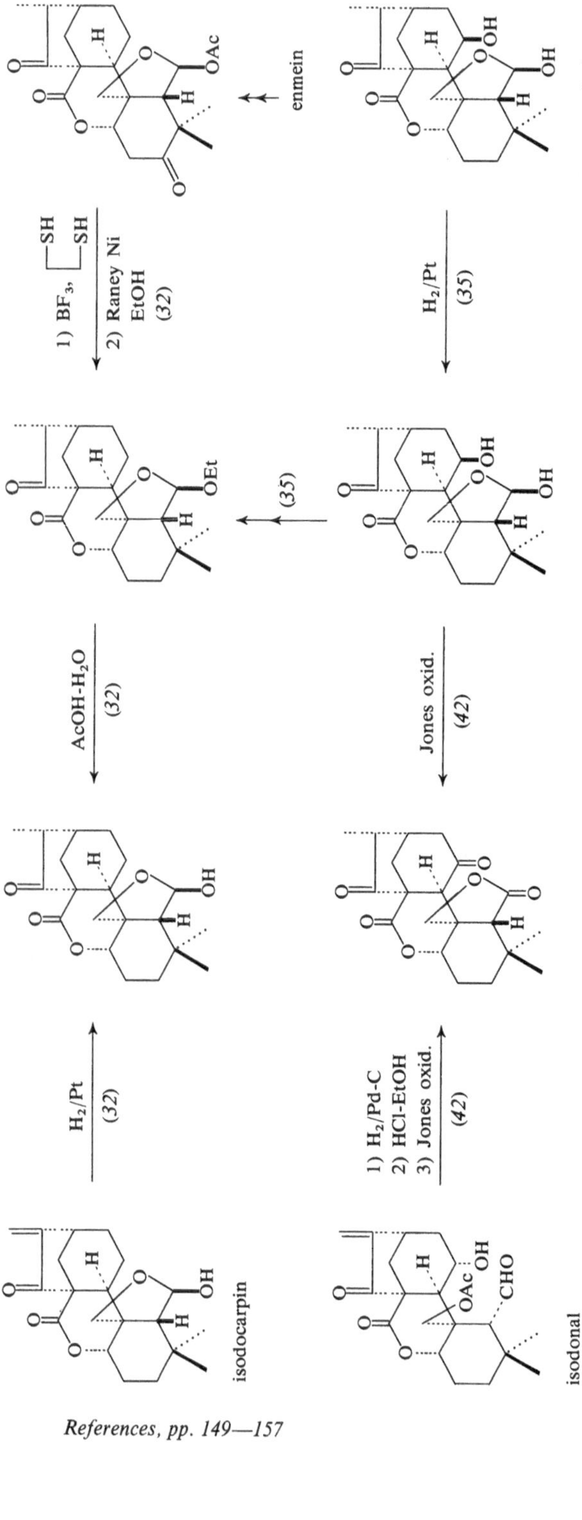

Scheme 6

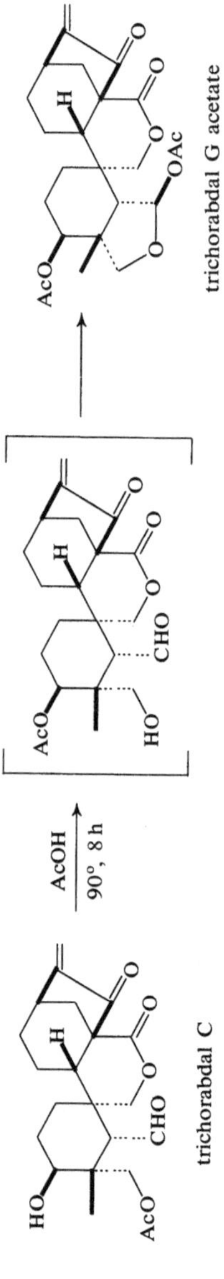

Scheme 7

trichorabdal C (**94**) was transformed to trichorabdal G acetate (**98**) as shown in Scheme 7. Hence this reaction is useful for establishing the location of a hydroxy group at C-3 in ring A (*79*).

Spirolactone-type diterpenoids having a 15-carbonyl group are generally unstable to alkali. For instance, trichorabdal B (**93**) on treatment with dilute alkali for a brief period undergoes rearrangement by way of a *retro*-Claisen reaction which is followed by an aldol condensation as shown in Scheme 8 (*79*). If the reaction time is extended isolation of the product is difficult, because a number of other products are formed.

Conformations of rings A and C in 6,7-secokaurane-type diterpenoids have been clarified by X-ray crystallographic analyses. Table IX summarizes the results. Enmein derivatives (**E-I**) and (**E-II**) display a difference in conformation of ring C. The conformation of ring A in O-methylshikodonin and trichorabdal B (**93**) differs from that in (**E-1**) and (**E-II**): the former has an equatorial 10-9 C-C bond, while the latter has an axial 10-9 C-C bond. The conformation of ring A in trichorabdal C (**94**) is different from that in trichorabdal B (**93**) but is the same as that found in (**E-I**) and (**E-II**). In solution ring A of trichorabdals A–D (**92–95**) may be an equilibrium mixture of those two chair forms; their NMR spectra generally show broad signals and hence measurement at elevated temperatures is necessary in order to obtain sharp signals (*79*).

d) Structure Determination of ent-8,9-Secokaurane-type Diterpenoids

Characteristic of this type of diterpenoid are spectral features indicating the presence of a cross conjugated dienone system in a five-membered ring: UV maxima at 244 ~ 248 nm and proton and ^{13}C NMR signals due to an exocyclic methylene, an olefinic methine, and an allylic bridge-head methine are observed (*71–73, 89*). As a representative example, the NMR data of shikoccin (**101**) are shown in Fig. 2.

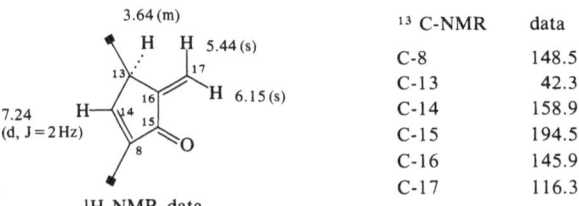

^{13}C-NMR	data
C-8	148.5
C-13	42.3
C-14	158.9
C-15	194.5
C-16	145.9
C-17	116.3

Fig. 2

The *ent*-8,9-secokaurane type skeleton is chemically derived from an *ent*-kaurane-type diterpene with a 9-hydroxy group. Thus shikoccidin (**15**) on acetylation with a mixture of acetic anhydride and pyridine gave shikoccin acetate as shown in Scheme 9 (*73*).

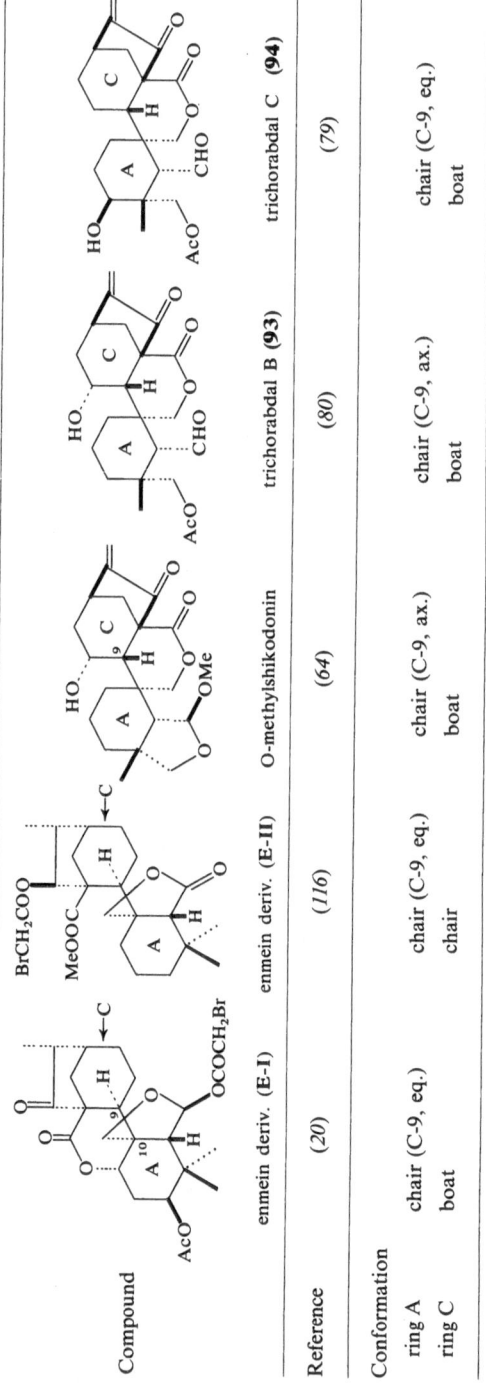

Scheme 8

Table IX. *Conformation of Rings A and C in 6,7-Secokaurane Type Diterpenoids*

Compound	enmein deriv. (E-I)	enmein deriv. (E-II)	O-methylshikodonin	trichorabdal B (93)	trichorabdal C (94)
Reference	(20)	(116)	(64)	(80)	(79)
Conformation					
ring A	chair (C-9, eq.)	chair (C-9, eq.)	chair (C-9, ax.)	chair (C-9, ax.)	chair (C-9, eq.)
ring C	boat	chair	boat	boat	boat

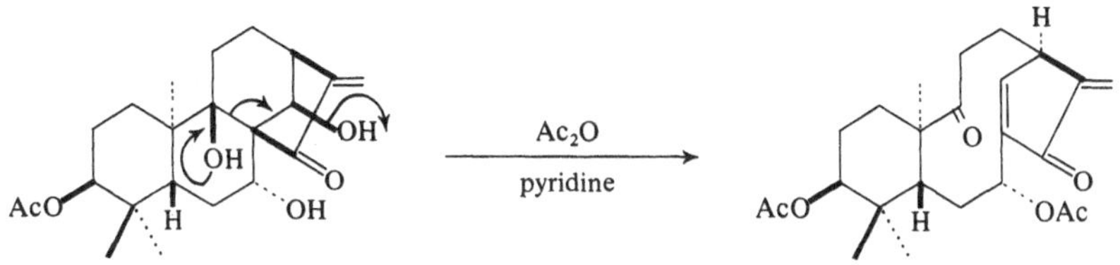

shikoccin acetate

Scheme 9

m-Chloroperbenzoic acid or Jones reagent selectively effects epoxidation of the *endo*-double bond of shikoccin (**101**) (*71, 72*). This is shown in Scheme 10. The 7-hydroxy is transformed to a methoxy group on treatment of shikoccin (**101**) with acid in methanol solution (*72*).

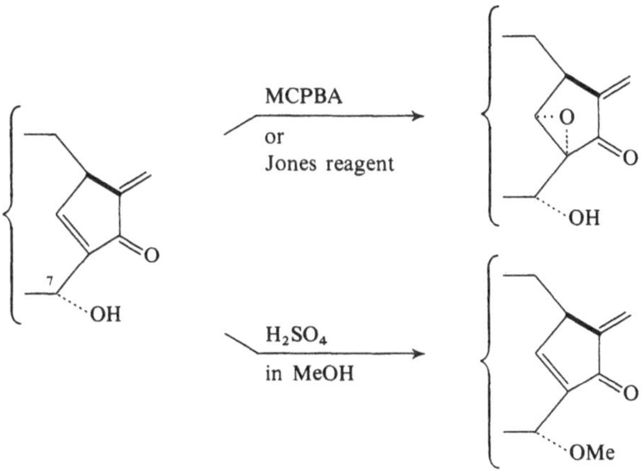

Scheme 10

X-ray crystallographic analyses were carried out on shikodomedin (**99**) (*71*) and shikoccin monoacetate (*74*).

e) Structure Determination of ent-7,20-Bonded Kaurane-type Diterpenoids

Only two members of this class have been reported so far. The structure determination (*58*) of one of these, phyllostachysin A (**106**), is shown in Scheme 11. Sodium periodate oxidation of the dihydro derivative of phyllostachysin A (**106**) selectively cleaved the 7,20-glycol to give a 7-oxo-20-aldehyde whose ring closure afforded a 7,20-dihemiacetal. On the other hand, the triacetate of dihydro-(**106**) was cleaved at the 6-7 C-C bond by lead tetraacetate to give a keto acid. These reactions proved that there is a C-C bond between the 7 and the 20 position.

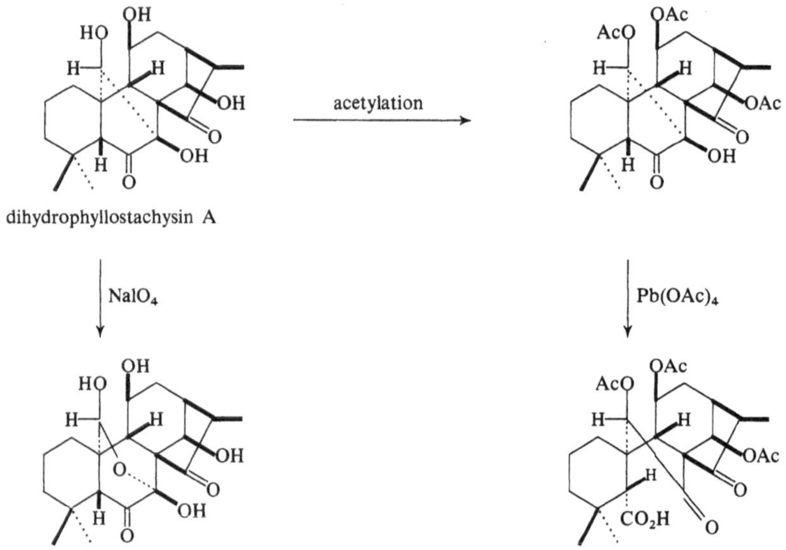
acetylation
dihydrophyllostachysin A
$NaIO_4$
$Pb(OAc)_4$

Scheme 11

f) Structure Determination of an ent-*Gibberellane-type Diterpenoid*

Only one such diterpenoid, rabdoepigibberellolide (**108**) has been reported (*66*). Comparison of its NMR spectrum with that of rabdosianin A (**50**) isolated from the same plant source led to assignment of structure (**108**), which was confirmed by X-ray analysis. Its biogenesis from rabdosianin A has been suggested (*66*).

III. Chemical Conversions

Because of their unique structures, their abundant occurrence in nature and their relative ease of isolation, the chemistry of *Rabdosia* diterpenoids has been studied thoroughly and many articles on their chemical behavior and transformations have been published.

1. Conversion of *Rabdosia* Diterpenoids to *ent*-Kaurane and *ent*-Kaurene

A chemical conversion of enmein (**62**) into *ent*-kaurane was attempted in order to provide chemical evidence for the structure and absolute configuration of enmein. In 1967, OKAMOTO and coworkers (*99*) and E. FUJITA and coworkers (*100*) independently of each other carried out this conversion successfully. Both groups utilized the acyloin condensation as the key reaction for ring B closure and used the same lactone-ester derivative (**109**) as the substrate. Compound (**109**) was prepared from enmein by the route shown in Scheme 12 by OKAMOTO's group and by the route shown in Scheme 13 by FUJITA's group. Both groups accomplished the conversion of (**109**) to *ent*-kaurane (**112**) by the route shown in Scheme 14: acyloin condensation of compound (**109**) gave (**110**) as the major product, which on Wolff-Kishner reduction followed by catalytic reduction gave alcohol (**111**). Oxidation of (**111**) followed by a second Wolff-Kishner reduction of the resulting aldehyde gave a hydrocarbon, which was identical with *ent*-kaurane (**112**). Thus the structure and absolute configuration of enmein were established chemically.

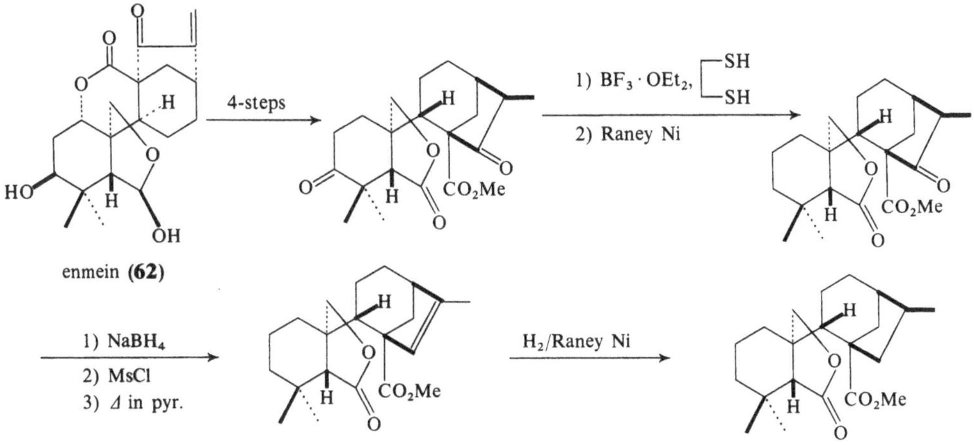

enmein (**62**)

Scheme 12   (**109**)

(**62**)

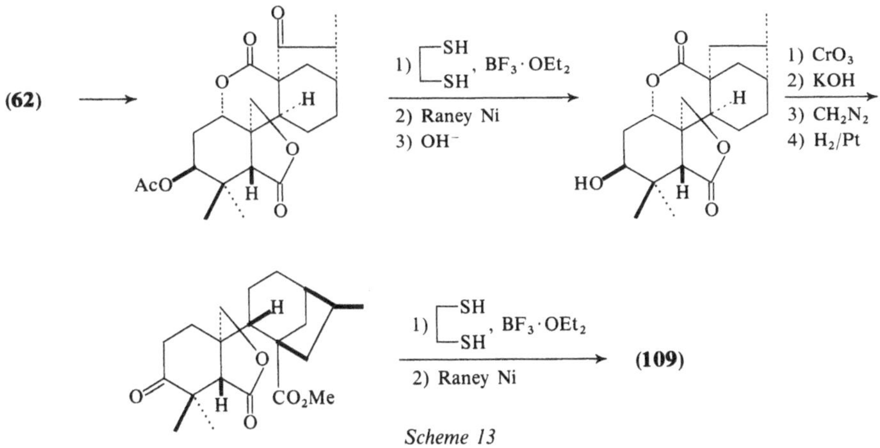

Scheme 13

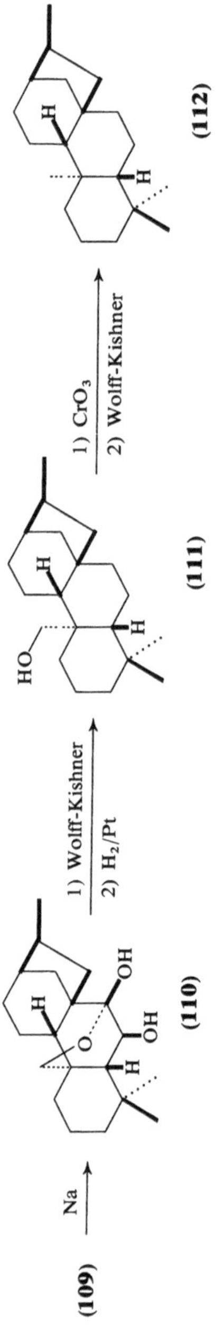

Scheme 14

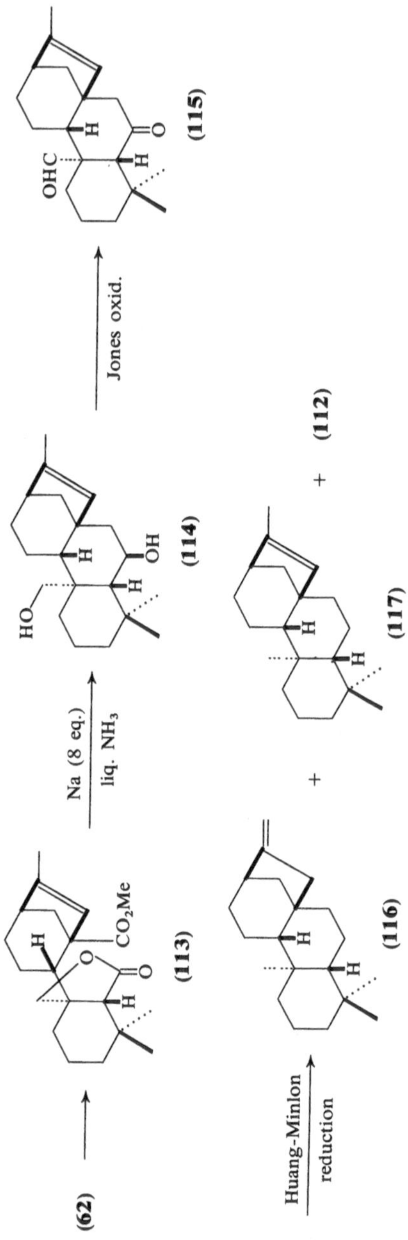

Scheme 15

The conversion of enmein (**62**) and trichokaurin (**56**) into *ent*-16-kaurene was achieved by E. FUJITA *et al.* (*101, 102*). This conversion also represents a formal conversion to the aconite alkaloids, hence will be described in III-3. A more direct conversion (*103*) of enmein (**62**) into *ent*-kaurenes carried out afterwards is described here and is shown in Scheme 15. Conditions for an acyloin reaction which gave diol (**114**) in good yield from (**113**), were found after lengthy investigation. Diol (**114**) gave the desired *ent*-16-kaurene (**116**), in addition to *ent*-15-kaurene (**117**), and *ent*-kaurane (**112**) (ratio of products 5 : 2 : 3), after oxidation to a keto aldehyde (**115**) followed by Huang-Minlon reduction of the latter. Under NAGATA's modification of the Wolff-Kishner reduction (**115**) gave exclusively *ent*-kaurane (**112**) (*104*).

Since *ent*-16-kaurene (**116**) has been converted to phyllocladene, atisirene, and neoatisirene, the foregoing conversion also implies a formal chemical conversion of enmein to these diterpenoid hydrocarbons.

2. Conversion of Enmein to *ent*-Abietane

E. FUJITA and coworkers were interested in conversion of a 6,7-secokaurane type diterpenoid such as enmein to *ent*-abietane (**123**), a hydrocarbon unknown at the time. Alcohol (**119**) [derived from enmein (**62**) as shown in Scheme 16] was cleaved with NaH to an aldehyde, a ring-D open derivative. The latter was converted into a lactone ester (**120**) which was subjected to the acyloin condensation to give isomeric products (**121**) and (**122**). Both compounds were deoxygenated in four steps to give a hydrocarbon (**123**). On the other hand, starting from (−)-abietic acid and/or dihydroabietic acid, hydrocarbon (**118**) was prepared by the route shown in Scheme 17. The products (**123**) and (**118**) were shown to be enantiomeric, hence they were *ent*-abietane and abietane, respectively (*105*).

3. Conversion of Enmein and Trichokaurin to Aconite Alkaloids

E. FUJITA and coworkers carried out chemical conversions of trichokaurin (**56**) to a keto acid (**127**) and of enmein (**62**) to triol (**126**). Compounds (**126**) and (**127**) had been converted into *ent*-16-kaurene (**116**) and three aconite alkaloids (atisine, garryine, and veatchine) by MASAMUNE and coworkers; hence these chemical conversions imply a formal chemical conversion of trichokaurin and enmein to *ent*-16-kaurene and three aconite alkaloids.

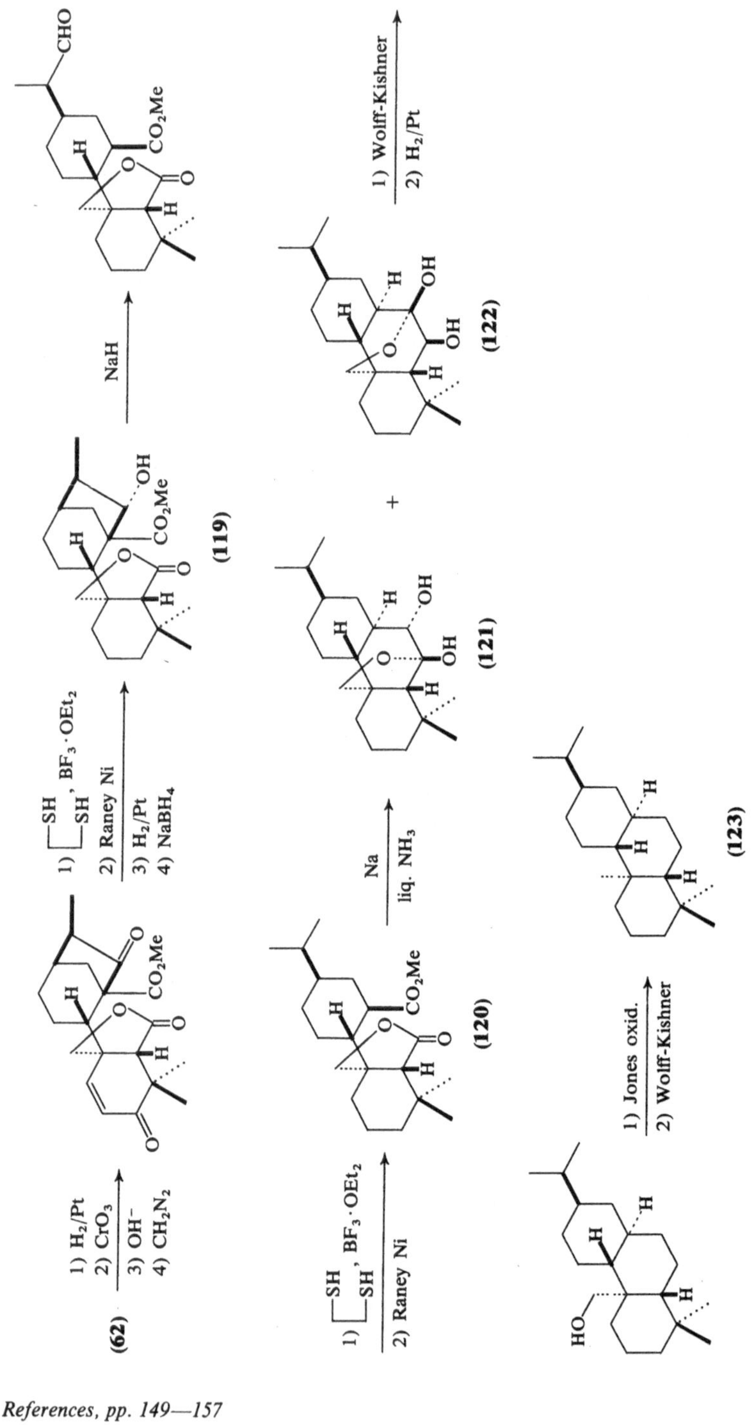
(62)
1) H_2/Pt
2) CrO_3
3) OH^-
4) CH_2N_2
CO_2Me
1) SH, SH, $BF_3 \cdot OEt_2$
2) Raney Ni
3) H_2/Pt
4) $NaBH_4$
OH
CO_2Me
(119)
NaH
CHO
CO_2Me
1) SH, SH, $BF_3 \cdot OEt_2$
2) Raney Ni
CO_2Me
(120)
Na
liq. NH_3
OH
OH
(121)
+
OH
OH
(122)
1) Wolff-Kishner
2) H_2/Pt
HO
1) Jones oxid.
2) Wolff-Kishner
(123)

Scheme 16

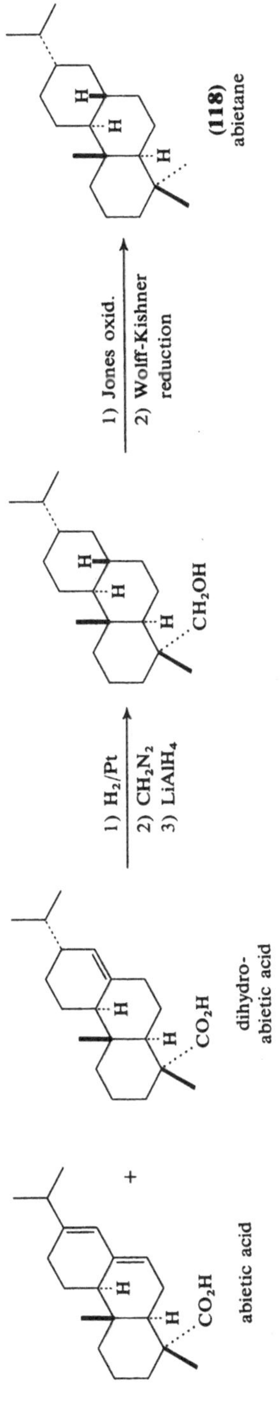

Scheme 17

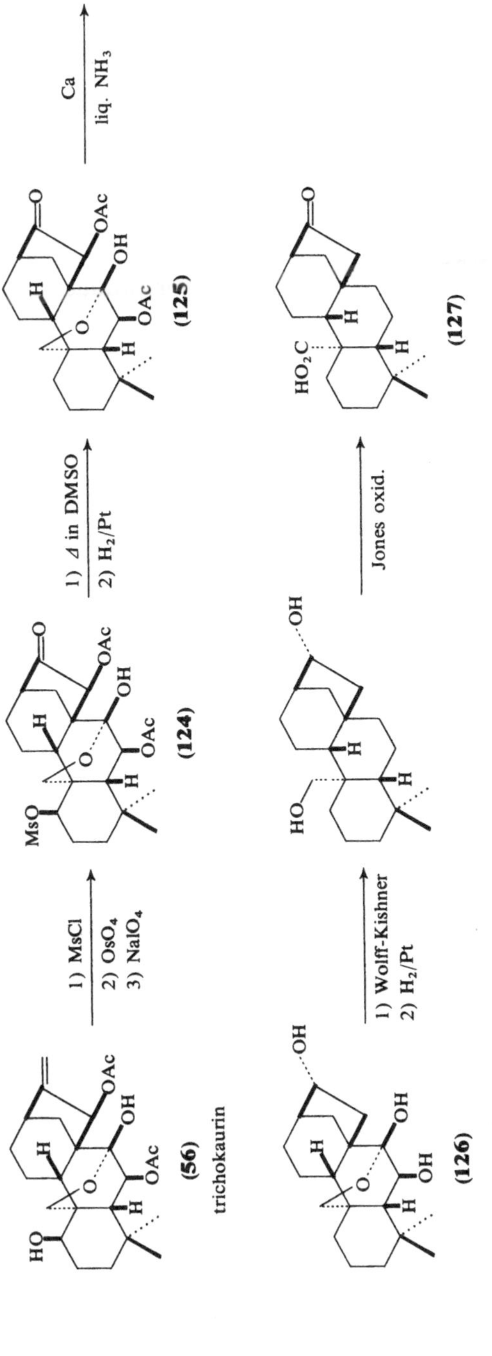

Scheme 18

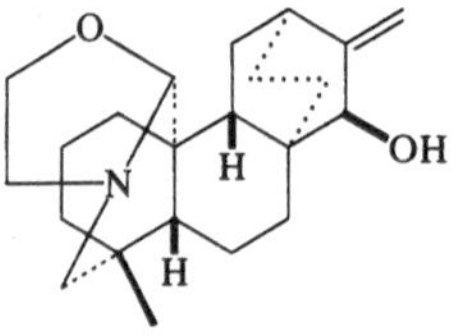

atisine 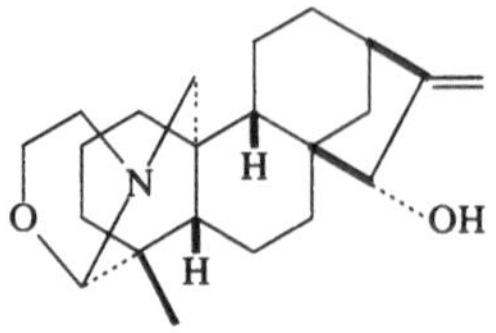   garryine 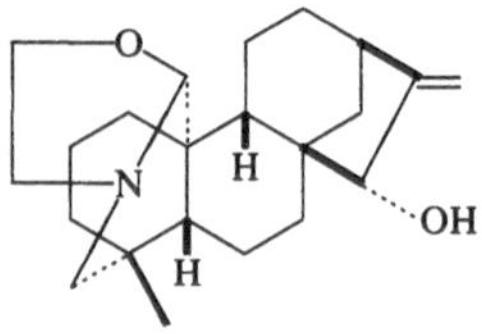   veatchine

The conversion of trichokaurin (56) to (127) was carried out by the route shown in Scheme 18 (78c, 102). Ketone (125), which was derived from trichokaurin (56) via (124) in four steps, was subjected to hydrogenolysis to give triol (126) as the major product. Reduction of (126) followed by oxidation gave the key keto acid (127).

The conversion of enmein (62) to (126) was accomplished through several steps of reactions as shown in Scheme 19 (101).

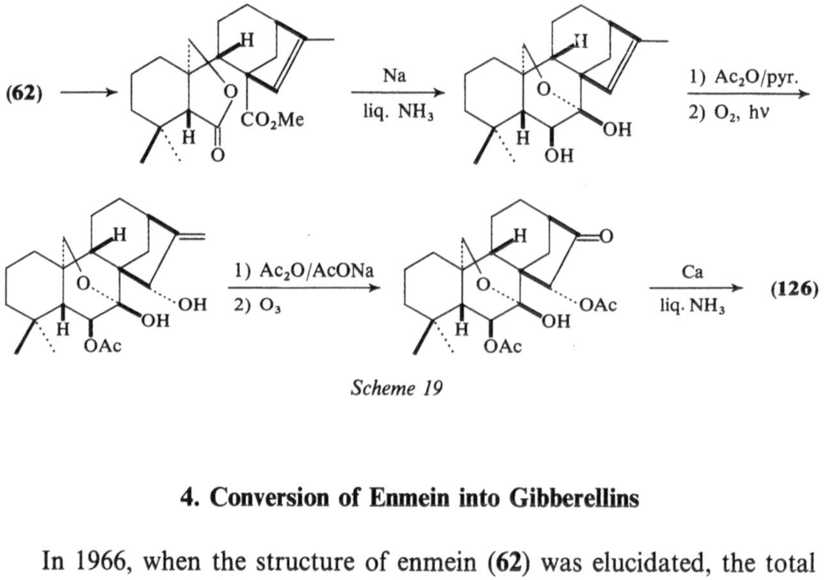

Scheme 19

4. Conversion of Enmein into Gibberellins

In 1966, when the structure of enmein (62) was elucidated, the total synthesis of natural gibberellins had not yet been reported. Hence, Okamoto and coworkers attempted the chemical transformation of enmein into a gibberellane derivative (106) and reported the formation of such compound by a benzilic acid rearrangement-like reaction of keto hemiacetal (128), an oxidation product of acyloin (110). This is shown in Scheme 20.

The chemical conversion of enmein into natural gibberellins was subsequently achieved by two groups: Okamoto and coworkers (107) converted enmein into gibberellin A_{15}, while E. Fujita and coworkers (108) converted enmein into gibberellins A_{15} and A_{37}.

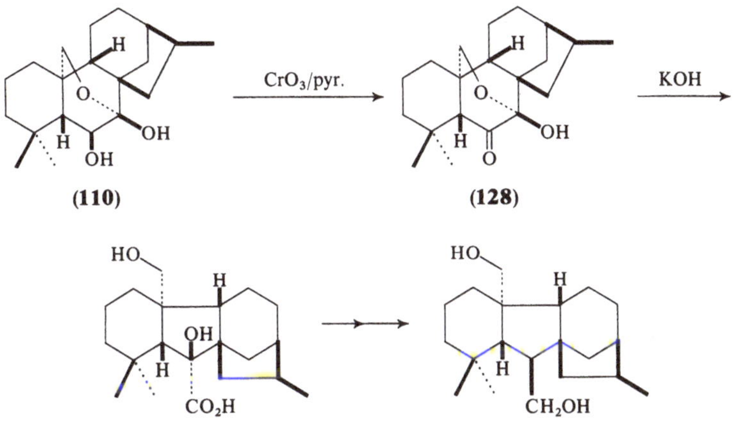

(110)   CrO₃/pyr. ⟶   (128)   KOH ⟶

Scheme 20

The first synthesis (Scheme 21) utilizes a new method, photolysis of a nitrone, for functionalization of the C-19 methyl groups. Hydrazone (**129**) was derived from enmein (**62**). It was then transformed into nitrone (**130**) by reaction with BrN_3. The nitrone on photolysis gave product (**131**) which had a nitrogen atom on C-19. Thermolysis of (**131**) converted it into imine (**132**), in which the C-6, N bond and the C-7, C-8 bond are in an anti-parallel relation because of the boat conformation of the ring B. The formation of diazonium salt from (**132**) by nitrous acid gave rise to the smooth rearrangement depicted in Fig. 3 to form the gibberellane skeleton. This rearrangement parallels the biosynthetic route because of the extrusion of C-7. Products (**133**) and (**134**) from the rearrangement were converted to gibberellin A_{15} (**135**) and its double bond isomer, as shown in Scheme 21.

The proposed mechanism for functionalization of the methyl group by photolysis of the nitrone is shown in Fig. 4.

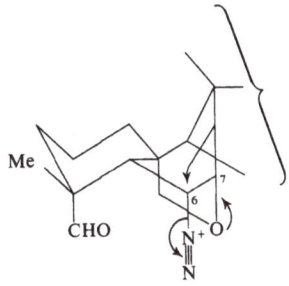

Fig. 3

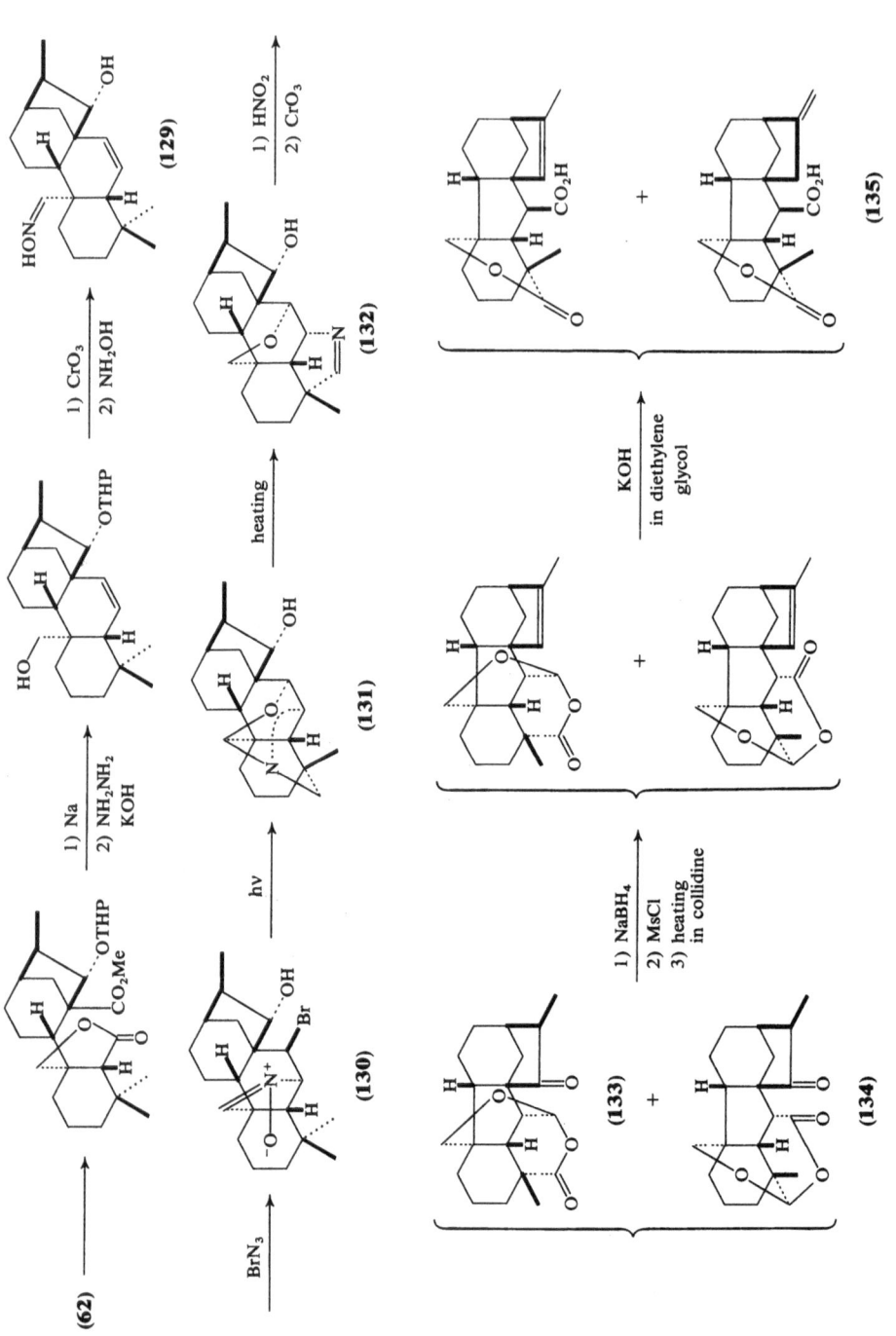

Scheme 21

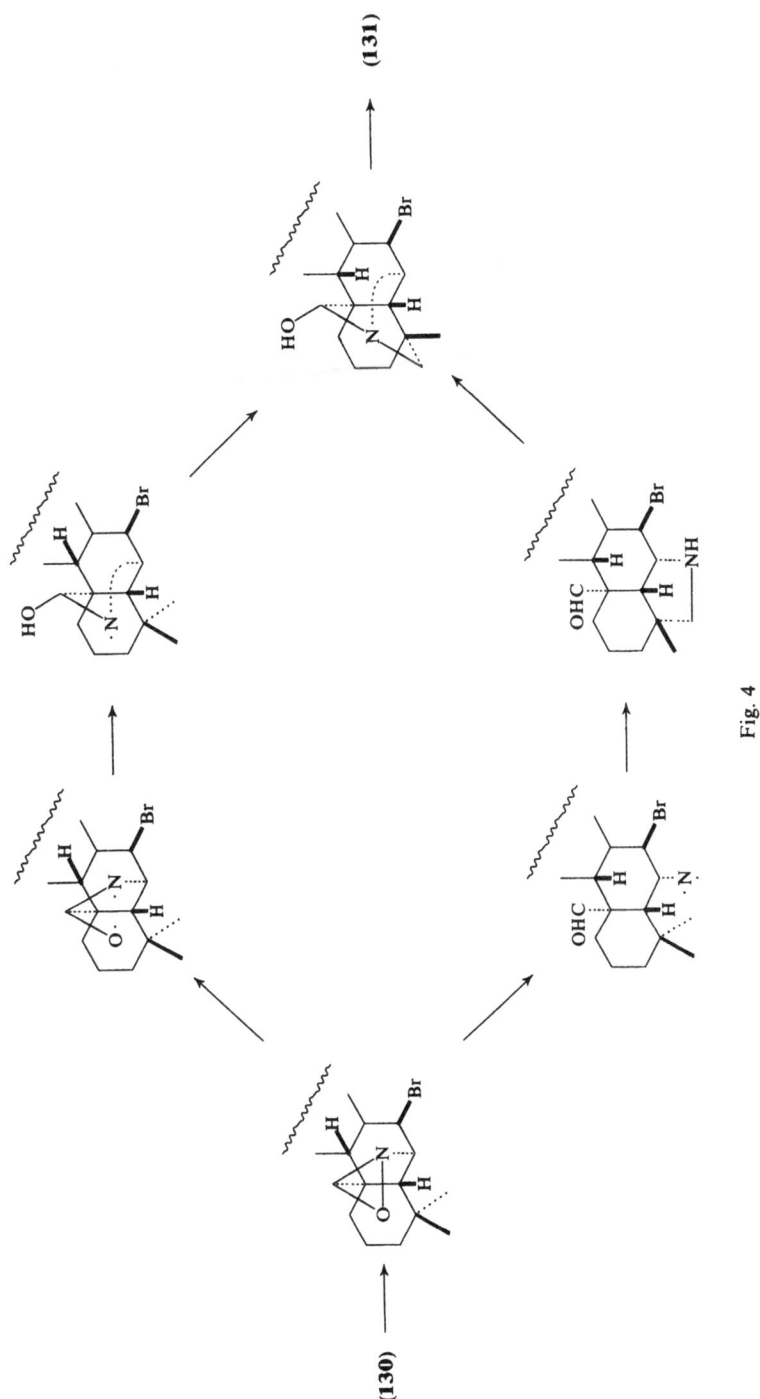

Fig. 4

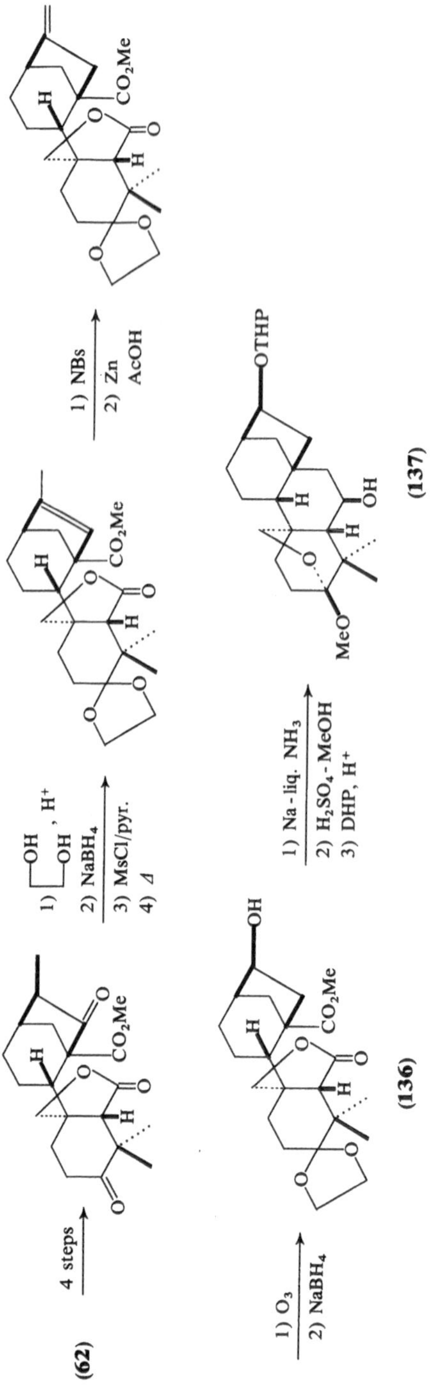

Scheme 22

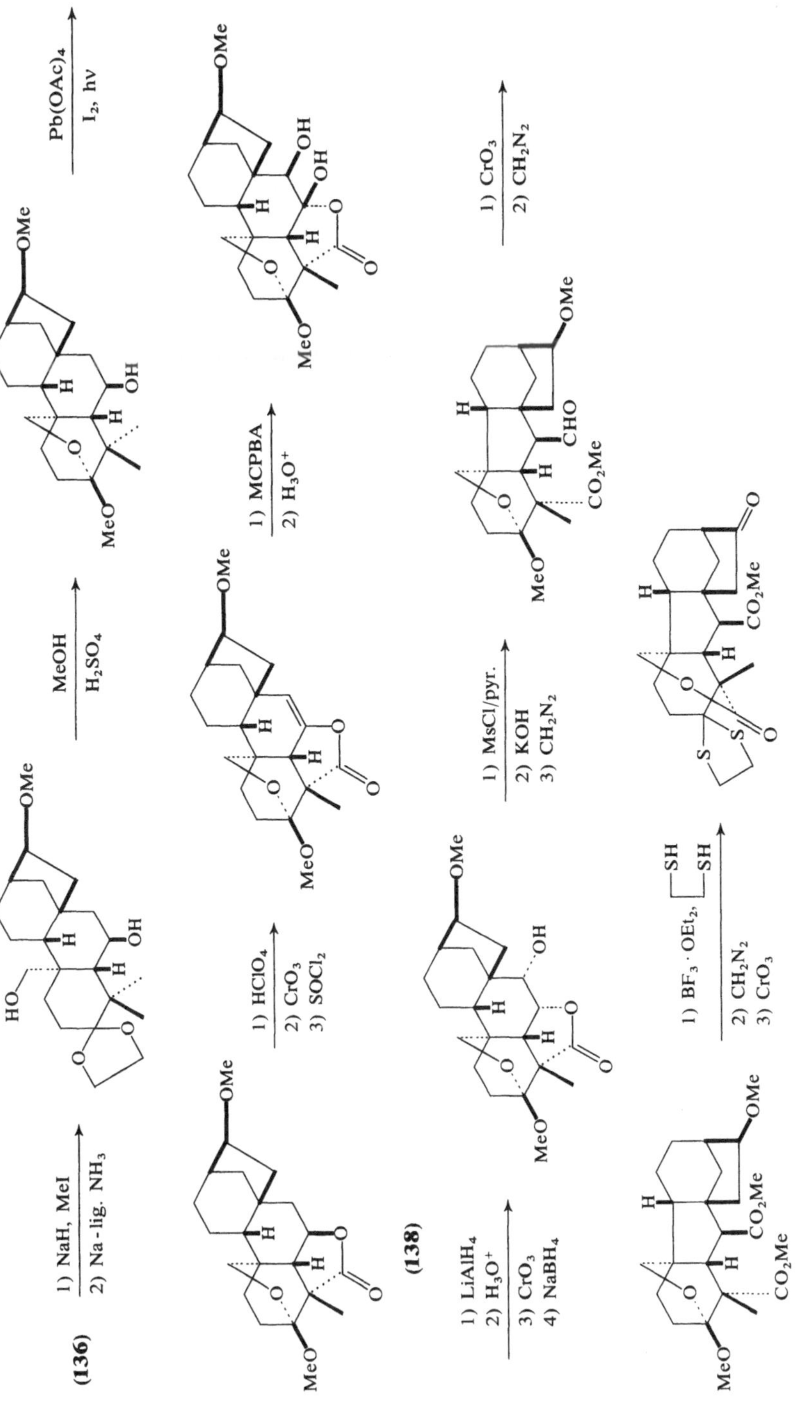
(136)
1) NaH, MeI
2) Na - lig. NH3
HO
OH
H
OMe
MeOH
H2SO4
MeO
Pb(OAc)4
I2, hν
(138)
1) HClO4
2) CrO3
3) SOCl2
1) MCPBA
2) H3O+
OH
OH
1) LiAlH4
2) H3O+
3) CrO3
4) NaBH4
1) MsCl/pyr.
2) KOH
3) CH2N2
CHO
CO2Me
1) CrO3
2) CH2N2
1) BF3 · OEt2, SH SH
2) CH2N2
3) CrO3
(139)
CO2Me
S
(140)

Scheme 23

Compound (**137**) used as a relay in a total synthesis of enmein was prepared *via* (**136**) from enmein as shown in Scheme 22. Compound (**136**) was also used as the starting material for the conversion of enmein to gibberellins by E. FUJITA's group (*108*). In this conversion, there were three problems to be solved. The first problem was introduction of the exomethylene group on ring D, which, because of its reactivity, must be formed in one of the final steps of the synthesis, preferably from a ketone. As a stable precursor for such a ketone, the methyl ether of a secondary alcohol was envisaged. Since no good reagent for demethylation of the methyl ether of a secondary alcohol was available, the reagent system $BF_3 \cdot OEt_2$-thiol (*109*) was developed and used with success [see (**139**→**140**) in Scheme 23]. The second problem was selective oxygenation of the C-19 methyl group. Preliminary experiments to this end are described in III-6(b), but in the event selective functionalization of C-19 was achieved by the hypoiodite reaction of C-6β-OH in a kaurane type compound, in which the ring A is fixed in the boat form. This reaction proceeded smoothly and gave a lactone in good yield [see the formation of compound (**138**) in Scheme 23]. The third problem was the proposed skeletal rearrangement of a kaurane into a gibberellane. Study of the rearrangement in the case of several kaurane derivatives led to the conclusion that compounds which had an oxygen function at C-6α and a leaving group at C-7α give rise to the rearrangement most efficiently (*111*), an observation which can be explained adequately on the basis of stereoelectronic considerations.

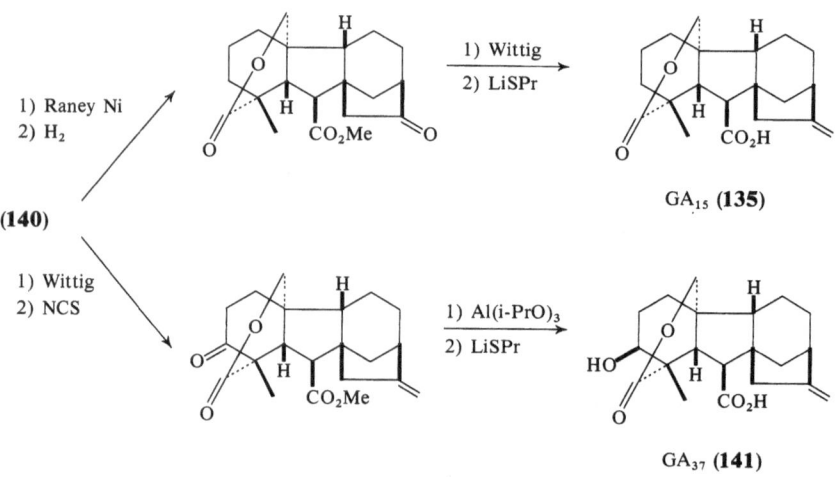

Scheme 24

Thus a common intermediate (**140**) for the synthesis of gibberellins A_{15} (**135**) and A_{37} (**141**) was derived from (**136**) as shown in Scheme 23. From this intermediate (**140**), gibberellins A_{15} (**135**) and A_{37} (**141**) were efficiently prepared by reduction, Wittig reaction, and demethylation or by Wittig reaction, dethioacetalization, reduction, and demethylation, respectively, as shown in Scheme 24 (*108*).

The foregoing chemical conversions of enmein into gibberellins A_{15} and A_{37} correspond to a formal total synthesis of the latter compounds, because the total synthesis of enmein has been accomplished. A more direct total synthesis was also carried out. Compound (**137**) was synthesized from 5-methoxy-2-tetralone and converted into (**138**) as shown in Scheme 25.

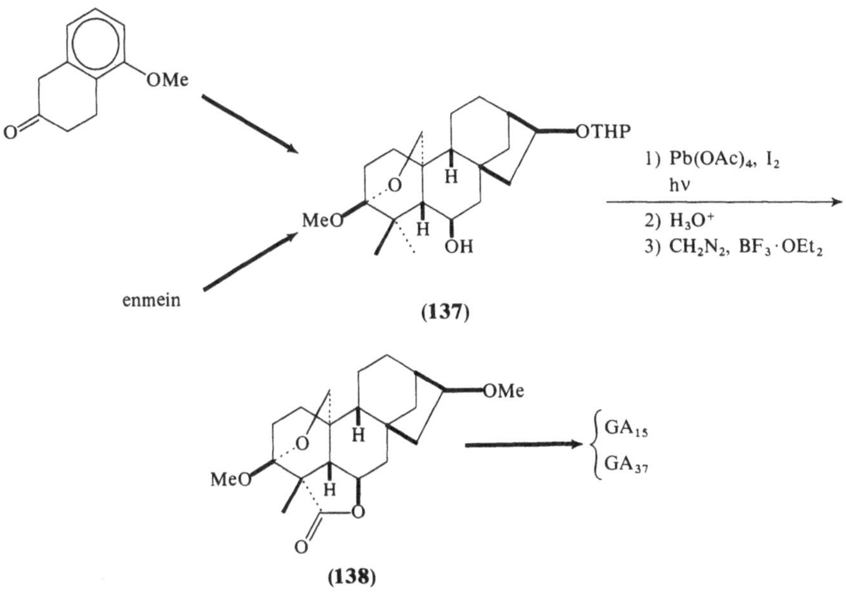

Scheme 25

5. Interconversions among *Rabdosia* Diterpenoids

Interconversions among *Rabdosia* diterpenoids have been frequently undertaken to establish the structures of new compounds, but were occasionally carried out as a matter of purely chemical interest. The following discussion is limited to interconversions between kaurene-type and 6,7-secokaurene-type diterpenes. The transformations are divided into biogenetic-type and retrobiogenetic-type chemical conversions.

(a) Biogenetic-type Conversion (Kaurene-type → 6,7-Secokaurene-type)

This type of conversion is easily achieved by periodate oxidation of a 6,20-dihydroxy-7-oxokaurene derivative. The 6,7-secokaurene-type product from this reaction is either of the enmein-type or the spirolactone-type: when the starting material has a hydroxy group at C-1, an enmein-type compound is produced; when it has no hydroxy group at C-1, a spirolactone-type product is formed (see Scheme 26).

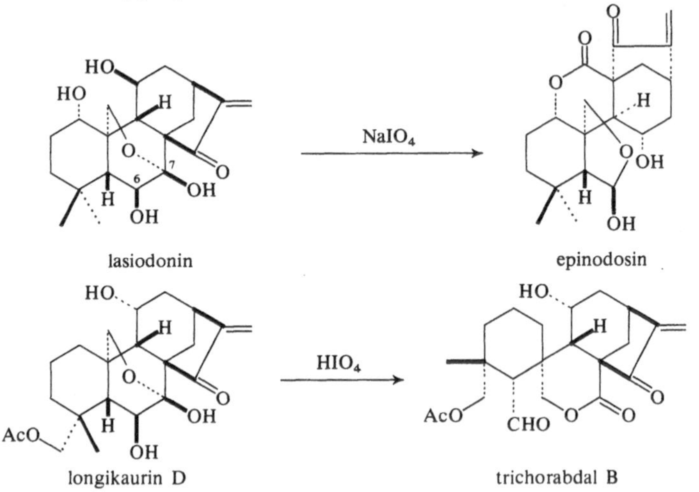

NaIO₄   (46)

lasiodonin   epinodosin

HIO₄   (73)

longikaurin D   trichorabdal B

Scheme 26

The chemical conversion of oridonin (32) into isodocarpin (64) was achieved through the route shown in Scheme 27 (*112*). Intermediate (142) was obtained from dihydroisodocarpin (143) by treatment with acidic methanol; as (143) is derived in turn from enmein (62) (*32*), nodosin (66) (*32, 35*), and trichokaurin (56) (*78*), this means the chemical conversion of enmein, nodosin, and trichokaurin into isodocarpin.

An improved chemical conversion of trichokaurin into isodocarpin was reported later on: it was achieved through a shorter route shown in Scheme 28 which is similar to the biosynthetic pathway (*113*).

(b) Retrobiogentic-type Conversions

The chemical conversion of enmein (62) into enmelol (58) (Scheme 29) constitutes a retrobiogenetic-type conversion (*114*). In this conversion, the acyloin reaction was used for the skeletal transformation, just as in the conversion of enmein (62) to *ent*-kaurane (112). Thus enmein (62) was converted into compound (145) by acyloin reaction of lactone ester (144). Compound (145) was further transformed into enmelol (58) by the route shown in Scheme 29.

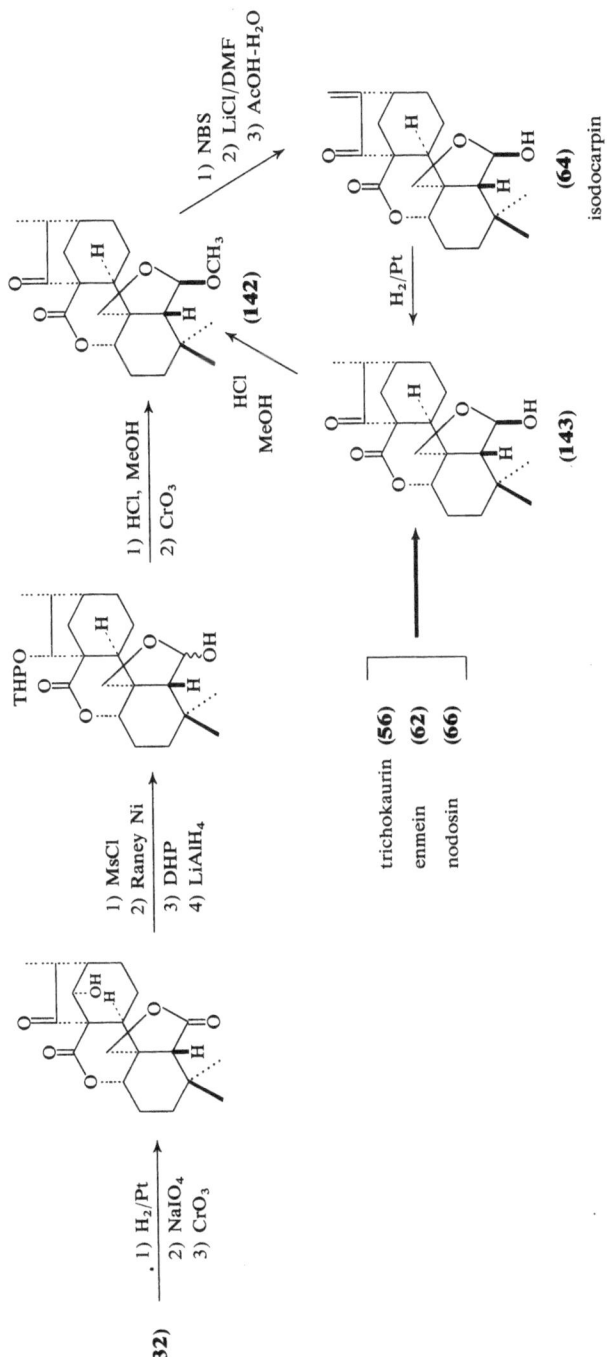

Scheme 27

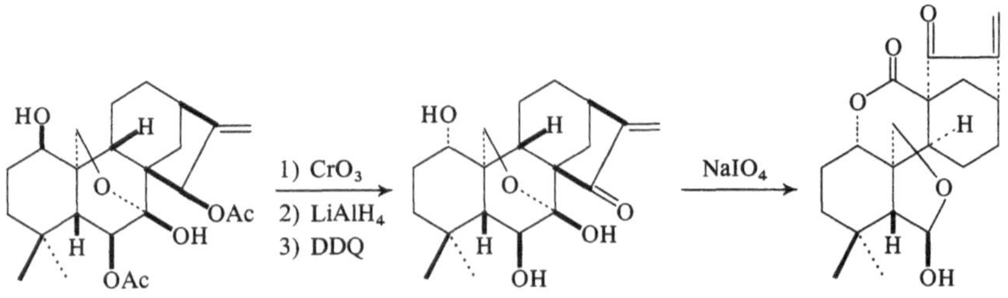

1) CrO$_3$
2) LiAlH$_4$
3) DDQ

NaIO$_4$

trichokaurin
(56)

isodocarpin
(64)

Scheme 28

6. Other Topics

(a) Stereochemistry of Ring D in Enmein Derivatives

In the course of various conversions, four epimeric alcohols **(147)**, **(148)**, **(149)**, and **(150)** (Scheme 30) were synthesized and their absolute configurations were elucidated (*115*). Keto-lactone ester **(146)** derived from enmein **(62)** was subjected to NaBH$_4$ reduction in aqueous methanol to give *trans*-β-alcohol **(147)**, while reduction of **(146)** with LiAl(t-BuO)$_3$H in anhydrous THF gave *cis*-α-alcohol **(148)**. Epimerization of alcohol **(148)** with mild alkali gave **(147)**. X-ray crystallographic analysis of **(147)**-bromoacetate established its absolute configuration (*116*). Thermolysis of the mesylate of **(147)** and epoxidation of the resulting olefin gave an epoxide which on hydrogenolysis with Adams catalyst gave **(147)** and *cis*-β-alcohol **(149)** in 4 : 3 ratio. Alcohol **(149)** was oxidized to ketone, which was reduced with LiAl(t-BuO)$_3$H to give the fourth epimer **(150)**. Direct epimerization of *cis*-alcohol **(149)** to the *trans*-alcohol **(150)** took place on heating with basic catalysts. The epimerizations [**(148)**→**(147)** and **(149)**→**(150)**] under the influence of bases are presumed to occur by a retroaldol reaction followed by recyclization so as to eliminate the *cis*-eclipsed interaction between the C-15 hydroxy and the C-16 methyl groups in **(148)** and **(149)**.

Evidence for this postulate was provided by epimerization of deuterio- or tritio-labeled derivatives of **(148)** (*117*). As the result, a retro-aldol type mechanism in which stereoelectronic requirements are satisfied in the transition state was suggested. The process is depicted in Figure 5. This concept was also applied to the retro-Dieckmann type cleavage of ketone **(151)** to carboxylic acid **(152)** (Scheme 31), the facile epimerization of the 3β-axial alcohol **(153)** to the 3α-equatorial alcohol **(154)** under the influence of dilute alkali (followed by esterification) and the cleavage of the 3-ketone **(155)** to methyl ester **(156)** on treatment with sodium hydroxide in methanol at 0 °C for one hour (Scheme 31).

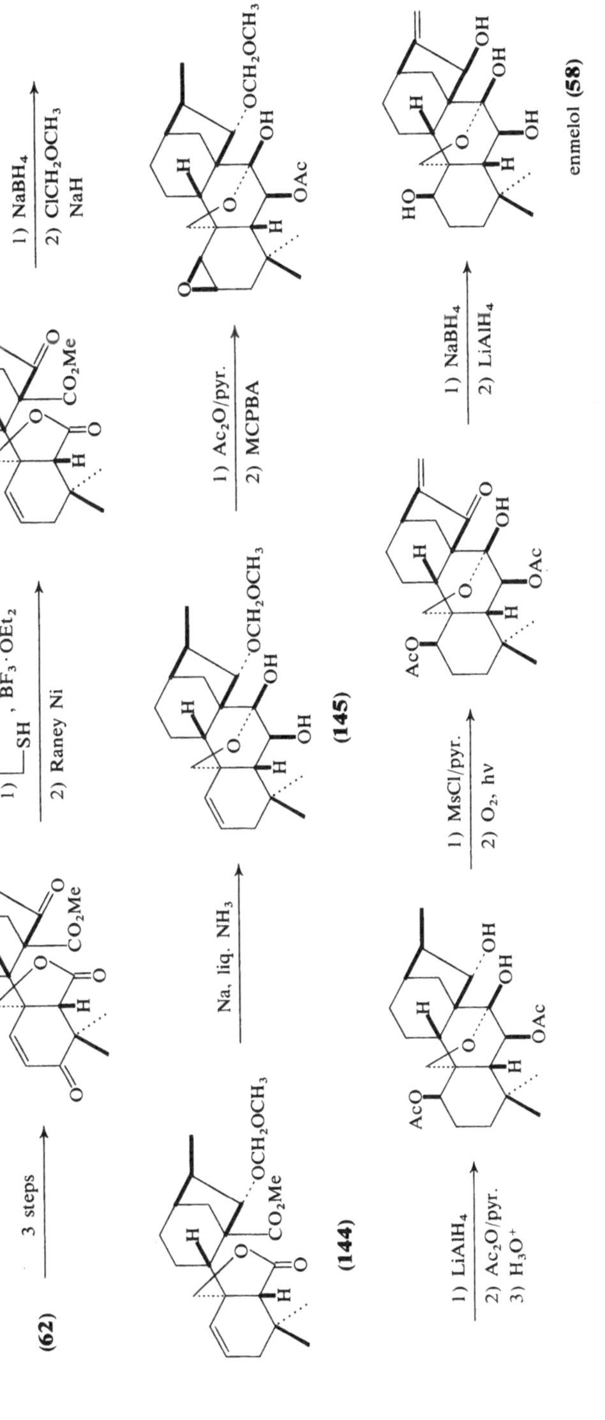
(62)
3 steps
1) HS–CH2CH2–SH, BF3·OEt2
2) Raney Ni
1) NaBH4
2) ClCH2OCH3
NaH
O
CO2Me
OCH2OCH3
(144)
Na, liq. NH3
OH
(145)
1) Ac2O/pyr.
2) MCPBA
OAc
1) LiAlH4
2) Ac2O/pyr.
3) H3O+
AcO
1) MsCl/pyr.
2) O2, hν
1) NaBH4
2) LiAlH4
HO
enmelol (58)

Scheme 29

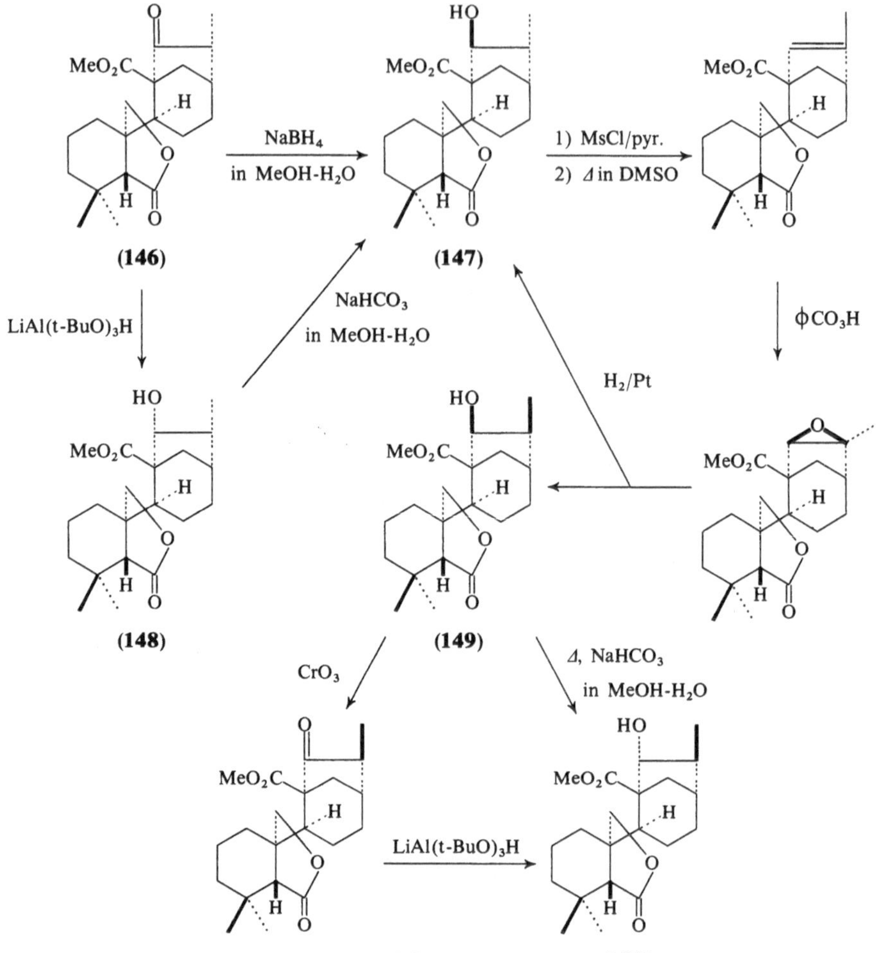

Scheme 30

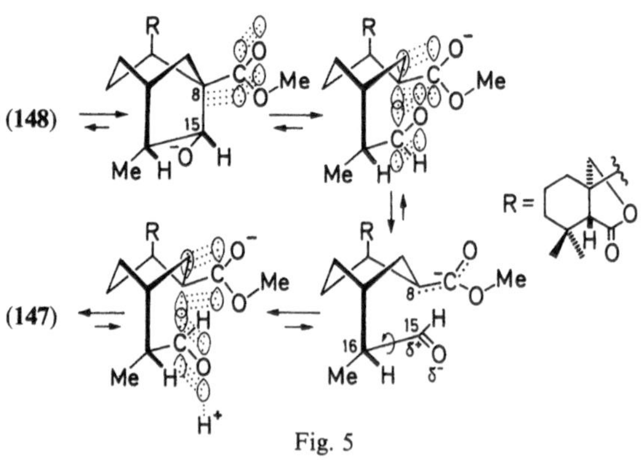

Fig. 5

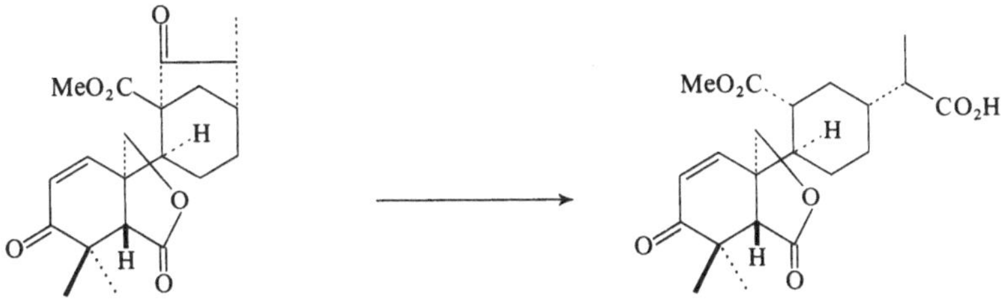

(151)   (152)

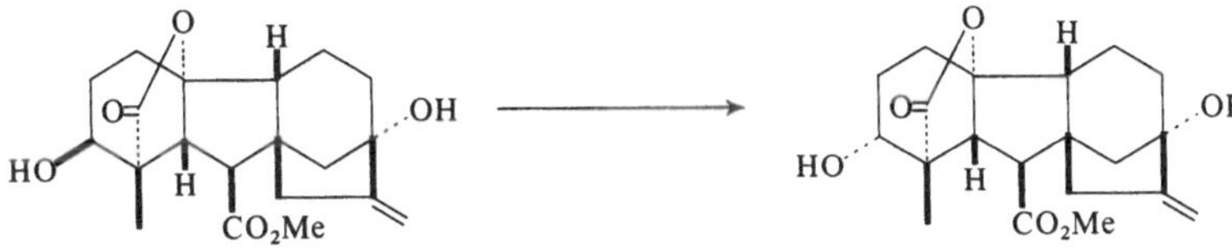

(153)   (154)

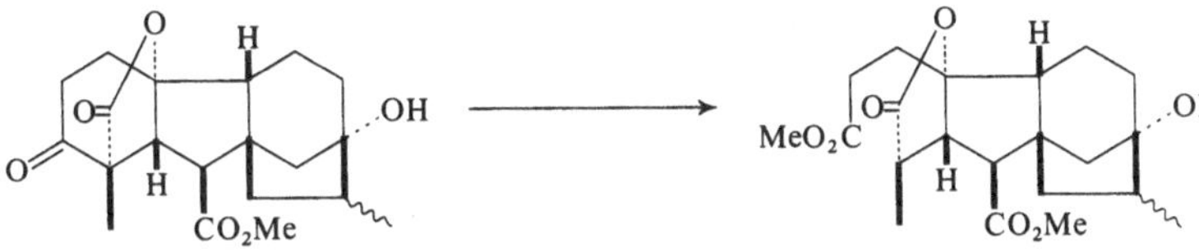

(155)   (156)

Scheme 31

(b) Hypoiodite Reaction of Enmein Derivatives

It has been suggested that nodosin (66) is biosynthesized from isodocarpin (64) *via* the isodoacetal-like intermediate (157). To prepare such an intermediate dihydroisodocarpin (158), dihydroenmein-3-acetate (159), and isodotricin-3-acetate (160), were subjected to the hypoiodite reaction (Scheme 32) but instead products of type (161) were obtained (*118*).

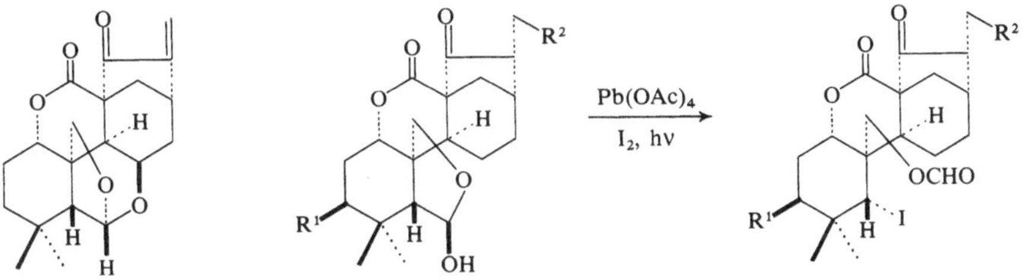

$\xrightarrow[\text{I}_2,\ hv]{\text{Pb(OAc)}_4}$

(157)   (158) R¹ = R² = H   (161)
              (159) R¹ = OAc, R² = H
              (160) R¹ = OAc, R² = OMe

Scheme 32

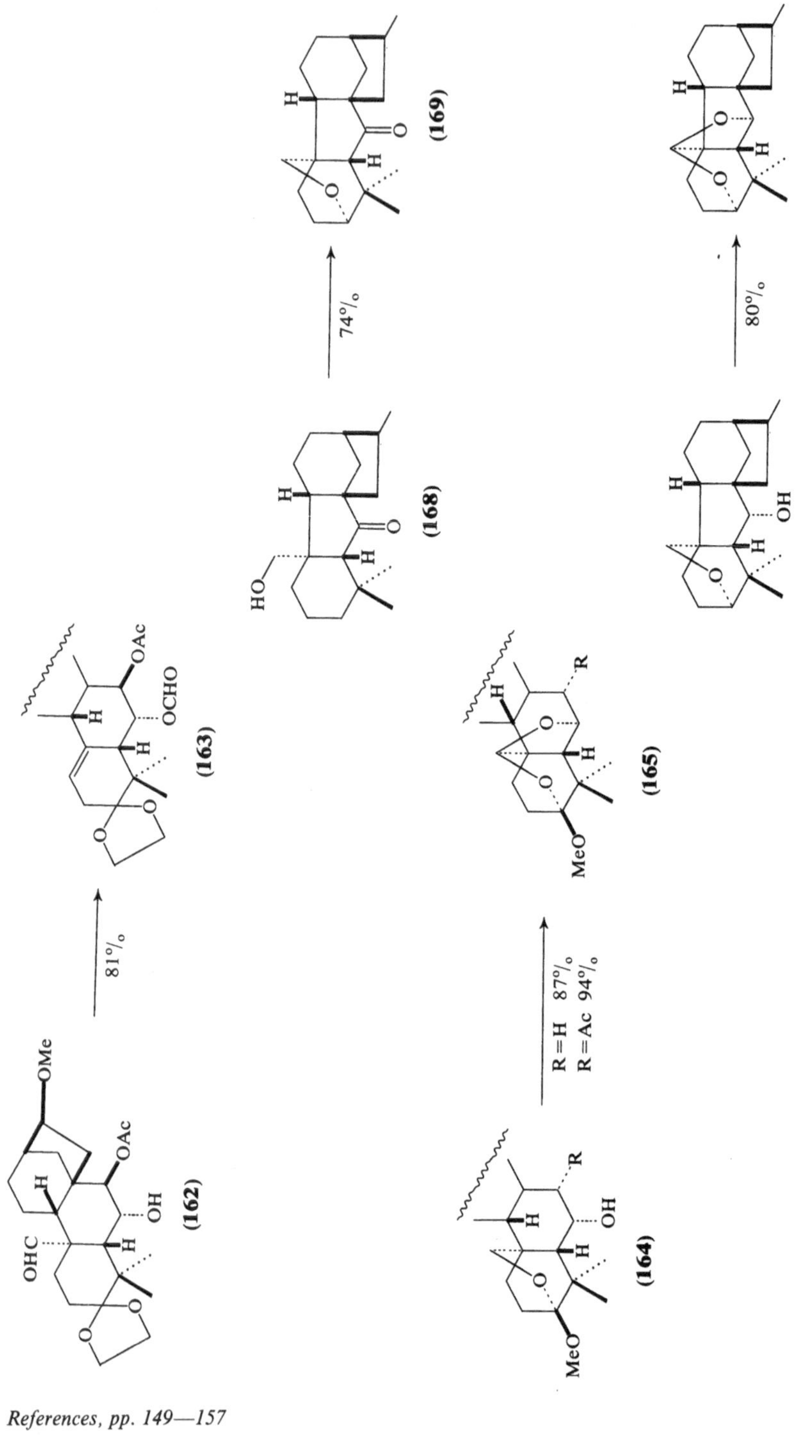

OHC
OMe
OAc
OH
(162)
81%
OAc
OCHO
(163)
MeO
R
OH
(164)
R = H 87%
R = Ac 94%
MeO
R
(165)
HO
(168)
74%
(169)
OH
(170)
80%
(171)

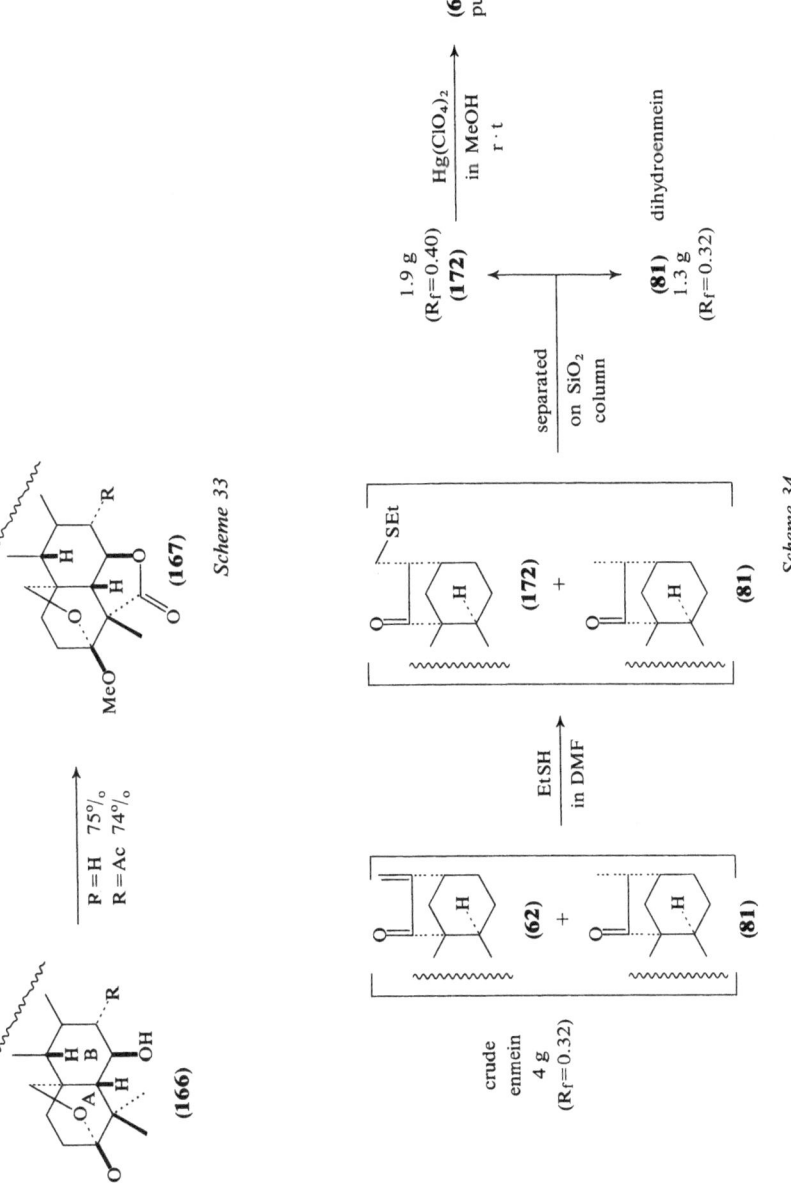

Scheme 33

Scheme 34

Subsequently, hypoiodite reactions of 17-norkauranol derivatives and 7-norgibberellanol derivatives were investigated in experiments aimed at exploring the possibility of functionalizing the C-19 methyl group for the conversion of enmein to gibberellins (*110*) (see III-4). Oxidation of 6α-hydroxy-17-norkaurane derivatives (**162**) and (**164**) (Scheme 33) occurred selectively at C-20 to give (**163**) and (**165**), respectively. On the other hand, oxidation of 6β-hydroxy-17-norkaurane derivative (**166**) occurred selectively at C-19 to give lactone (**167**). Oxidation of 20-hydroxy- (**168**) and 6α-hydroxy-7-norgibberellane (**170**) occurred at C-3 and at C-20 to give (**169**) and (**171**), respectively. In conclusion, the O-functionalization of the C-19 methyl group by means of the hypoiodite reaction could be achieved in kaurane-6β-ol derivatives where ring A is in a rigid boat conformation.

(c) Isolation of Dihydroenmein and Its Conversion to Enmein

While enmein (**62**) can be isolated from *Rabdosia japonica* in the pure state, enmein isolated from *R. trichocarpa* is contaminated by a large quantity (about half of the amount of enmein) of dihydroenmein and their separation is very difficult. Enmein has antitumor and antibacterial activities, while dihydroenmein is inactive. In order to obtain relatively large amounts of pure enmein for biological tests, development of a good method for separating dihydroenmein from the crude enmein mixture was required. Taking advantage of the large difference between the Rf values of the EtSH adduct of enmein and dihydroenmein E. Fujita and coworkers (*130*) chromatographed the mixture obtained by treatment of crude enmein with EtSH in DMF on a SiO_2 column. The isolated EtSH adduct (**172**) was treated with $Hg(ClO_4)_2 \cdot 3 H_2O$ to afford pure enmein in 83% yield (see Scheme 34).

The chemical transformation of dihydroenmein into enmein was achieved by two groups (*119, 120*) (see Scheme 35). Dihydroenmein diacetate (**173**) was converted to the α,β-unsaturated ketone (**175**) by bromination followed by dehydrobromination of the resulting bromo derivative (**174a**). Kubota and coworkers (*120*) studied the dehydrobromination of bromo derivatives (**174a**), (**174b**), and (**174c**); (**174b**) and (**174c**) gave the expected products (**175a**) and (**175b**), respectively, while (**174a**) gave dimer (**176a**). The formation of dimer (**176a**) is due to formation of HBr in the dehydrobromination, because dimer (**176b**) is formed from (**173**) by treatment with LiCl-EtBr in DMF and (**176a**) is also formed from (**175c**) by HBr gas in DMF.

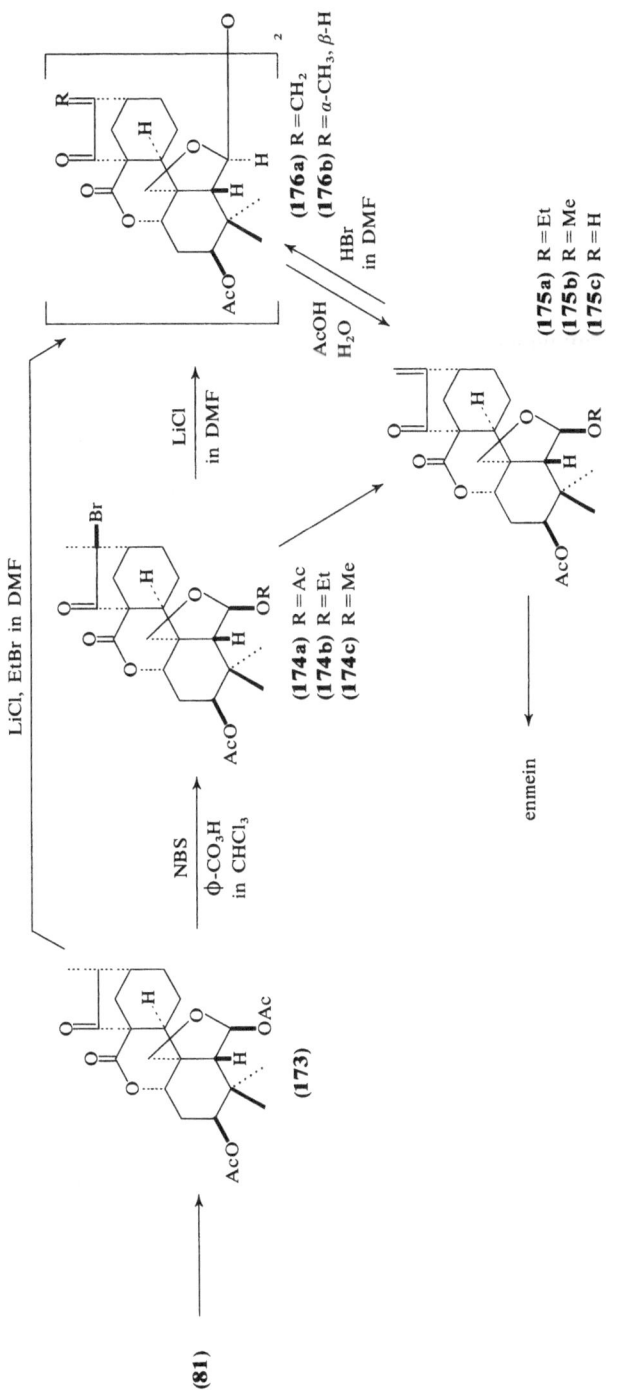

Scheme 35

IV. Total Synthesis of Enmein

Enmein (62) is a very attractive target for total synthesis because of its complicated and highly oxygenated structure and because of its antitumor and antibacterial activities. This challenge has been met successfully by E. Fujita and coworkers who utilized the norkaurane derivative (178), which was derived in good yield from enmein (121), as an important relay compound. After the racemic compound (178) had been synthesized from a phenanthrene derivative (177), the optically active compound (178) derived from enmein (62) was transformed into the 6,7-secokaurene derivative (179) by a pathway which included as one step cleavage of the C-6, C-7 bond. Compound (179) was then converted to enmein (62) by stereoselective lactonization to C-1 and by stereoselective introduction of two hydroxy groups at C-3 and C-6 followed by modification of ring D. The approach is outlined in Scheme 36 (122).

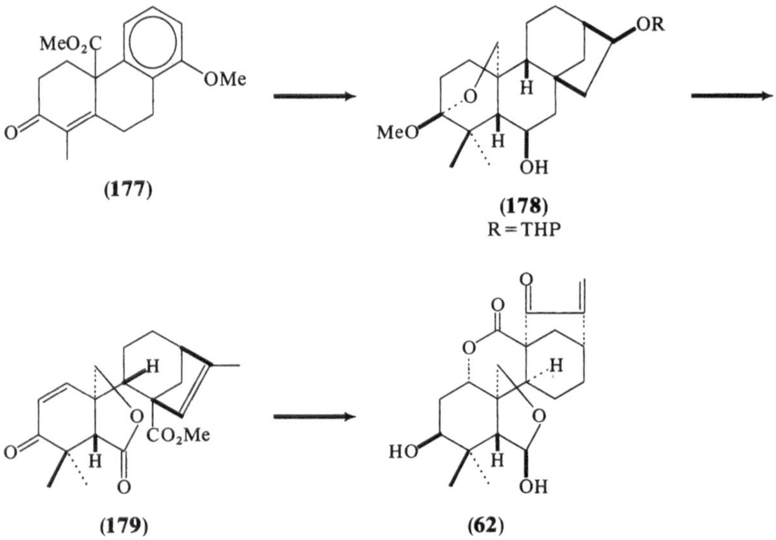

Scheme 36

Synthesis of racemic (178) is shown in Scheme 37. Ketoester (177), synthesized from naphthalene-1,6-diol *via* 5-methoxy-2-tetralone, was converted into compound (180) by successive methylation at C-4, acetalization at C-3, reduction of the ester group to hydroxymethyl group, epoxidation of the C-5, C-6 double bond, and ring opening of the epoxide. Birch reduction of diol (180) with 18 equivalents of lithium in liquid ammonia followed by acid hydrolysis and subsequent methyl acetalization gave ketone (**181**) with the desired *trans, anti,* and *trans* junctions of rings A, B, and C, respectively, in high yield. The extraordinary reactivity of compound (**180**) towards the Birch reduction was attributed to the suitable location and stereochemistry of the hydroxy groups. After the reactive C-13 methylene group of compound (**181**) was protected, allylation at C-8 followed by elimination of the protecting group at C-13 gave allyl ketone (**182**), whose ozonolysis gave a keto diacetal (**183**) in addition to (**184**). Partial hydrolysis of (**183**) under acidic conditions gave aldehyde (**184**), which on exposure to 0.2% sodium methoxide in methanol for 0.5 hour at room temperature gave 16β-hydroxyderivative (**185**), the product of a kinetically controlled aldol reaction. The tetrahydropyranyl derivative of (**185**) on Huang-Minlon reduction followed by dehydration gave 14-deoxo-5-ene derivative (**186**), which on hydroboration afforded the desired relay compound (**178**).

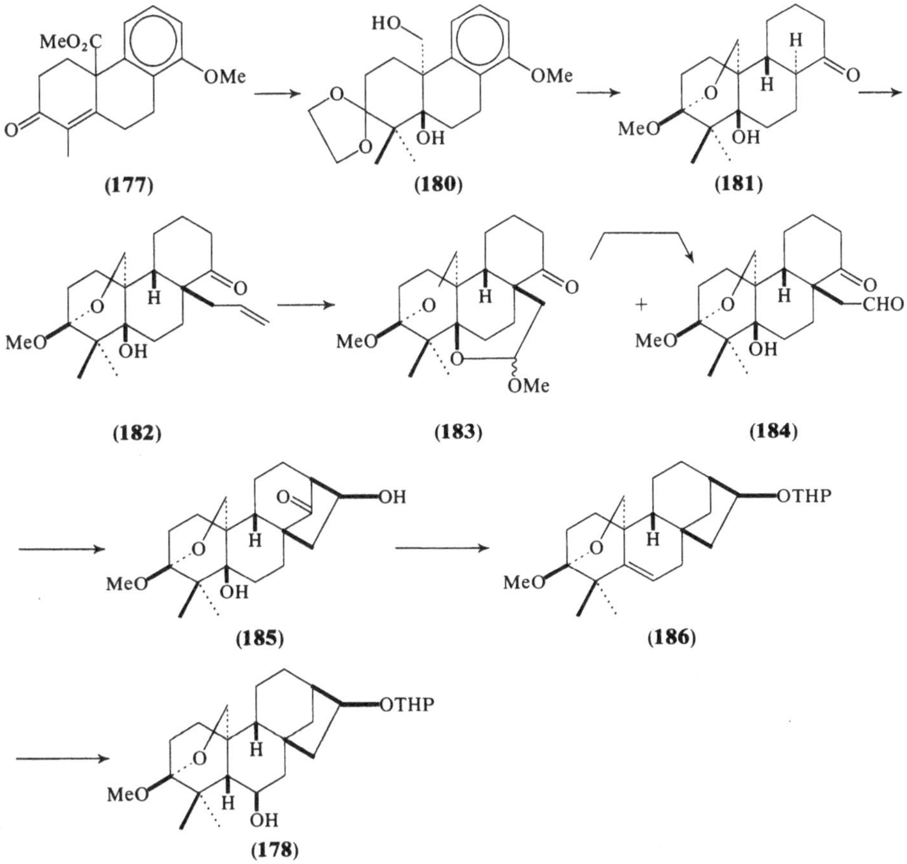

Scheme 37

Dehydration of the optically active relay (**178**) derived from enmein (**62**) gave a 1 : 2 mixture of 5,6-ene and 6,7-ene. Separation could be achieved by means of the ethylene acetal (**187**), whose ozonolysis product was subjected to successive Jones oxidation, methylation, Wittig reaction, and treatment with dilute hydrochloric acid to afford the 3-on-16-ol derivative (**188**). Bromination of (**188**) followed by dehydrobromination and subsequent dehydration afforded (**189**). The purified compound (**189**), after conversion into the acetal, was hydrolyzed to carboxylic acid (**190**), which was transformed into the desired lactone (**191**) by treatment with boron trifluoride etherate. The reaction produced a single product uncontaminated by the C-1 epimer, because of easy formation of a favored transition state which satisfied the stereoelectronic requirements.

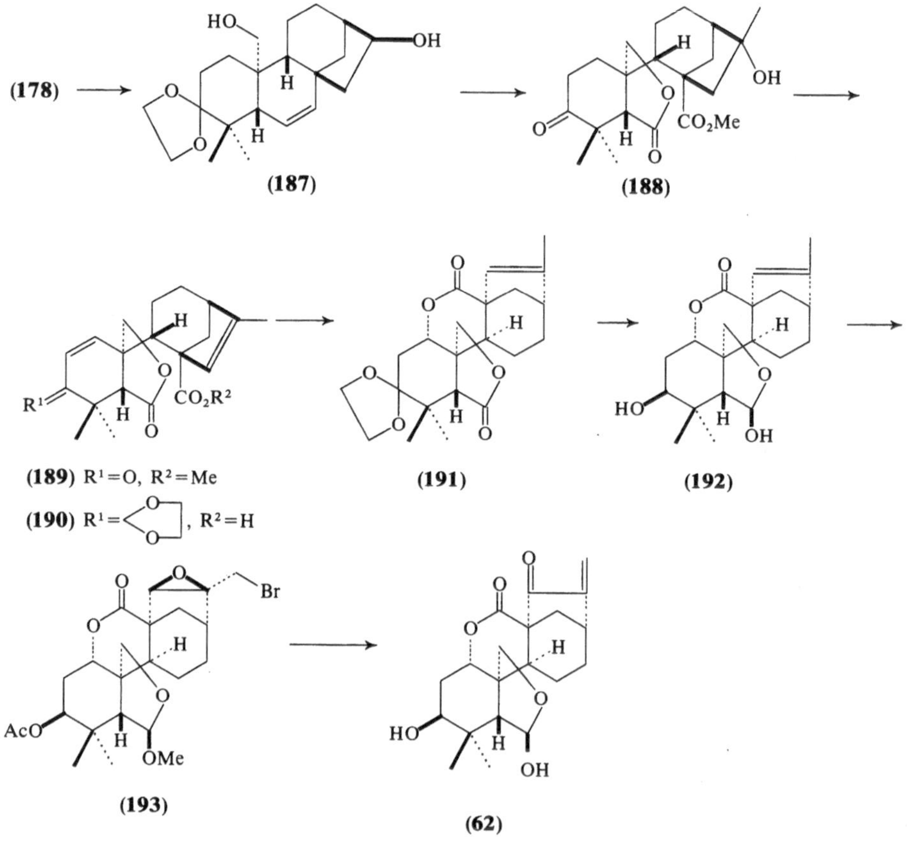

(**178**) ⟶

(**187**)

(**188**)

(**189**) R¹ = O, R² = Me

(**190**) R¹ = <structure>, R² = H

(**191**)

(**192**)

(**193**)

(**62**)

Scheme 38

Meerwein-Ponndorf reduction of the ketone obtained by deprotection of compound (191) gave the desired 3β-axial hydroxy product. Selective reduction of the γ-lactone ring of this substance was achieved by LiAlH₄ in THF at −30 °C to give hemiacetal (192). Subsequent acetalization at C-6, acetylation at C-3, bromination at allylic C-17, and epoxidation converted (192) to (193), which on treatment with zinc dust in ethanol at reflux gave the Δ¹⁶-allylic 15β-ol. Oxidation of the latter, deacetylation, and hydrolysis produced the desired enmein (62). Thus, a relay total synthesis of enmein was achieved (see Scheme 38).

As the chemical conversion of enmein into several other naturally-occurring diterpenoids has been reported (see III and Scheme 39), this work also constituted a formal total syntheses of these natural products.

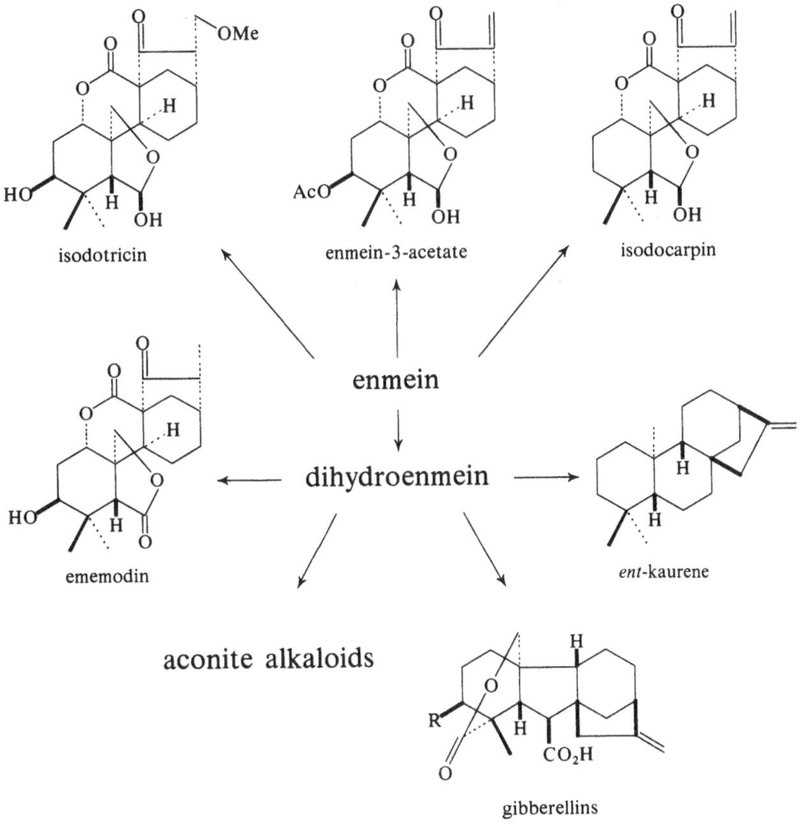

isodotricin

enmein-3-acetate

isodocarpin

enmein

dihydroenmein

ememodin

ent-kaurene

aconite alkaloids

gibberellins

Scheme 39

V. Biosynthesis

The biosynthesis of enmein (**62**) and oridonin (**32**), the major bitter principles in the leaves of *Rabdosia japonica,* has been investigated by E. Fujita and coworkers. It is thought to proceed by a pathway similar to that for cyclic diterpenoids in general, with *ent*-16-kaurene (**116**) as an important precursor. A biogenetic pathway and a classification of the *Rabdosia* diterpenoids based on the biogenesis were proposed (*123*).

As the result of feeding experiments with growing plants, [17-^{14}C]-*ent*-16-kaurene (**194**) was proved to be incorporated into enmein (**62**) and oridonin (**32**); specific incorporation of the isotope into C-17 of both diterpenes was demonstrated (*124, 125*). Subsequently, functionalization at the allylic C-15 position of *ent*-16-kaurene (**116**) was suggested to proceed through direct oxygenation with triplet oxygen, on the basis of the fact that [17-^{14}C]-*ent*-16-kaurene (**194**) and [17-^{14}C]-*ent*-16-kauren-15-one (**195a**) were incorporated into enmein and oridonin, while [17-^{14}C]-*ent*-15-kaurene (**196**), [17-^{14}C]-*ent*-15β,16β-epoxykaurane (**197**), [17-^{14}C]-*ent*-16-kauren-15β-ol (**198**), and [17-^{14}C]-*ent*-kauran-15-one (**199**) were not incorporated at all into these diterpenoids (*125, 126*).

To discover whether 7-oxygenated *ent*-kaurene derivatives are potential precursors of enmein and oridonin, the 17-labeled 7-oxygenated kaurenes (**200, 201a, 202**, and **203**) were prepared (*127*) and administered to growing plants. The results showed that 7-oxygenated kaurenes as well as 15-oxo-kaurene (**195**) are potential precursors of *Rabdosia* diterpenes (*128*).

The relative efficiency of oxygenated kaurenes as precursors of enmein and oridonin was examined using variously labeled compounds (*128*). First, the relative efficiencies of *ent*-16-kauren-15-one (**195**) and 7-oxygenated kaurenes (**200–202**) were examined. Thus, ^{3}H-labeled 7-oxo-16-kaurene (**201b**) was prepared (*127*), and mixtures of (**201b**) and (**195a**), of (**201b**) and (**200**), and of (**201b**) and (**202**) were administered. The incorporations of 7β-ol (**200**) and 7-one (**201**) into enmein were very similar, but 7α-ol (**202**) was incorporated with an efficiency less than 1/5 of that of 7-one (**201**). A comparison of the incorporation into oridonin shows the order 7-one (**201**) > 7β-ol (**200**) > 7α-ol (**202**). Comparison of 7-one (**201**) and 15-one (**195**) suggests that incorporation of 7-one (**201**) into enmein and oridonin is preferred to incorporations of 15-one (**195**).

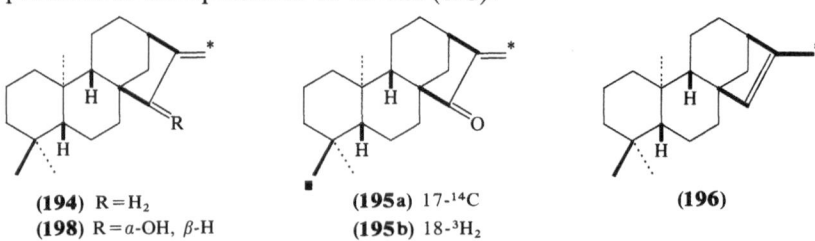

(**194**) R = H₂
(**198**) R = a-OH, β-H

(**195a**) 17-[14]C
(**195b**) 18-[3]H₂

(**196**)

(197)   (199)   (200) R = H₂
                                                    (203) R = O

(201a) 17-¹⁴C   (202)   (204a) 17-¹⁴C
(201b) 18-³H₁   (204b) 18-³H₁

Secondly, the relative efficiencies of 15- and 7-mono-oxygenated kaurenes (**195, 201**) and 7,15-dioxygenated kaurene (**204**) as precursors were examined by use of the double-isotope technique (*128*). Thus, [17-¹⁴C]- and [18-³H₁]-labeled 7,15-dioxokaurene (**204a, 204b**) were prepared (*127*), and mixtures of (**204b**) and (**201a**), of (**204a**) and (**201b**), and of (**204b**) and (**195a**) were fed to *Rabdosia japonica* plants. The mono-oxo kaurenes (**195, 201**) were found to be more effectively incorporated into enmein and oridonin than the dioxo compound (**204**).

Finally, [17-¹⁴C]-labeled 20-oxygenated kaurene (**205**) and [17-¹⁴C]-labeled 3-oxygenated kaurenes (**206–208**) were synthesized (*129*). As a reference for comparing the relative incorporation, (**195b**) was synthesized and mixtures of (**205**) + (**195b**), (**206**) + (**195b**), (**207**) + (**195b**), and (**208**) + (**195b**) were administered to *Rabdosia* plants. In order to see the fates of ³H and ¹⁴C of the same compound in the plant, a mixture of (**195b**) and (**195a**) was fed; there was no major difference between the retentions of ³H and ¹⁴C in oridonin. The 20-hydroxykaurene (**205**) was incorporated into enmein to almost the same degree as 15-oxokaurene (**195**), while its incorporation into oridonin was very low. This result suggests that oxygenation at the C-20 atom of *ent*-6,7-secokaurenes proceeds earlier than C-20-oxygenation of *ent*-kaurene-type diterpenoids. The relative order of incorporation of 3-oxygenated kaurenes (**206–208**) into enmein was 3α-ol (**206**) > 3-one (**207**) > 3β-ol (**208**) ≃ 15-one (**195**).

In summary, *ent*-16-kaurene (**116**) was proved to be a precursor of enmein and oridonin. Substrate specificity of the oxygenase in this biosynthesis seems low; all of the 3, 7, 15, or 20-oxygenated kaurenes were incorporated. Much work remain to be done to elucidate the complete biosynthetic pathway.

(205)

(206) R = α-OH, β-H
(207) R = O
(208) R = α-H, β-OH

* ^{14}C
■ ^3H

VI. Physiological Activity

In Japan, R. japonica and R. trichocarpa have been used as bitter stomachics and anthelminthics, while in China R. ternifolia has been used as an antiphlogistic, germicide, antipyretic, and toxicide and R. rubescens is used in cancer of the esophagus. Initial studies to track down the compounds responsible for the activity of Rabdosia species were carried out in 1954. Subsequent studies dealt with cytotoxicity, antitumor activity, inhibition of oxidative phosphorylation, anti-feeding activity, growth inhibitory activity, and bitterness. All of the foregoing activities have been attributed to the diterpenoid constituents.

1. Antibacterial Activity

The first report on antibacterial activity of Rabdosia constituents appeared in 1954, when it was discovered that a crystalline substance isolated from the ethanolic extract of Rabdosia japonica inhibited the growth of Gram-positive bacteria (3). Later two groups found that an extract of R. trichocarpa also had antibacterial and antitumor activities (131, 132). It was shown (133) that the activity was due to enmein. Dihydroenmein was inactive, while O,O-diacetylenmein remained active. Hence it was assumed that the biologically active site of enmein was the α-methylenecyclopentanone moiety (133).

KUBOTA and coworkers reported high antibacterial activity for enmein (62), isodonal (71), nodosin (66), oridonin (32), and enmein-3-acetate (63) against Gram-positive bacteria (134) and showed that trichodonin (70), shikokianin (24), umbrosin A (1), and umbrosin B (2) inhibited the growth of Bacillus subtilis (134, 135). Shikokianidin (49) and the dihydro derivatives of the active diterpenoids were inactive, while their acetates retained the activity. It was therefore concluded that the α-methylenecyclopentanone moiety was essential for antibacterial activity, the activity being attributed to a Michael-type addition of a sulfhydryl enzyme to this function (Scheme 40) (134).

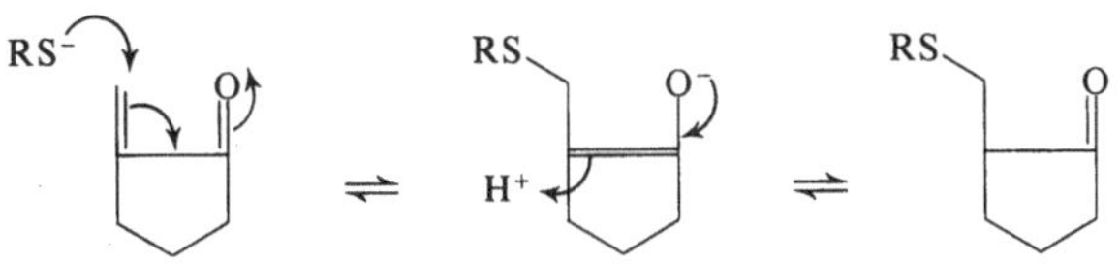

Scheme 40

E. FUJITA and coworkers (*136*) also examined the antibacterial activity of the natural diterpenoids, oridonin (**32**), lasiokaurin (**37**), enmein (**62**), enmein-3-acetate (**63**), trichokaurin (**56**), dihydroenmein (**81**) and their derivatives 14-deoxyoridonin (**25**) (effusanin A), the dihydro oridonin derivative (**209**), and oridonin butanethiol adduct (**210**). Oridonin (**32**), lasiokaurin (**37**), and 14-deoxyoridonin (**25**) had especially high activity against *Sarcina lutea,* while trichokaurin (**56**), dihydroenmein (**81**), (**209**), and (**210**) were inactive. On the basis of these results and the addition reaction of alkanethiols to C-17, it was again concluded that the α-methylenecyclopentanone system was an active center; an important role in enhancing the activity was attributed to hydrogen-bonding between C-6-OH and the C-15 carbonyl group.

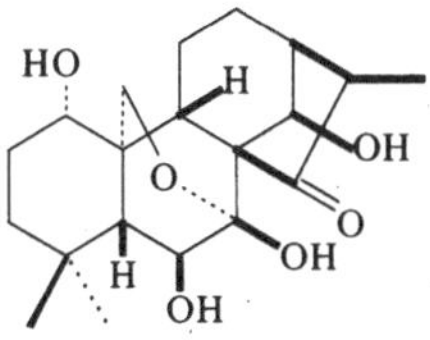

(209)

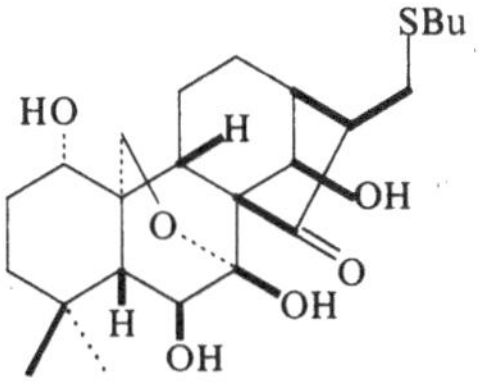

(210)

Reports on the antibacterial activities of longikaurins A (**42**) and B (**43**) (*52*), longikaurins C–F (**44**–**47**) (*53*), effusanins A–E (**25**–**29**) (*27*), kamebanin (**7**) (*91*), umbrosins A (**1**) and B (**2**) (*88*), isodomedin (**12**) (*69*), and amethystoidin A (**23**) (*24*) have also appeared.

2. Antitumor Activity

The first report dealing with antitumor activity of *Rabdosia* constituents referred to crude crystalline material isolated from *R. trichocarpa* (*132, 137–139*). Subsequently, enmein (**62**) and enmein diacetate (**195**) were shown to have antitumor activity against Ehrlich ascites carcinoma, while dihydroenmein (**81**) was inactive (*133*). Although at the time the structure of enmein (**62**) had not been established it was assumed that the α-methylenecyclopentanone whose presence had been recognized was essential for activity (*133*).

E. Fujita and coworkers (*136, 140*) examined the *in vivo* antitumor activity of oridonin (**32**), lasiokaurin (**37**), enmein (**62**), enmein-3-acetate (**63**) and related compounds against Ehrlich ascites carcinoma in the mouse, and proved that all compounds which had the α-methylenecyclopentanone moiety had antitumor activity. On the other hand, trichokaurin (**56**), the thiol adduct (**210**), and the dihydro derivatives (**81 and 209**) were inactive. The facile addition of soft nucleophiles such as alkanethiols and L-cysteine in a 1,4-fashion to the α,β-unsaturated ketone moiety of oridonin (**32**) or enmein (**62**) supported the hypothesis that the physiological activity of *Rabdosia* diterpenoids may be attributed to deactivation of SH-enzymes (or SH-coenzymes). The greater activity of oridonin (**32**) and lasiokaurin (**37**) was attributed to hydrogen-bonding between the C-6-OH and the C-15 carbonyl groups; the hydroxyl groups at C-7 and C-14 were assumed to play a role as binding sites to specific enzymes in the tumor cells (*136, 140*) (Fig. 6).

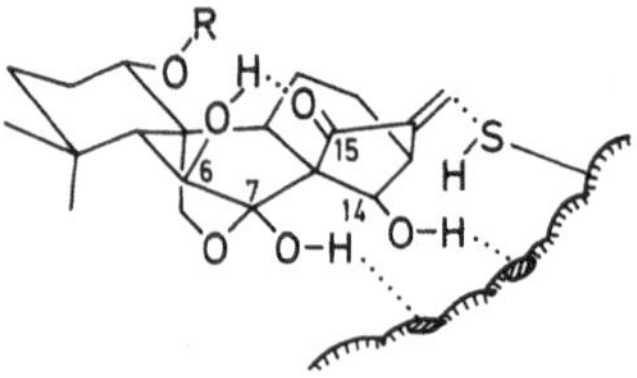

Fig. 6

Various acylated oridonins have also been examined (*141*). An increase in the chain length of 14-acyloridonins tended to increase antitumor activity; thus the activities of 14-dodecanoyl-, tetradecanoyl-, and hexadecanoyl oridonin were higher than that of oridonin itself. On the other hand, acylation of the 6-OH group resulted in deactivation, an observation which supported the important role of this hydroxyl in hydrogen-bonding to the C-15 carbonyl mentioned earlier (*141*).

Acylation of enmein (**62**) led to an increase in antitumor activity, but also to an increase in toxicity; the 6-propanoate, benzoate, and hexadecanoate had increased activity (*130*).

The antitumor activity of spirolactone-type diterpenoids has also been studied. Shikodonin (**88**) had antitumor activity (*64*). Very recently *in vivo* antitumor activity has been reported for trichorabdals A–C (**92–94**) and related diterpenoids against Ehrlich ascites carcinoma in the mouse. The trichorabdals exhibited greater activity than oridonin and while the activity of their dihydro derivatives was greatly reduced, they were still active, in contrast to dihydroenmein and its analogs which are inactive. Hence, the trichorabdals may possess a second active site, presumably the spirolactone aldehyde part, and it was assumed that the enhanced activity of the trichorabdals is due to synergism between the two active sites (*82, 142*).

Similar observations were made by another group which compared the antitumor activity against Ehrlich ascites carcinoma in the mouse of sculponeatin A (**78**), an enmein-type diterpenoid, with that of sculponeatin C (**87**), a spirolactone-type diterpenoid. The *in vivo* activity of sculponeatin C (**87**) was higher than that of sculponeatin A (**78**), although the *in vitro* activity was almost the same and hence it was assumed that the C-6 acetal and spirolactone functions might play an important role in causing the *in vivo* activity (*143*). The cytotoxicities of 24 other *Rabdosia* diterpenoids and *in vivo* antitumor activities of 11 other diterpenoids against Ehrlich ascites carcinoma were also examined; the polyhydroxylated 15-oxokaurenes leukamenin B (**18**), kamebanin (**7**), kamebakaurin (**31**), and effusanin A (**25**) exhibited activity greter than that of oridonin (**32**) (*143*).

Relatively weak antitumor activity is present in the *ent*-8,9-secokaurene-type diterpenoids (**101**), (**103**), and (**104**) but not in *O*-methylshikoccin (**102**). The oxidation product (**211**) of shikoccin exhibited greatly increased activity compared with that of shikoccin (**101**), a result also attributed to a synergistic effect of the plural active centers (*82*).

(**211**)

An ethanolic extract of *R. japonica* had high activity against Sarcoma 180 ascites (*144*), and a crude crystalline substance from *R. trichocarpa* (*137*) was also shown to be active. The anticancer activity of *Rabdosia* diterpenoids against P-388 lymphocytic leukemia is generally low; oridonin (**32**), enmein (**62**), enmein-3-acetate (**63**), nodosin (**66**), and shikoccin (**101**) showed low activity (*145*), but trichorabdal G acetate (**98**) and F acetate (**97**) were considerably more active (*146*). Rabdosins A (**73**) and B (**74**) had *in vivo* antileukemic activity (*44*). Anticancer activity has been explored for rubescensin C (**48**) (*147*), eriocalyxin B (**30**) (*28*), and ponicidin (rubescensin B) (**33**) (*60*) as well as *in vivo* activity of kamebanin (**7**) against Walker intramuscular carcinosarcoma (*91*).

Cytotoxicity has been reported for shikodomedin (**99**) (*71*), shikokiamedin (**100**) (*71*), isodomedin (**12**) (*69*), shikokianin (**24**) (*64*), inflexinol (**11**) (*31*), inflexin (**10**) (*30*), and longikaurin A (**42**) (*147*). Finally clinical tests have been carried out recently in China with oridonin (**32**) and ponicidin (**33**) (*60*).

3. Insect Growth Inhibitory Activity

As a result of tests aimed at evaluating antifeeding activity of the bitter *Rabdosia* diterpenoids, oridonin (**32**), enmein (**62**), nodosin (**66**), isodonal (**71**), and epinodosin (**72**) against the African army worm *(Spodoptera exempta)*, it was found that these diterpenoids were growth inhibitors or biological poisons rather than antifeedants. As the activity disappeared when the methylene group was hydrogenated, it was also attributed to the α-methylenecyclopentanone moiety *(148, 149)*.

The effect of isodonal (**71**), nodosin (**66**), enmein (**62**), oridonin (**32**), and umbrosin A (**1**) on oxidative phosphorylation in mitochondria isolated from rat liver, silkworm midgut, and termite ovaries was investigated. All compounds exhibited an inhibitory effect, except for umbrosin A (**1**) which had no activity in mitochondria isolated from insect. The active site was also associated with the α-methylenecyclopentanone system *(135, 148, 149)*.

4. Bitterness

A qualitative theory on the relationship between bitterness and chemical structures of bitter *Rabdosia* diterpenoids has been proposed *(143)*. To be bitter a substance must have at least one bitter unit; it consists of a hard acid and a hard base which are located within 1.5 Å of each other so that intramolecular hydrogen-bonding is possible. Cleavage of this hydrogen bond and concomitant formation of a new hydrogen bond to the receptor site are responsible for bitterness *(150)*. For instance, isodonal (**71**) which possesses an α-orientated 11-OH is very bitter, while trichodonin (**70**), its 11β-epimer, is not. In bitter isodonal, the distance between the 11-hydroxy proton, the donor proton, and the 6-aldehydic oxygen, the proton acceptor, is ca. 1 Å, while in tasteless trichodonin it is ca. 3 Å.

The bitter unit of enmein (**62**) is presumed to be associated with the 20-hydroxy group and the 6-aldehyde in a formula in which the hemiacetal ring is opened; the one in oridonin (**32**) is presumed to be associated with the 6β-hydroxyl and the 15-carbonyl groups. In the kaurenes, umbrosins A (**1**) and B (**2**), kamebanin (**7**), isodomedin (**12**), and mebadonin (**16**), the bitter unit is assumed to involve the 7α- and the 14β-hydroxy groups. These hydroxyls form an intramolecular hydrogen bond as demonstrated by X-ray analysis, and their acetylation or transformation into acetonide leads to tasteless derivatives. However, diol (**212**) which has only this bitter unit and no other functional group was not as bitter as the more highly substituted natural product, mebadonin (**16**). The decrease is attributed to the change in physical properties brought about by removal of functional group and the resultant influence on transport to the receptor site rather than to the binding itself (*151*).

(212)

References

1. HARA, H.: On the Asiatic Species of the Genus *Rabdosia* (Labiatae). J. Japan Bot. (Shokubutsu Kenkyu Sasshi) **47**, 198 (1972).
2. YAGI, S.: On "Plectranthin", a Bitter Principle Derived from *Plectranthus glaucocalyx* Maxim. *var. japonicus* Maxim. J. Kyoto Med. Soc. **7**, 30 (1910).
3. TANABE, S., and H. NISHIKAWA: Screening Tests for Antibiotic Action of Plant Extracts Jpn. J. Bact. **9**, 475 (1954).
4. FUJITA, E.: The Chemistry on Diterpenoids of *Isodon* Species. Bull. Inst. Chem. Res., Kyoto Univ. **46**, 161 (1968).
5. — Diterpenoids of Enmei-so. Kagaku no Ryoiki, Zokan 86 (Chemistry of Natural Products '68), **1968**, 173.
6. FUJITA, E., and T. FUJITA: The Constituents of Enmei-so (*Isodon japonicus* Hara and *I. trichocarpus* Kudo). Proc. Symp. Wakan-Yaku **9**, 1 (1975).
7. FUJITA, E., Y. NAGAO, and M. NODE: Diterpenoids of *Isodon* and *Teucrium* Plants. Heterocycles **5**, 793 (1976).
8. a) FUJITA, E., T. FUJITA, and M. SHIBUYA: Isolation of Enmein and its 3-Acetate from *Isodon japonicus*. Chem. Commun. **1966**, 297; b) Isolation of Enmein and its 3-Acetate from *Isodon japonicus* Hara. Yakugaku Zasshi **87**, 1076 (1967).
9. IKEDA, T., and S. KANATOMO: Study of Bitter Principles of *Isodon trichocarpus*. I. Yakugaku Zasshi **78**, 1128 (1958).
10. KANATOMO, S.: Alkaline Decomposition of Enmein. Chem. Pharm. Bull. (Japan) **6**, 680 (1958).
11. NAYA, K.: Studies on Bitter Principles in Plants (1st Report). A Bitter Principle in *Isodon trichocarpus* "Isodonin". Nippon Kagaku Kaishi **79**, 885 (1958).
12. TAKAHASHI, M., T. FUJITA, and Y. KOYAMA: Studies on the Components of *Isodon japonicus* (Burm.) Hara and *I. trichocarpus* (Maxim.) Kudo. Yakugaku Zasshi **78**, 699 (1958).
13. — — — Components of *Isodon japonicus* (Burm.) Hara and *I. trichocarpus* (Maxim.) Kudo. II. On the Reduction of Enmein and its Derivatives. Yakugaku Zasshi **80**, 594 (1960).
14. — — — Studies on the Components of *Isodon japonicus* (Burm.) Hara and *I. trichocarpus* (Maxim.) Kudo. III. On the Optical Rotatory Dispersion of Dihydroenmein. Yakugaku Zasshi **80**, 696 (1960).
15. KANATOMO, S.: Studies on Bitter Principles of *Isodon trichocarpus*. III. Baryta Distillation of Enmein. Yakugaku Zasshi **81**, 1437 (1961).
16. — On the Structure of Enmein. Yakugaku Zasshi **81**, 1049 (1961).
17. KUBOTA, T., T. MATSUURA, T. TSUTSUI, and K. NAYA: Chemical Constitution of Enmein,

a Bitter Principle from *Isodon trichocarpus* Kudo. Bull. Chem. Soc. Japan **34**, 1737 (1961). *Idem*, Nippon Kagaku Kaishi **84**, 353 (1963).

18. KUBOTA, T., T. MATSUURA, T. TSUTSUI, S. UYEO, M. TAKAHASHI, H. IRIE, A. NUMATA, T. FUJITA, T. OKAMOTO, M. NATSUME, Y. KAWAZOE, K. SUDO, T. IKEDA, M. TOMOEDA, S. KANATOMO, T. KOSUGE, and K. ADACHI: The Constitution and Stereochemistry of Enmein. Tetrahedron Letters **1964**, 1243.

19. KUBOTA, T., T. MATSUURA, T. TSUTSUI, S. UYEO, H. IRIE, A. NUMATA, T. FUJITA, and T. SUZUKI: Constitution and Stereochemistry of Enmein, A Diterpene from *Isodon trichocarpus* Kudo, Tetrahedron **22**, 1659 (1966).

20. a) IITAKA, Y., and M. NATSUME: The X-ray Study of Acetyl-bromoacetyldihydroenmein. Tetrahedron Letters **1964**, 1257; b) The Chrystal and Molecular Structure of Acetylbromoacetyldihydroenmein. Acta Crystallogr. **20**, 197 (1966).

21. UYEO, S.: The Structure & Stereochemistry of the Diterpenoid Enmein. J. Sci. Indust. Res. (India) **26**, 386 (1967).

22. LI, G., Y. WANG, Z. XU, P. ZHANG, and W. ZHAO: Studies on Chemical Constituents of *Isodon amethystoides* (Benth) C Y Wu et Hsuan. Yaoxue Xuebao **16**, 667 (1981) [Chem. Abstr. **96**, 31688s (1982)].

23. WANG, X., Z. WANG, P. C. SHI, and B. ZHOU: Wangzaozine A, a Novel Diterpenoid from *Isodon amethystoides*. Zhongcaoyao **13**, 12 (1982) [Chem. Abstr. **98**, 50318d (1983)].

24. CHANG, T., M. XU, Y. LIN, and J. SHI: Study on Antitumor Constituents of *Rabdosia amethystoides* Hara. Yaoxue Tongbao **16**, 57 (1981) [Chem. Abstr. **96**, 118987n (1982)].

25. ISOBE, T., T. KAMIKAWA, and T. KUBOTA: Bitter Components from the Leaves of *Isodon* Species. Nippon Kagaku Kaishi **1972**, 2143.

26. KUBO, I., T. KAMIKAWA, T. ISOBE, and T. KUBOTA: Structure of Effusin. Chem. Commun. **1980**, 1206.

27. FUJITA, T., Y. TAKEDA, T. SHINGU, and A. UENO: Structures of Effusanins, Antibacterial Diterpenoids from *Rabdosia effusa*. Chemistry Letters **1980**, 1635.

28. WANG, Z., and Y. XU: New Diterpenoid Constituents of *Rabdosia eriocalyx*. I. Yunnan Zhiwu Yanjiu **4**, 407 (1982) [Chem. Abstr. **98**, 104285m (1983)].

29. SUN, H., X. SUN, Z. LIN, Y. XU, Y. MINAMI, T. MARUNAKA, and T. FUJITA: Excisanin A and B, New Diterpenoids from *Rabdosia excisa*. Chemistry Letters **1981**, 753.

30. KUBO, I., K. NAKANISHI, T. KAMIKAWA, T. ISOBE, and T. KUBOTA: The Structure of Inflexin. Chemistry Letters **1977**, 99.

31. FUJITA, T., Y. TAKEDA, E. YUASA, A. OKUMURA, T. SHINGU, and T. YOKOI: Structure of Inflexinol, a New Cytotoxic Diterpene from *Rabdosia inflexa*. Phytochem. **21**, 903 (1982).

32. FUJITA, E., T. FUJITA, and M. SHIBUYA: Diterpenoids Constituents of *Isodon trichocarpus* and *Isodon japonicus*. Tetrahedron Letters **1977**, 3153.

33. — — — Terpenoids. IX. The Structure and Absolute Configuration of Isodocarpin, a New Diterpenoid from *Isodon trichocarpus* Kudo and *I. japonicus* Hara. Chem. Pharm. Bull. (Japan) **16**, 1573 (1968).

34. FUJITA, E., T. FUJITA, Y. OKADA, S. NAKAMURA, and M. SHIBUYA: Terpenoids. XXI. The Structure and Stereochemistry of Isodotricin, a Diterpenoid of *Isodon trichocarpus* and *I. japonicus*. Chem. Pharm. Bull. (Japan) **20**, 2377 (1972).

35. FUJITA, E., T. FUJITA, and M. SHIBUYA: Terpenoids. VII. The Structure and Absolute Configuration of Nodosin, a New Diterpenoid of *Isodon* Species. Chem. Pharm. Bull. (Japan) **16**, 509 (1968).

36. a) FUJITA, E., T. FUJITA, H. KATAYAMA, and M. SHIBUYA: Oridonin, a New Diterpenoid from *Isodon* Species. Chem. Commun. **1967**, 252; b) FUJITA, E., T. FUJITA, H. KATAYAMA, M. SHIBUYA, and T. SHINGU: Terpenoids. Part XV. Structure and Absolute Configuration of Oridonin Isolated from *Isodon japonicus* and *Isodon trichocarpus*. J. Chem. Soc. (C) **1970**, 1674.

37. FUJITA, E., M. TAOKA, M. SHIBUYA, T. FUJITA, and T. SHINGU: Terpenoids. Part XXVII. Structure and Stereochemistry of Ponicidin, a Diterpenoid of *Isodon japonicus*. J. Chem. Soc., Perkin Trans. I **1973**, 2277.

38. a) FUJITA, E., T. FUJITA, M. TAOKA, H. KATAYAMA, and M. SHIBUYA: The Structure and Absolute Configuration of Sodoponin and Epinodosinol, New Minor Diterpenoids of *Isodon japonicus*. Tetrahedron Letters **1970**, 421; b) Terpenoids. XXIV. Isolation of Isodonal and Epinodosin from *Isodon japonicus* and Structure Elucidation of Sodoponin and Epinodosinol, Novel Diterpenoids of the Same Plant. Chem. Pharm. Bull. (Japan) **21**, 1357 (1973).

39. FUJITA, E., M. TAOKA, Y. NAGAO, and T. FUJITA: Terpenoids. Part XXV. Structures and Absolute Configurations of Isodoacetal, Nodosinin, and Odonicin, Novel Diterpenoids of *Isodon japonicus*. J. Chem. Soc., Perkin Trans. I **1973**, 1760.

40. FUJITA, E., T. FUJITA, and M. SHIBUYA: The Structures of Isodocarpin, Nodosin, Oridonin and Other Constituents of *Isodon Trichocarpus* and *Isodon Japonicus*. Abstract Papers of the 10th Symposium on the Chemistry of Natural Products. (Tokyo) P. 224 (1966).

41. KUBOTA, T., and I. KUBO: The Structures of Trichodonin and Epinodosin. Chem. Commun. **1968**, 763.

42. KUBO, I., T. KAMIKAWA, and T. KUBOTA: Studies on Constituents of *Isodon japonicus* Hara. The Structures and Absolute Stereochemistry of Isodonal, Trichodonin and Epinodosin. Tetrahedron **30**, 615 (1974).

43. KUBOTA, T., and I. KUBO: A New Bitter Principle of *Isodon japonicus* Hara. Tetrahedron Letters **1967**, 3781.

44. LI, J., C. LIU, X. AN, M. WANG, T. ZHAO, S. YU, and G. ZHAO: Studies on the Antitumor Constituents of *Rabdosia japonica* (Burm. f) Hara. I. Structures of rabdosin A and B. Yaoxue Xuebao **17**, 682 (1982) [Chem. Abstr. **98**, 59801 p (1983)].

45. LIU, C., J. LI, X. AN, R. CHENG, F. SHEN, Y. XU, and D. WANG: Studies on the Antitumor Constituents of *Rabdosia japonica* (Burm. f) Hara. II. Structure of Rabdosin C. Yaoxue Xuebao **17**, 750 (1982) [Chem. Abstr. **98**, 77989k (1983)].

46. ISOBE, T., Y. NODA, and T. KUBOTA: The Constituents from *Rabdosia japonica*. Nippon Kagaku Kaishi **1982**, 883.

47. SEO, S., Y. TOMITA, and K. TORI: Biosynthesis of Oleanene- and Ursene-Type Triterpenes from [4-^{13}C]Mevalonolactone and [1,2-^{13}C$_2$]Acetate in Tissue Cultures of *Isodon japonicus* Hara. J. Am. Chem. Soc. **103**, 2075 (1981).

48. XU, Y., X. SUN, H. SUN, Z. LIN, and D. WANG: Chemical Structures of glaucocalyxin A and B. Yun Nan Zhi Wu Yan Jiu **3**, 283 (1981) [Chem. Abstr. **96**, 100893u (1982)].

49. FUJITA, E., and M. TAOKA: Terpenoids. XX. The Structure and Absolute Configuration of Lasiokaurin and Lasiodonin, New Diterpenoids from *Isodon lasiocarpus* (Hayata) Kudo. Chem. Pharm. Bull. (Japan) **20**, 1752 (1972).

50. FUJITA, E., M. TAOKA, and T. FUJITA: Terpenoids. XXVI. Structures of Lasiokaurinol and Lasiokaurinin, Two Novel Diterpenoids of *Isodon lasiocarpus* (Hayata) Kudo. Chem. Pharm. Bull. (Japan) **22**, 280 (1974).

51. TAKEDA, Y., T. FUJITA, and C.-C. CHEN: Structures of Lasiocarpanin, Rabdolasional, and Carpalasionin: New Diterpenoids from *Rabdosia lasiocarpa*. Chemistry Letters **1982**, 833.

52. FUJITA, T., Y. TAKEDA, and T. SHINGU: Longikaurin A and B; New, Biologically Active Diterpenoids from *Rabdosia longituba*. Chem. Commun. **1980**, 205.

53. — — — Longikaurin C, D, E, and F; New Antibacterial Diterpenoids from *Rabdosia longituba*. Heterocycles **16**, 227 (1981).

54. CHENG, P.-Y., M.-C. HSU, Y.-T. LIN, and C.-C. SHIH: Studies on the Antitumor Components of *Rabdosia mactophylla* (I.). Yao Hsueh T'ung Pao **15**, 43 (1980) [Chem. Abstr. **95**, 49257q (1981)].

55. CHENG, P., M. XU, Y. LIN, and J. SHI: Antitumor Constituents of *Rabdosia macrophylla* (Migo). Yaoxue Xuebao **16**, 796 (1981) [Chem. Abstr. **96**, 48973d (1982)].

56. — — — — Structure of Rabdophyllin G, an Antitumor Constituent of *Rabdosia macrophylla*. Yaoxue Tongbao **17**, 174 (1982) [Chem. Abstr. **97**, 115212r (1982)].

57. — — — — Studies on Rabdophyllin G, an Antitumor Constituent of *Rabdosia macrophylla* (II). Yaoxue Xuebao **17**, 917 (1982) [Chem. Abstr. **98**, 95653a (1983)].

58. SUN, H., Z. LIN, Y. MINAMI, T. MARUNAKA, T. FUJITA, and Y. TAKEDA: Structures of New Diterpenoids isolated from *Rabdosia* Plants, *R. phyllostachys, R. rubescens* and *R. sculponeata,* in China. Abstract Papers of the 25 th Symposium on The Chemistry of Natural Products (Japan) p. 266 (1982).

59. SUN, H.-D., Z.-W. LIN, C.-Q. QIN, J.-H. CHAO, and Q.-Z. ZHAO: Studies on the Chemical Constituents of Antitumor Plant *Rabdosia rubescens* (Hemsl.) Hara. Yun-nan Chih Wu Yen Chiu **3**, 95 (1981) [Chem. Abstr. **95**, 121032g (1981)].

60. CHANG, T. L., C.-Y. CHEN, C.-H. MIAU, C.-T. CHAO, H.-T. SUN, and C.-W. LIU: Rubescensin B — Another Effective Antitumor Agent in *Rabdosia rubescens* Hemsl. Kô Hsueh Túng Pao **25**, 1051 (1980) [Chem. Abstr. **94**, 145226d (1981)].

61. SUN, H., J. CHAO, Z. LIN, T. MARUNAKA, Y. MINAMI, and T. FUJITA: The Structure of Rubescensin C: A New Minor Diterpenoid isolated from *Rabdosia rubescens*. Chem. Pharm. Bull. (Japan) **30**, 341 (1982).

62. KUBOTA, T., and I. KUBO: The Two Bitter Principles of *Isodon shikokianus* Kudo. Bull. Chem. Soc. Japan **42**, 1778 (1969).

63. ISOBE, T., T. KAMIKAWA, I. KUBO, and T. KUBOTA: The Structure of Shikokianidin, a Minor Component of *Isodon shikokianus* Hara. Bull. Chem. Soc. Japan **46**, 583 (1973).

64. KUBO, I., M. J. PETTEI, K. HIROTSU, H. TSUJI, and T. KUBOTA: Structure of Shikodonin, a Unique Anti-Tumor Spiro-secokaurene Diterpenoid. J. Am. Chem. Soc. **100**, 628 (1978).

65. OCHI, M., M. OKAMURA, H. KOTSUKI, I. MIURA, I. KUBO, and T. KUBOTA: Bitter Diterpenoids from *Rabdosia shikokiana* (Makino) Hara. Bull. Chem. Soc. Japan **55**, 2208 (1982).

66. OCHI, M., K. HIROTSU, I. MIURA, and T. KUBOTA: Rabdoepigibberellolide, a Novel Diterpenoid from *Rabdosia shikokiana* (Makino) Hara. J. Chem. Soc., Chem. Commun. **1982**, 810.

67. OCHI, M., M. OKAMURA, H. KOTSUKI, I. MIURA, K. KUBO, and T. KUBOTA: The Structures of Rabdosianin A, B, and C, Three New Diterpenoids from *Rabdosia shikokianus* Hara. Bull. Chem. Soc. Japan **54**, 2786 (1981).

68. ISOBE, T., Y. NODA, K. SHIBATA, and T. KUBOTA: The Structures of Diterpene Glycoside, Shikokiaside A and B. Chemistry Letters **1981**, 1225.

69. KUBO, I., I. MIURA, K. NAKANISHI, T. KAMIKAWA, T. ISOBE, and T. KUBOTA: Structure of Isodomedin, a Novel *ent*-Kaurenoid Diterpene. J. Chem. Soc., Chem. Commun. **1977**, 555.

70. FUJITA, T., Y. TAKEDA, and T. SHINGU: Bitter Diterpenoids from *Isodon shikokianus var. intermedius*. Phytochem. **18**, 299 (1979).

71. FUJITA, T., Y. TAKEDA, T. SHINGU, M. KIDO, and Z. TAIRA: Structures of Shikodomedin (X-Ray Analysis) and Shikokiamedin: New Cytotoxic 8,9-seco-*ent*-Kaurenoids from *Rabdosia shikokiana*. J. Chem. Soc., Chem. Commun. **1982**, 162.

72. NODE, M., N. ITO, K. FUJI, and E. FUJITA: Three New 8,9-seco-*ent*-Kaurane Diterpenoids from *Rabdosia shikokiana* (Labiatae). Chem. Pharm. Bull. (Japan) **30**, 2639 (1982).

73. FUJITA, E., N. ITO, I. UCHIDA, and K. FUJI: Structures of Shikoccin, a Unique 8,9-seco-*ent*-Kaurene Diterpenoid, and Shikoccidin (X-Ray Crystallography), a New Penta-oxygenated *ent*-Kaurene Diterpenoid. Chem. Commun. **1979**, 806.

74. TAGA, T., K. OSAKI, N. ITO, and E. FUJITA: Structure of *ent*-9,15-Dioxo-8,9-seco-14,16-

Diterpenoids of *Rabdosia* Species 153

kauradiene-3α,7β-diol Diacetate (Shikoccin Moncacetate). Acta Crystal. **B 38**, 2941 (1982).
75. SUN, H., Z. LIN, Y. MINAMI, Y. TAKEDA, and T. FUJITA: On the Constituents of *Rabdosia ternifolia* (D. Don) Hara: The structure of a New Diterpenoid, Isodonoic Acid. Yakugaku Zasshi **102**, 887 (1982).
76. FUJITA, T., Y. TAKEDA, H. SUN, and Y. MINAMI: The Chemical Structure of a New Diterpene of *Rabdosia ternifolia*. Abstract Paper of the 103rd Annual Meeting of Japan Pharmaceutical Society, Tokyo, p. 242 (1983).
77. KANATOMO, S., and S. SAKAI: Studies on Bitter Principles of *Isodon trichocarpus*. IV. Yakugaku Zasshi **81**, 1807 (1961).
78. a) FUJITA, E., T. FUJITA, and M. SHIBUYA: The structure and Stereochemistry of Trichokaurin, a New Diterpenoid from *Isodon trichocarpus* Kudo. Chem. Commun. **1967**, 148; b) On the Stereochemistry of the Acetoxy-group at C-15 in Trichokaurin. Chem. Commun. **1967**, 466; c) The Structure and Absolute Configuration of Trichokaurin and its Chemical Conversion into (−)-Kaurene and Diterpene Alkaloids. Tetrahedron **25**, 2517 (1969).
79. NODE, M., M. SAI, K. FUJI, E. FUJITA, T. SHINGU, W. H. WATSON, and D. GROSSIE: Three New Anti-tumor Diterpenoids, Trichorabdal A, C, and D. Chemistry Letters **1982**, 2023.
80. FUJITA, E., K. FUJI, M. SAI, and M. NODE: The Structure of Trichorabdal B and its Transformation into a Novel Skeleton; X- Ray Crystal Structures. Chem. Commun. **1981**, 899.
81. NODE, M., M. SAI, E. FUJITA, and K. FUJI: Structure of New Diterpenoids, Trichorabdals E, F, and G from *Rabdosia trichocarpa*. Abstract papers of the 27th Symposium on the Chemistry of Terpenes, Essential Oils, and Aromatics (Japan), p. 70 (1983).
82. NODE, M., M. SAI, N. ITO, K. FUJI, E. FUJITA, T. SHINGU, S. TAKEDA, and N. UNEMI: Antitumor Active Diterpenoids from Labiatae Plant. Abstract Papers of 4th Symposium on the Development and Application of Naturally Occurring Drug Materials (Japan), p. 37 (1982).
83. MORI, S., K. SHUDO, T. AGETA, T. KOIZUMI, and T. OKAMOTO: Studies on the Constituents of *Isodon trichocarpus* Kudo. I. Isolation of the Constituent and the Structures of Isodonol, Enmedol, and Enmenol. Chem. Pharm. Bull. (Japan) **18**, 871 (1970).
84. MORI, S., T. KOIZUMI, K. SHUDO, and T. OKAMOTO: Studies on the Constituents of *Isodon trichocarpus* Kudo. II. The Structures of Enmenin, Enmelol, and Ememodin. Chem. Pharm. Bull. (Japan) **18**, 884 (1970).
85. a) FUJITA, T., Y. TAKEDA, and K. TAKEDA: The Chemical Structures of New Minor Diterpenes of *Rabdosia trichocarpa*. Abstract Papers of the 25th Symposium on the Chemistry of Terpenes, Essential Oils, and Aromatics (Japan), p. 104 (1981); b) The Chemical Structure of Minor Diterpenoid from *Rabdosia trichocarpa*. Abstract Paper of the 29th Meeting of Japan Pharmacognosy Society, p. 71 (1982).
86. FUJITA, E., Y. NAGAO, S. NAKANO, Y. MASADA, K. HASHIMOTO, and T. INOUE: Change of Quantity of Each Major Diterpenoid During Growth of *Isodon trichocarpus* Kudo, its Exploration by GC and GC-MS. Yakugaku Zasshi **92**, 1400 (1972).
87. FUJITA, E., T. FUJITA, and N. ITO: Studies on the Constituents of the Stems of *Isodon trichocarpus* Kudo. Yakugaku Zasshi **87**, 1150 (1967).
88. KUBO, I., T. KAMIKAWA, T. ISOBE, and T. KUBOTA: Bitter Principles of *Isodon umbrosus* Hara. The Structures of Umbrosin A and B. Bull. Chem. Soc. Japan **47**, 1277 (1974).
89. TAKEDA, Y., T. FUJITA, and A. UENO: A New Diterpenoid from *Rabdosia umbrosa var. latifolia*. Abstract Paper of the 21st Chugoku-Shikoku Regional Meeting of Pharmaceutical Society of Japan, p. 53 (1982).
90. HIROTSU, K., T. KAMIKAWA, T. KUBOTA, and A. SHIMADA: Mebadonin, a New *ent-*Kaurane Diterpenoid from *Isodon kameba* Okuyama. Chemistry Letters **1973**, 255.

91. Kubo, I., I. Miura, T. Kamikawa, T. Isobe, and T. Kubota: The Structure of Kamebanin, a New Antitumor *ent*-Kaurenoid from *Isodon kameba* Okuyama. Chemistry Letters **1977**, 1289.
92. Takeda, Y., T. Fujita, and A. Ueno: Structures of Leukamenins. Chemistry Letters **1981**, 1229.
93. Fujita, T., T. Ichihara, Y. Takeda, Y. Takaishi, Y. Minato, and T. Shingu: The Chemical Structure of a New Diterpene, Kamebakaurin from *Isodon kameba*. Abstract papers of the 21st Symposium on the Chemistry of Terpenes, Essential Oils, and Aromatics (Japan), p. 24 (1977).
94. Fujita, T., T. Ichihara, Y. Takeda, Y. Takaishi, and T. Shingu: Study on the Constituents of *Isodon kameba;* Structure of A New Diterpene, Kamebakaurinin. Abstract paper of the 98th Annual Meeting of Pharmaceutical Society of Japan, p. 331 (1978).
95. — — — — Chemical Structures of New Diterpenes, Kamebacetal A and B from *Isodon kameba*. Abstract papers of the 23rd Symposium on the Chemistry of Terpenes, Essential Oils, and Aromatics (Japan), p. 279 (1979).
96. Kubo, I., I. Ganjian, and T. Kubota: Chemotaxonomic Significance of *ent*-Kaurene Derivatives in *Rabdosia umbrosus* Varieties. Phytochemistry **21**, 81 (1982).
97. Ganjian, I., I. Kubo, and T. Kubota: Rapid and Micro Identification of Biologically Active Diterpenes in *Rabdosia umbrosus var. excisinflexus* (Labiatae) by Reversed-phase High-performance Liquid Chromatography. J. Chromatogr. **200**, 250 (1980).
98. Triveda, G. K., I. Kubo, and T. Kubota: High Performance Liquid Chromatography of Isodon Diterpenoids. J. Chromatogr. **179**, 219 (1979).
99. Shudo, K., M. Natsume, and T. Okamoto: Conversion of Enmein to (−)-Kaurane. The Absolute Stereochemistry of Enmein. Chem. Pharm. Bull. (Japan) **13**, 1019 (1965).
100. *a)* Fujita, E., T. Fujita, K. Fuji, and N. Ito: The Absolute Configuration of Enmein. Transformation of Enmein into (−)-Kaurane. Chem. Pharm. Bull. (Japan) **13**, 1023 (1965); *b)* Terpenoids – II. The Chemical Conversion of Enmein into (−)-Kaurane. The Absolute Configuration of Enmein. Tetrahedron **22**, 3423 (1966).
101. *a)* Fujita, E., T. Fujita, and H. Katayama: Formal Chemical Conversion of Enmein into *ent*-Kaurene, Atisine, Garryine and Veatchine. Tetrahedron Letters **1969**, 2577; *b)* Terpenoids – XIV. Formal Chemical Conversion of Enmein into *ent*-Kaurene, Atisine and Veatchine. Tetrahedron **26**, 1009 (1970).
102. Fujita, E., T. Fujita, and M. Shibuya: The Chemical Conversion of Trichokaurin into (−)-Kaurene, Atisine, Garryine, and Veatchine. Chem. Commun. **1967**, 468.
103. Fujita, E., T. Fujita, and Y. Nagao: Terpenoids XIX. Chemical Conversion of Enmein into *ent*-15-Kaurene and *ent*-16-Kaurene. Tetrahedron **28**, 555 (1972).
104. Fujita, E., and Y. Nagao: Terpenoids XXIII. Reduction of Kaurene with Hydrazine and Hydrazine Hydrochloride. Yakugaku Zasshi **92**, 1405 (1972).
105. *a)* Fujita, E., T. Fujita, and H. Katayama: Synthesis of Abietane and Transformation of Enmein into *enantio*-Abietane. Chem. Commun. **1967**, 968; *b)* Fujita, E., T. Fujita, H. Katayama, and Y. Nagao: Terpenoids – X. Chemical Conversion of Enmein into *enantio*-Abietane and Total Synthesis of Abietane. Tetrahedron **25**, 1335 (1969).
106. Shudo, K., M. Natsume, and T. Okamoto: Synthesis of Gibbane Derivatives from Enmein. Chem. Pharm. Bull. (Japan) **14**, 311 (1966).
107. *a)* Somei, M., and T. Okamoto: A Novel Method for Attacking Non-activated C – H Bond and its Application to the Synthesis of Gibberellin-A_{15} from Enmein. Chem. Pharm. Bull. (Japan) **18**, 2135 (1970); *b)* A Novel Method for Functionalization of Non-activated C – H Bond and its Application to the Synthesis of Gibberellin-A_{15} (Chemical Studies on Diterpenoids. I). Yakugaku Zasshi **92**, 397 (1972).
108. *a)* Node, M., H. Hori, and E. Fujita: Syntheses of Methyl Esters of Gibberellin A_{15} and Gibberellin A_{37}. Chem. Commun. **1975**, 898; *b)* Fujita, E., M. Node, and H. Hori:

Terpenoids. Part 39. Total Synthesis of Gibberellins A_{15} and A_{37}. J. Chem. Soc. Perkin Trans. I **1977**, 611; *c)* FUJITA, E., and M. NODE: Synthesis of Gibberellins. Heterocycles **7**, 709 (1977).

109. NODE, M., H. HORI, and E. FUJITA: Demethylation of Aliphatic Methyl Ethers with a Thiol and Boron Trifluoride. J. Chem. Soc. Perkin Trans. I **1976**, 2237.

110. — — — — Terpenoids. XXXVII. Hypoiodite Reactions with 6-Hydroxy-17-norkaurane- and 7-Norgibberellane-derivatives. Chem. Pharm. Bull. (Japan) **24**, 2149 (1976).

111. — — — — Terpenoids. Part XXXVIII. Ring B Contraction of Kauranolides and Related Compounds into Gibberellane-type Compounds. J. Chem. Soc. Perkin Trans. I **1976**, 2144.

112. FUJITA, E., T. FUJITA, and H. KATAYAMA: Terpenoids. Part XVI. Chemical Conversion of Oridonin into Isodocarpin. J. Chem. Soc. (C) **1970**, 1681.

113. FUJITA, E., T. FUJITA, and Y. NAGAO: Terpenoids. XVII. Chemical Conversion of Trichokaurin into Isodocarpin *via* a Direct Pathway. Chem. Pharm. Bull. (Japan) **18**, 2343 (1970).

114. FUJITA, E., and S. NAKAMURA: Terpenoids. XXXIII. Chemical Conversion of Enmein into Enmelol. Chem. Pharm. Bull. (Japan) **23**, 858 (1975).

115. *a)* FUJITA, E., T. FUJITA, Y. NAGAO, P. COGGAN, and G. A. SIM: On the Stereoisomers of Some Enmein Derivatives. Tetrahedron Letters **1968**, 4191; *b)* FUJITA, E., T. FUJITA, and Y. NAGAO: Terpenoids – XII. The Stereochemistry of Some Alcohols Derived from Enmein. Tetrahedron **25**, 3717 (1969).

116. COGGON, P., and G. A. SIM: Molecular Conformations. IX. Bicyclo[3.2.1]octene System: X-Ray Analysis of Enmein Derivative. J. Chem. Soc. (B) **1969**, 413.

117. *a)* FUJITA, E., and Y. NAGAO: A Common Stereoelectronic Requirement in Epimerisations with Some Diterpene Alcohols. Chem. Commun. **1970**, 1211; *b)* Terpenoids. Part XVIII. A Common Stereoelectronic Requirement in Epimerizations and Retro-Dieckmann-type Cleavages of Some Diterpene Alcohol and Ketones. J. Chem. Soc. (C) **1971**, 2902.

118. FUJITA, E., I. UCHIDA, and T. FUJITA: Reactions of Enmein-type Compounds with Lead Tetraacetate and Iodine under Irradiation. Chem. Pharm. Bull. (Japan) **22**, 1656 (1974).

119. KATAYAMA, H.: Studies on Inter-conversion of Kaurane Type and B-Secokaurane Type Diterpenoids and on the Structure of Oridonin. Dissertation (Kyoto Univ.), p. 21 (1969).

120. ISOBE, T., and T. KUBOTA: Dehydrobromination of 16-Bromodihydroenmein-type Compounds. Bull. Chem. Soc. Japan **48**, 949 (1975).

121. SHIBUYA, M., and E. FUJITA: Terpenoids. Part XXX. Chemical Conversion of Enmein into an Important Relay Compound for its Total Synthesis. J. Chem. Soc. Perkin Trans. I **1974**, 178.

122. *a)* FUJITA, E., M. SHIBUYA, S. NAKAMURA, Y. OKADA, and T. FUJITA: Total Synthesis of Enmein. Chem. Commun. **1972**, 1107; *b)* Terpenoids. Part XXVIII. Total Synthesis of Enmein. J. Chem. Soc. Perkin Trans. I **1974**, 165.

123. FUJITA, E., M. NODE, Y. NAGAO, and T. FUJITA: Terpenoids. XXXI. Biogenetic Classification of *Isodon* Diterpenoids. Yakugaku Zasshi **94**, 788 (1974).

124. FUJITA, T., S. TAKAO, and E. FUJITA: Biosynthesis of the Diterpenes Enmein and Oridonin from *ent*-16-Kaurene. Chem. Commun. **1973**, 434.

125. FUJITA, T., I. MASUDA, S. TAKAO, and E. FUJITA: Biosynthesis of Natural Products. Part 1. Incorporations of *ent*-Kaur-16-ene and *ent*-Kaur-16-en-15-one into Enmein and Oridonin. J. Chem. Soc. Perkin Trans. I **1976**, 2098.

126. FUJITA, T., S. TAKAO, Y. NAGAO, and E. FUJITA: Biosynthesis of Enmein and Oridonin from 15-Oxygenated Kaurenoids and 14-Deoxyoridonin. J. Chem. Soc. Chem. Commun. **1974**, 666.

127. FUJITA, T., S. TAKAO, and E. FUJITA: Biosynthesis of Natural Products. Part 2. Syntheses of ^{14}C- or ^{3}H-Labelled ent-Kaur-16-ene Derivatives Oxygenated at C-7, or at C-7 and C-15, from Epicandicandiol. J. Chem. Soc. Perkin Trans. I **1979**, 910.

128. — — — Biosynthesis of Natural Products. Part 4. Biosynthesis of Enmein and Oridonin from Mono- or Di-oxygenated Kaurenoids. J. Chem. Soc. Perkin Trans. I **1979**, 2468.

129. FUJITA, T., I. MASUDA, S. TAKAO, and E. FUJITA: Biosynthesis of Natural Products. Part 3. Syntheses of ent-[17-^{14}C] Kaur-16-en-20-ol from Enmein and ent-[17-^{14}C] Kaur-16-ene Derivatives oxygenated at C-3 from ent-Kaur-16-ene-3β, 19-diol. J. Chem. Soc. Perkin Trans. I **1979**, 915.

130. NAGAO, Y., and E. FUJITA: Studies for Exploitation of the Biologically Active Components in "Enmei-so" (Rabdosia japonica Hara and R. trichocarpa Kudo) and Related Plants. Proc. Symp. Wakan-Yaku **13**, 79 (1980).

131. OHYAMA, A., Y. FURUMOTO, S. CHIKUGO, A. ISHIGA, Y. TAKAMI, M. NOMURA, S. YOKOTA, F. LIAO, T. AWAI, A. FUWA, M. YAMADA, Y. UEDA, N. ISHII, T. KURIMURA, and S. TANAMA: Antimicrobial activity of extracts of Amethystanthus trichocarpus Nakai (Action of Enmein and C_1 Upon Some Bacteria and Fungi). Bull. Kobe Med. College **13**, 253 (1958).

132. ARAI, T., Y. KOYAMA, T. SUENAGA, and H. KAJI: Studies on antitumor substance in I. trichocarpus. II. Chemotherapy **9**, 404 (1961).

133. ARAI, T., Y. KOYAMA, T. SUENAGA, and T. MORITA: Antitumor Activity of the Components of Isodon trichocarpus and Related Plants. J. Antibiotics Ser. A **16**, 132 (1963).

134. KUBO, I., M. TANIGUCHI, Y. SATOMURA, and T. KUBOTA: Antibacterial Activity and Chemical Structure of Diterpenoids. Agr. Biol. Chem. **38**, 1261 (1974).

135. YAMAGUCHI, M., M. TANIGUCHI, I. KUBO, and T. KUBOTA: Inhibitory Effect of Antibacterial and Antitumor Diterpenoids on Oxidative Phosphorylation in Mitochondria Isolated from Rat Liver. Agr. Biol. Chem. **41**, 2475 (1977).

136. FUJITA, E., Y. NAGAO, K. KANEKO, S. NAKAZAWA, and H. KURODA: The Antitumor and Antibacterial Activity of the Isodon Diterpenoids. Chem. Pharm. Bull. (Japan) **24**, 2118 (1976).

137. ARAI, T., Y. KOYAMA, T. MORITA, and H. KAJI: Studies on Antitumor Substance in I. trichocarpus. I. Chemotherapy **9**, 403 (1961).

138. AISO, K., T. ARAI, and Y. KOYAMA: Studies on Antitumor Substance in I. trichocarpus. III. Chemotherapy **9**, 404 (1961).

139. ARAI, T., Y. KOYAMA, T. SUENAGA, and T. MORITA: Studies on Antitumor Substance in I. trichocarpus. IV. Chemotherapy **10**, 197 (1962).

140. FUJITA, E., Y. NAGAO, M. NODE, K. KANEKO, S. NAKAZAWA, and H. KURODA: Antitumor Activity of the Isodon Diterpenoids: Structural Requirements for the Activity. Experientia **32**, 203 (1976).

141. FUJITA, E., Y. NAGAO, T. KOHNO, M. MATSUDA, and M. OZAKI: Antitumor Activity of Acylated Oridonin. Chem. Pharm. Bull. (Japan) **29**, 3208 (1981).

142. NODE, M., M. SAI, K. FUJI, E. FUJITA: S. TAKEDA, and N. UNEMI: Antitumor Activity of Diterpenoids, Trichorabdal A, B, and C, and the Related Compounds: Synergism of Two Active Sites. Chem. Pharm. Bull. (Japan) **31**, 1433 (1983).

143. FUJITA, T., Y. TAKEDA, H.-D. SUN, Y. MINAMI, T. MARUNAKA, S. TAKEDA, Y. YAMADA, and T. TOGO: Structures and Antitumor Activity of Diterpenoids from Rabdosia (Labiatae) Plants. Abstract papers of 4th Symposium on the Development and Application of Naturally Occurring Drug Materials (Japan), p. 34 (1982).

144. ITOKAWA, H., K. WATANABE, and S. MIHASHI: Screening Test for Antitumor Activity of Crude Drugs (I). Shoyakugaku Zasshi **33**, 95 (1979).

145. NAGAO, Y., N. ITO, T. KOHNO, H. KURODA, and E. FUJITA: Antitumor Activity of

Rabdosia and *Teucrium* Diterpenoids against P. 388 Lymphocytic Leukemia in Mice. Chem. Pharm. Bull. (Japan) **30**, 727 (1982).

146. FUJITA, E., K. FUJI, M. NODE, and M. SAI: unpublished results.
147. FUJITA, T., Y. TAKEDA, and T. SHINGU: Biologically Active Diterpenoids from *I. longitubus*. Abstract papers of 2nd Symposium on the Development and Application of Naturally Occurring Drug Materials (Japan), p. 16 (1978).
148. TANIGUCHI, M., M. YAMAGUCHI, I. KUBO, and T. KUBOTA: Inhibitory Effect of *Isodon* Diterpenoids on Growth and Mitochondrial Oxidative Phosphorylation in Lepidopterous Insects. Agr. Biol. Chem. **43**, 71 (1979).
149. KUBO, I., M. TANIGUCHI, and T. KUBOTA: The Biological Activities of the *Isodon* Diterpenoids. Rev. Latinoamer. Quim. **9**, 157 (1978).
150. KUBOTA, T., and I. KUBO: Bitterness and Chemical Structure. Nature **223**, 97 (1969).
151. KUBO, I., and T. KUBOTA: The Structural Basis for Bitterness in the *Isodon* Diterpenes. J. Food Chem. **4**, 235 (1979).

(Received July 10, 1983)

The Quinazoline Alkaloids

By S. JOHNE, Institute of Plant Biochemistry, The Academy of Sciences of the German Democratic Republic, Halle (Saale), German Democratic Republic

Contents

I. Introduction

Quinazoline (1) is a building block for approximately eighty naturally occurring alkaloids isolated from a number of families of the plant kingdom, from microorganisms and from animals. The first known quinazoline alkaloid was vasicine (peganine), isolated in 1888 from *Adhatoda vasica,* and later from other species. This plant has been used in Indian indigenous medicine for centuries. The antimalarial activity of febrifugine (9) provided a strong stimulus for the synthesis and biological screening of a vast number of quinazoline derivatives. A number of interesting new alkaloids have been discovered in a variety of sources, and study of the chemistry and pharmacology of quinazoline alkaloids has intensified in the last twenty years. In many cases, structures of the quinazolines have been confirmed by synthesis. The biosynthesis of this type of alkaloid has received intensive study also.

(1)

Interestingly, quinazoline alkaloids occur in the following taxonomically unrelated plant families: Acanthaceae *(Adhatoda, Anisotes)*, Araliaceae *(Mackinlaya)*, Leguminosae-Papilionatae *(Galega)*, Malvaceae *(Sida)*, Palmae *(Daemonorops)*, Rutaceae *(Euxylophora, Evodia, Glycosmis, Hortia, Vepris, Zanthoxylum)*, Saxifragaceae *(Dichroa, Hydrangea)*, Scrophulariaceae *(Linaria)* and Zygophyllaceae *(Peganum)*.

A previous chapter in this series by PRICE dealing with this subject appeared in 1956 (*169*). The reader is also referred to the standard work in alkaloid chemistry, *The Alkaloids* [MANSKE, ed. (*16a, 133—135, 147*)], and to the authoritative books by BOIT (*25*), ARMAREGO (*6*), and DÖPKE (*60*). A comprehensive review on the quinazoline alkaloids by JOHNE and GRÖGER appeared in 1970 (*94*). The chemistry of quinazoline alkaloids is being continuously updated in *The Alkaloids*, a Specialist Periodical Report published by The Chemical Society (London) (*78, 196*). The present review summarizes current knowledge of the occurrence, properties, synthesis, biosynthesis, and biological activity of the quinazoline alkaloids and covers the literature up to the middle of 1983. Table 2 at the end of the chapter lists all presently known quinazoline alkaloids and their sources.

II. Simple Substituted Quinazolin-4-ones

1. General Characteristics

Physical methods have been found to be particularly useful in structure elucidation which chemical degradations and even synthesis failed to accomplish (*34, 155, 157*). Several generalizations can be made.

The simple quinazolines (Chart 1) almost invariably exist in tautomeric forms with the keto form generally predominating (*34, 63, 83, 157*). Their chemical behavior depends upon the nature of the substituents at positions 1 and 2. PAKRASHI *et al.*, and other authors have observed that quinazolin-4-ones with or without substitution at position 2, when N-1-methylated undergo facile reduction under mild conditions, whereas those compounds with a free NH group are extremely resistant to reducing agents (*155*). PAKRASHI and BHATTACHARYYA (*154*) have studied some interrelationships among the simple substituted quinazoline alkaloids. The hydrochlorides of glycorine (**3**) and arborine (**2**), on vacuum sublimation above their melting points, yield quinazolin-4-one and glycosminine (**5**), respectively. On remethylation with alkaline methyl iodide, quinazolin-4-one yields almost exclusively 3-methylquinazolin-4-one while glycosminine (**5**) gives a mixture of arborine and the 3-methyl derivative (**10**). Oxidation of arborine with chromic oxide in acetic acid yields glycorine and glycosmicine (**4**). The extreme reluctance of glycosmicine (**4**) to form salts precluded its de-methylation to the corresponding benzoylene urea (**11**). The latter, on direct methylation with alkaline methyl iodide, is known (*218*) to yield (**12**) and then the 1,3-dimethyl derivative, and not (**4**). It has also been shown (*215*) that even during the condensation of N-methylanthranilamide and urea in the preparation of glycosmicine, the methyl group migrates from position 1 to position 3 to give a mixture of (**4**) and (**12**) in a ratio of 1 : 1.4.

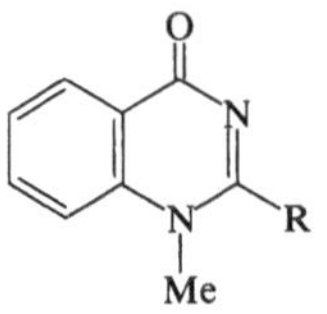

(**2**) R = CH₂Ph

(**3**) R = H

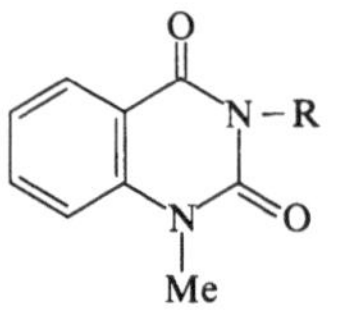

(**4**) R = H

(**7**) R = CH₂CH₂Ph

(**8**) R = CH₂CH₂PhOMe-(p)

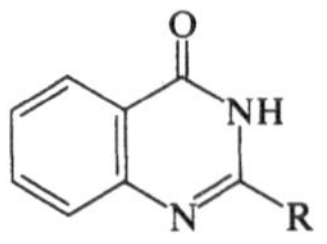

(**5**) R = CH₂Ph

(**6**) R = CH₂CH₂CH₂OH

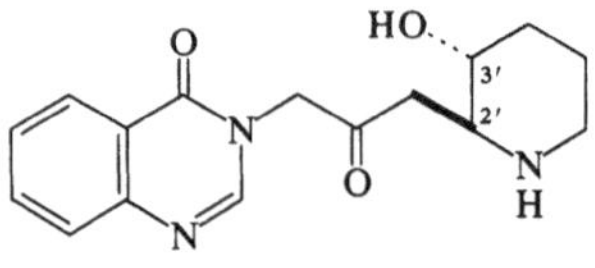

(**9**)

Chart 1. Simple substituted quinazolin-4-ones

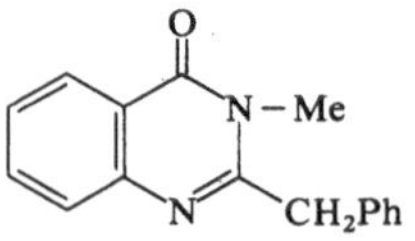

(**10**)

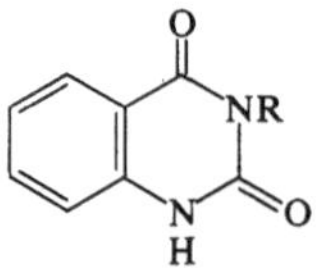

(**11**) R = H

(**12**) R = Me

Although position 3 remains the preferred site for chemical methylation in this type of compound (*83, 218*), the naturally occurring quinazolinones, more often than not, contain alkyl groups at position 1. Furthermore, while one of the lactam forms (**13a**) and (**13b**) is less favored (*83, 157*) in N-1 unsubstituted quinazolin-4-ones [quinazolin-4-one and (**5**)], it has been shown (*157*) to be the preferred one in the corresponding N-1 methylated derivatives [(**3**) and (**2**)], irrespective of the substituent at position 2 (*154*). HAGIWARA et al. (*81*) have studied the intramolecular alkyl rearrangements and tautomerism of quinazolinone derivatives. On the basis of ir, uv, and nmr spectra, the authors concluded that the quinazolin-4-one structure (**13a**) is the thermodynamically most stable one among the three possible tautomeric structures (**13a—13c**). In structure (**13a**), the C=N double bond is located in the β,γ-nonconjugated position relative to the carbonyl group, and the latter is stabilized by amide resonance. Thus, at elevated temperatures the methyl group of 1-methyl-2-phenylquinazolin-4-one, undergoes intramolecular rearrangement to give 3-methyl-2-phenyl-quinazolin-4-one.

(13a)   **(13b)**   **(13c)**

BHATTACHARYYA and PAKRASHI (*21*) reported ^{13}C-nmr spectra of some quinazolin-4-ones which give clues to the predominant tautomeric form and are useful for resolving controversial structural features. Characteristic of the spectra of quinazolin-4-ones are the chemical shifts of C-4 and C-8a. In compounds with a double bond between C-2 and N-3 the resonance of C-4 is about 7 PPM downfield from the C-4 signal in compounds with a double bond between N-1 and C-2. There is also a significant difference in the resonances of C-8a. In the uv spectra quinazolinones show maxima at 225, 270, 300 and 310 nm, the absorption around 225 nm being very intense. Additional details on uv spectral characteristics of these compounds can be found in (*34, 83, 155*). As regards ir absorption, CULBERTSON et al., (*47*) have reported infrared spectra of a number of quinazolinones. The ring system shows several characteristic absorptions in the region 1500—1700 cm^{-1}. According to these authors, the two bands at 1605 cm^{-1} and 1484 cm^{-1} are typical of a quinazolin-2,4-dione system. PAKRASHI et al. (*157*) have demonstrated that two strong bands at 1626—1676 cm^{-1} and 1593—1625 cm^{-1} may reasonably be regarded as characteristic of quinazolin-4-ones. Further details on infrared spectra of these compounds can be found in (*34*).

In the mass spectra the main fragmentation occurs in the pyrimidone moiety, the 3-4 bond being most readily broken (Scheme 1). The first fragmentation step seems to be the loss of C-2 and N-3, together with their substituents, by a concerted cleavage. Further fragmentation depends upon the nature of the substituent at N-1. For a more detailed discussion of this topic, the reader is referred to (29, 155, 157, 181).

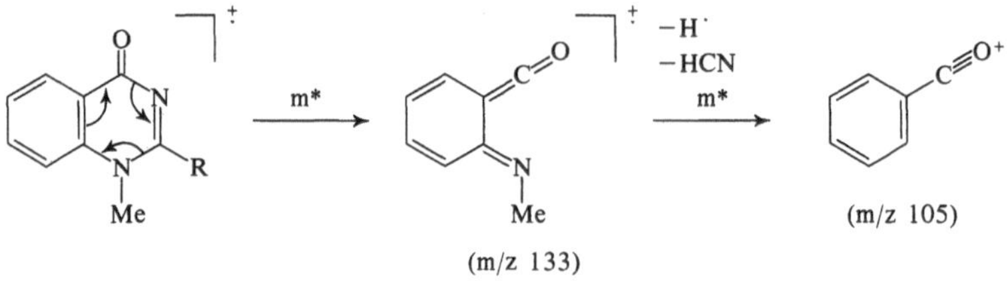

Scheme 1. Mass spectrometric fragmentation of simple substituted quinazolines

2. Arborine

Much controversy has centered around the structure and source of arborine (36, 41). The elucidation of structure (2) (Chart 1) by degradation and synthesis was described in part in a previous chapter in this series (169). CHAKRAVARTI et al. (36) proposed the benzylquinazolinone structure (2), whereas CHATTERJEE and GHOSH MAJUMDAR (42) preferred formula (14) because ozonolysis or periodic acid oxidation of arborine yields benzaldehyde. The yield of benzaldehyde obtained by these methods, or by oxidation with hydrogen peroxide, is extremely low (34) whereas phenylacetic acid may be obtained in almost quantitative yield. The controversy was resolved by CHAKRAVARTI et al. (37) on the basis of the physical properties of arborine. Ultraviolet absorption studies and a detailed critical study of the infrared absorption spectra of arborine, dihydroarborine and model compounds of unambiguously defined structure including N-methylanthranilamide, 1,2-dimethylquinazolin-4-one, 2,3-dihydro-1,2-dimethylquinazolin-4-one, 2-ethyl-1-methylquinazolin-4-one, 2,3-dihydro-2-ethyl-1-methylquinazolin-4-one strongly supported structure (2) for arborine.

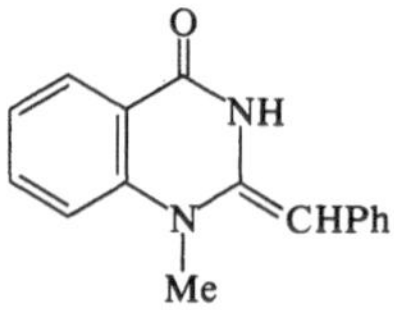

(14)

The ^{1}H-nmr spectrum of (2) showed the absence of vinyl protons, confirming that this alkaloid does not exist in the benzylidene form in appreciable amount. The formation of benzaldehyde during oxidative degradation could be rationalized as being due to the production of a small portion of the highly reactive tautomeric form (**14**) which undergoes rapid oxidative fission while further conversion of (**2**) to this form occurs. There are many well-known analogies (*34*). Mass spectral studies on (**2**) have been performed by PAKRASHI *et al.* (*157*). For further information on the structure elucidation of arborine, see (*33, 37, 155*).

It has been conclusively established (*34, 35, 38*) that the alkaloid named glycosine (*41*) from *Glycosmia pentaphylla* (Retz) DC is identical with (**2**) (*36*) isolated from *Glycosmis arborea* (Roxb.) DC. (Rutaceae) (*156*), *G. pentaphylla* is believed to be more correctly named *G. arborea* (Roxb.) DC. (*22, 157*).

The first syntheses of (**2**) were accomplished by CHAKRAVARTI *et al.* (*36*) and CHATTERJEE and GHOSH MAJUMDAR (*42*). PAKRASHI *et al.* (*158*) realized the dehydrocyclization of the intermediate N-methyl-N-phenylacetyl-anthranilamide either by passage through a column of alumina, by brief heating with dilute sodium bicarbonate solution (pH 9.3), or by treatment with sodium hydroxide (pH 12.5) at room temperature.

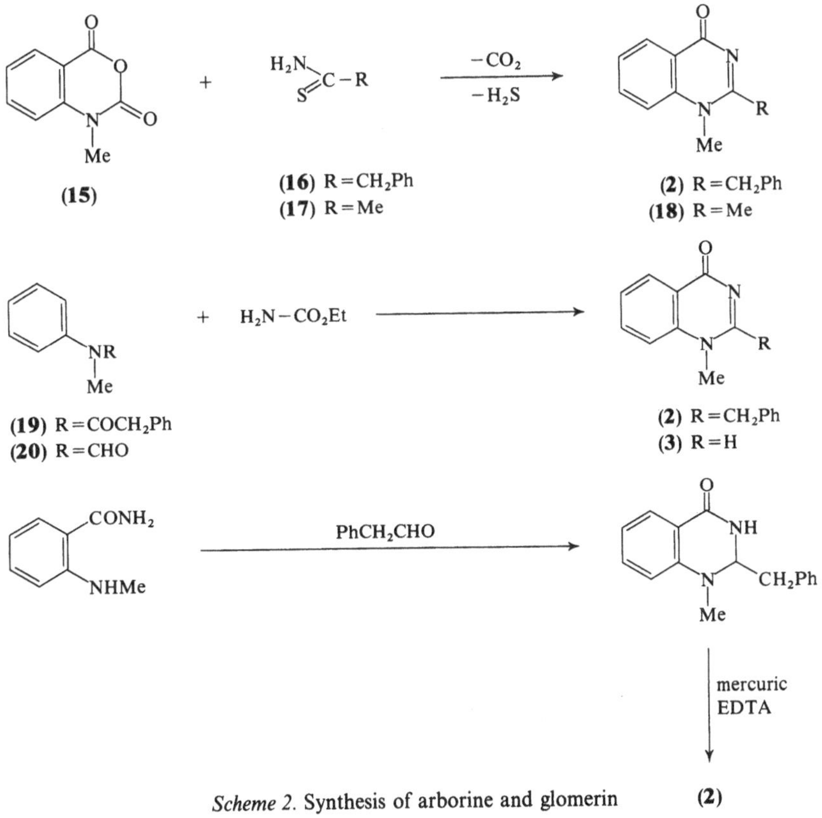
(15)
(16) R=CH$_2$Ph
(17) R=Me
(2) R=CH$_2$Ph
(18) R=Me
(19) R=COCH$_2$Ph
(20) R=CHO
(2) R=CH$_2$Ph
(3) R=H
−CO$_2$
−H$_2$S
H$_2$N−CO$_2$Et
CONH$_2$
NHMe
PhCH$_2$CHO
mercuric
EDTA
(2)

Scheme 2. Synthesis of arborine and glomerin

A new short route to (2) has been reported by ZIEGLER et al. (231), in which phenylacetic acid thioamide (16) is treated with N-methylisatoic anhydride (15) according to Scheme 2. BHATTACHARYYA et al. (23) described another synthesis which is also shown in Scheme 2. N-methylphenyl-acetanilide (20) was cyclized with ethylurethane in the presence of phosphoric oxide in xylene. MÖHRLE and SEIDEL (140) have prepared (2) and a large number of other 2-substituted quinazolines by intramolecular Mannich reactions of N-methylanthranilamide with aldehydes, and dehy-drogenation of the resultant tetrahydroquinazolines (Scheme 2). KAMETANI et al. have developed a convenient synthesis of quinazolines. The procedure depends upon the formation of sulfinamide anhydrides by reaction of anthranilic acids with thionyl chloride, and the subsequent in situ gene-ration of the corresponding iminoketene. Addition of the latter to an imine or a primary or secondary amide affords a quinazoline derivative. (2) was synthesized in this way (104, 105) by condensation of the N-methyl sulfinamide anhydride (prepared from N-methylanthranilic acid) with phenylacetamide in 54.5% yield (Scheme 3). The procedure has been improved by reacting the sulfinamide anhydrides with thioamides; in this way (2) and other quinazoline alkaloids have been prepared in high yields (106, 107).

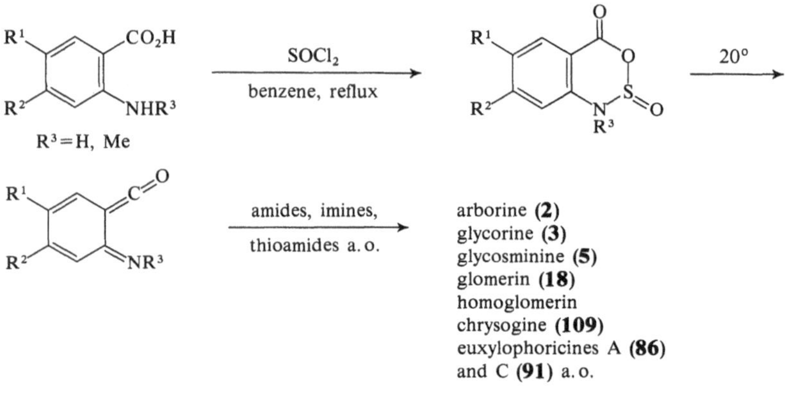

Scheme 3. Synthesis of quinazoline alkaloids according to KAMETANI et al. (105—107)

3. Glycorine

Glycorine (3) (Chart 1) was isolated together with glycosminine and glycosmicine as minor alkaloids of the leaves of Glycosmis arborea (Roxb.) DC. (153, 156). The structure elucidation of these alkaloids was described by PAKRASHI and BHATTACHARYYA (157), using physical data.

Glycorine is a hygroscopic tertiary base, the hydrochloride of which, on vacuum sublimation above its melting point, yields quinazolin-4-one (*154*). The uv spectrum (λ_{max} 269, 278, 306 and 317 nm) is essentially the same as that reported for 1-methylquinazolin-4-one. In 0.01 N HCl, the maxima ·shift to 282, 295 and 304 nm. As regards their spectrum, it is interesting that the acyl carbonyl of the free base absorbs at a frequency (1635 cm^{-1}) in KBr, 69 cm^{-1} lower than that of the hydrochloride. According to (*47*), this shift must be attributed to the direct conjugation of the carbonyl with an imino function. The large shift of the carbonyl absorption in the hydrochloride may thus be explained by assuming that the base exists as the zwitterionic forms (**21**) and (**22**), which are stabilized by salt formation to (**23**) and (**24**). The ^{1}H-nmr spectrum of the base in dimethylsulfoxide is consistent with the presence of eight protons on a quinazoline skeleton and is listed below. Signals centered at δ 8.08 are due to the four phenyl ring protons. The N-methyl resonance is observed at δ 3.83. A signal at δ 8.25 can be assigned to the C-2 proton. The ^{13}C-nmr spectrum indicates that structure (**23**) is preferred for glycorine hydrochloride (see above).

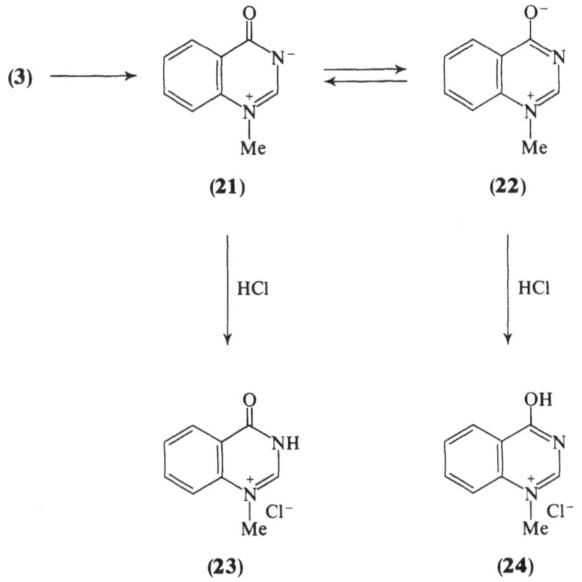

(3) ⟶

(**21**)   (**22**)

| HCl | HCl

(**23**)   (**24**)

That (**3**) represents the structure of glycorine was confirmed by its synthesis from N-methylanthranilamide and ethyl orthoformate (*157*). Oxidation of (**2**) with chromic oxide in acetic acid yields glycorine and glycosmicine (*154*). BHATTACHARYYA *et al.* (*23*) have synthesized (**3**) from N-methylformanilide (**20**) and ethylurethane in the presence of phosphoric oxide in xylene (Scheme 2), while KAMETANI *et al.* (*103*) have synthesized glycorine by treating the sulfinamide anhydride, prepared from N-methylanthranilic acid and thionyl chloride in hot dry benzene, with formamide in dry dioxane at room temperature (Scheme 3).

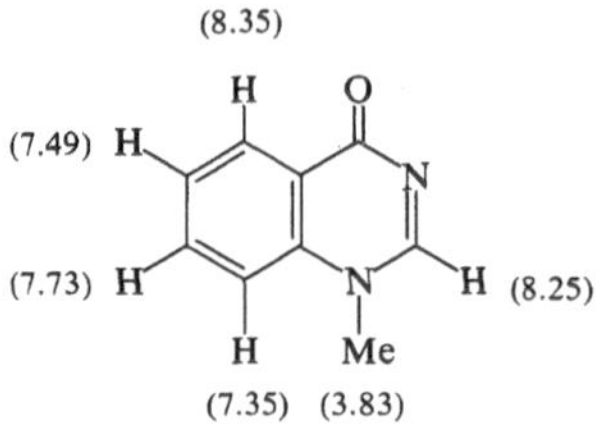

(8.35)

(7.49) H

(7.73) H

H (8.25)

H   Me
(7.35) (3.83)

(3)

4. Glycosmicine

Glycosmicine (**4**) (Chart 1) was isolated in 0.001% yield together with arborine, glycorine and glycosminine from the leaves of *Glycosmis arborea* (Roxb.) DC. (*153, 156*). The mass spectrum and infrared spectral characteristics of (**4**) have been discussed in detail by PAKRASHI *et al.* (*157*). The ir spectrum exhibits, among others, two strong absorptions at 1701 cm^{-1} and 1661 cm^{-1}, corresponding to two carbonyl functions, one at the 4-position and the other at the 2-position, along with two bands at 1605 cm^{-1} and 1484 cm^{-1}, typical of a quinazolindione system. The absorption at 1399 cm^{-1} is attributed to the N-methyl group. Proton nmr data support the presence of one N-methyl group (δ 3.58), four aromatic protons (on one ring), one of which must be *peri*- to a carbonyl group (doublet at δ 8.22), and a hydrogen-bonded NH (broad signal at δ 8.59). The structure was confirmed by direct comparison with a synthetic product prepared from N-methylanthranilamide and ethyl chloroformate. (**4**) was also obtained by oxidation of (**2**) with periodate or permanganate (*41*) or with chromic oxide (*154*).

5. Glycosminine

This alkaloid is identical with glycosmine (*156*), the structure elucidation being accomplished by PAKRASHI *et al.* (*157*). The uv spectrum has absorption maxima at 225, 265, 303 and 312 nm, while in the ir spectrum, the absorption bands at 1676 cm^{-1} and 1613 cm^{-1} suggest that the structure is a quinazolin-4-one substituted at position 2 and/or 3. An intense absorption at 770 cm^{-1} is due to an ortho substituted aromatic ring, while bands at 713 cm^{-1} and 748 cm^{-1} are attributed to a mono-substituted phenyl ring. A sharp absorption band at 3356 cm^{-1} suggests the presence of either an NH or OH group. The nmr spectrum contains signals for a hydrogen-bonded NH (δ 10.25) and a proton on a phenyl ring *peri-* to a carbonyl (δ 8.26). A signal at δ 4.08 is assigned to the two protons of a benzylic methylene group attached to one additional deshielding group. A complex group of signals centered at δ 7.5 accounts for the remaining hydrogens. These data lead to structures (5) (Chart 1) or (25) for glycosminine. Nmr studies and formation of glycosminine by N-demethylation of (2) (*154*) favor (5) as the predominating structure for this alkaloid (*155*). Methylation of (5) with alkaline methyl iodide yields a mixture of the 1-methyl and 3-methyl derivatives. The structure has been confirmed by synthesis from anthranilamide and phenylacetyl chloride or phenylacetic acid (*157*).

(25)

Analogous to the synthesis of arborine, dehydrocyclization of the intermediate 2-phenylacetylanthranilamide to (5) in good yield can be achieved either by passage through a column of neutral or acid-washed alumina, by brief heating in a dilute solution of sodium bicarbonate, or treatment with sodium hydroxide at room temperature. Cyclization at physiological pH (7.4) is also significant biogenetically (*158*). KAMETANI *et al.* (*104, 106, 107*) have synthesized (5) in a manner analogous to the synthesis of arborine (2) (Scheme 3) and it has also been obtained by condensation of anthranilamide with phenylacetamide in the presence of polyphosphoric acid (*159*). RHEE and WHITE (*172*) have synthesized glycosminine by condensation of anthranilamide (26) with phenylpyruvic acid (27). The intermediate quinazolinone (28) is then converted to its methyl ester (29). Treatment of (29) with piperidine furnishes (30) which, on heating in the presence of acetic acid, affords (5) (Scheme 4).

An alkaloid named glycophymine from the flower heads of so-called *Glycosmis pentaphylla* (*177*) is optically inactive. The ir and uv spectra of this compound suggested the presence of a quinazolinone system as in arborine or glycosminine. The mass spectrum of glycophymine, like that of other quinazolinone derivatives, contained intense peaks at m/z 235 (M$^+$ − 1), 119, and 91. Analytical data, the mass fragmentation pattern and formation of "glycophymine" by demethylation of arborine suggested that

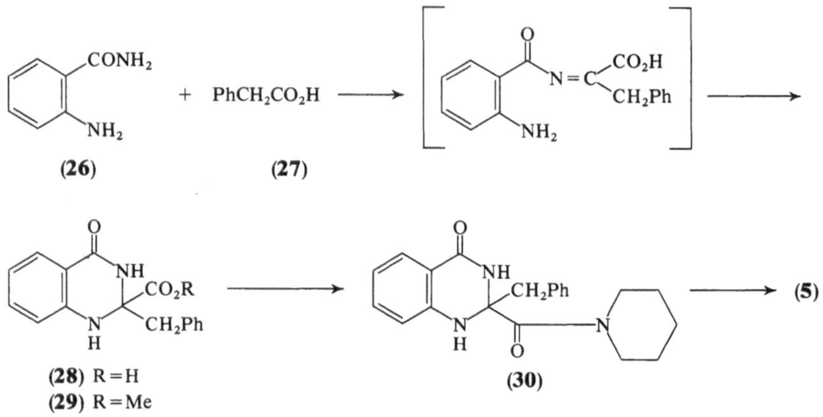

(28) R = H
(29) R = Me

(30)

Scheme 4. Synthesis of glycosminine according to RHEE and WHITE (*172*)

"glycophymine" was either identical with glycosminine or with the tautomeric 2-benzylquinazolin-4-one. The discoverers preferred the latter structure. "Glycophymine" has been synthesized by BHATTACHARYYA and PAKRASHI (*22*) by the procedure originally used by the discoverers (*177*) (from phenylacetanilide and ethyl carbamate), and was found to be identical in all respects with (**5**) synthesized from o-aminobenzamide and phenylacetic acid. The prevalence of the thermodynamically more stable isomer (**5**) rather than its tautomer was also confirmed (*63, 81*) by extensive uv, ir, ^{1}H-nmr, and chemical investigations.

6. Glycophymoline

This alkaloid was also isolated from the flower heads of *Glycosmis pentaphylla* and has the following properties. Ir: 1545 cm^{-1} (quinazoline system), 1618, 1610 cm^{-1} (aromatic system), and 1208 cm^{-1} (OMe group). Uv: λ_{max} 230, 268, 276, 302, 312 nm (log ε 4.35, 3.85, 3.78, 3.50, 3.45). The mass spectrum of glycophymoline, like that of other quinazolines, shows an intense peak at m/z 249 (M${}^+$ −1) and other significant fragments at m/z 235, 119 and 91. ^{1}H-nmr: δ 3.7 (aromatic OMe), δ 4.23 (benzylic methylene group), δ 7.25 (isolated phenyl nucleus), δ 8.1 (for the deshielded aromatic proton in the *peri*-position to the C=O group of the quinazolinone system), δ 7.5—7.8 (other aromatic protons). These data are in agreement with the structure of glycophymoline as an enol ether of glycosminine. Treatment of glycosminine with dimethyl sulfate leads to glycophymoline (*177a*).

7. Pegamine

The [1]H-nmr spectrum of pegamine (6) (Chart 1) contains the following signals: Four aromatic protons in the region between δ 7.5—8.3, of which the signal at δ 8.15 (1 H, d, J = 7 Hz) is attributed to the proton in the position *peri*- to the carbonyl group (C-5), and a triplet centered at δ 7.47 is assigned to the proton at C-7. Signals for the protons of C-9, C-10 and C-11 appear at δ 3.0 (t), 2.14 (m) and 4.2 (t). The [1]H-nmr spectrum of acetylpegamine exhibits a singlet at δ 1.72 for the three protons of the acetyl group. The uv maxima at 226, 266, 306 and 318 nm (log ε 4.27, 3.74, 3.35, 3.21, respectively) are in agreement with the proposed structure (*114*).

8. Quinazolines of *Zanthoxylum arborescens*

DREYER and BRENNER (*62*) isolated two new quinazoline alkaloids from seed husks of *Zanthoxylum arborescens*. The major compound, $C_{17}H_{16}N_2O_2$ exhibits a blue fluorescence on TLC. In the [1]H-nmr spectrum the downfield aromatic quartet at δ 8.12 (J = 7.1 Hz) and a one-proton triplet at δ 7.8 (J = 7 Hz), each further split (J = 1 Hz) suggest the presence of four adjacent aromatic protons on an ortho-substituted benzene ring. The aromatic quartet farthest downfield at δ 8.12 indicates the presence of a *peri*-related carbonyl group. Two two-proton multiplets (δ 2.93—4.40) are consistent with the presence of a 2-phenylethyl system. Other nmr data, in light of the mass spectrum, suggest the presence of an N-methylanthranilic acid system. These characteristics could be assembled in two different ways, leading to structures (7) (Chart 1) and (31). ^{13}C-spectroscopy and synthesis from N-methylisatoic anhydride and 2-phenylethylamine finally allowed elucidation of the structure. The synthetic product, 1-methyl-3-(2'-phenylethyl)-1H,3H-quinazolin-2,4-dione (7) was identical with the alkaloid.

(31)

A second alkaloid from the mother liquors of the major constituent, had prominent infrared absorptions at 1700, 1660 and 1620 cm^{-1}; its [1]H-nmr spectrum exhibited many similarities to that of the major component. The aromatic region contained an A_2B_2 system (δ 6.90 and 7.20) as well as signals associated with an N-methylanthranilic acid system. Two three-proton singlets, δ 3.58 and 3.77, suggest the presence of both N-methyl and methoxyl groups. These and other data supported structure (8) (Chart 1), an assignment which was confirmed by synthesis by the route used also for the major alkaloid.

9. Febrifugine and Isofebrifugine

These interconvertible alkaloids were isolated from *Dichroa febrifuga* (Saxifragaceae) and from *Hydrangea umbellata* (Saxifragaceae) in low yields. *Dichroa febrifuga* is an ingredient in a traditional Chinese herbal remedy effective against malaria and febrifugine (9) (Chart 1) appears to be primarily responsible for the antimalarial activity of this preparation. Febrifugine was the first alkaloid other than those of the *Cinchona* group to possess marked antimalarial activity and therefore it is perhaps the most interesting quinazoline alkaloid of all. Significant biological activity and extensive efforts toward the structure elucidation of (9) were strong stimuli for the synthesis and biological screening of vast numbers of quinazoline derivatives, especially following World War II.

Isofebrifugine differs from febrifugine in not giving a semicarbazone or an oxime. Both bases give 3,4-dihydroquinazolin-4-one on oxidation with alkaline permanganate, and with 2.5 N sodium hydroxide both yield anthranilic acid. Oxidation with neutral permanganate yields (32), while periodate oxidation of (9) and isofebrifugine both produces the same optically inactive product (33). Heating with semicarbazide converts (33) to (34), a key derivative for structure elucidation. Heating with zinc dust yields (35). Further degradation studies, inlcuding direct comparison with synthetic models at each step, led to proposal of structure (9), except for the location of the hydroxyl group in the piperidine ring and the stereochemistry. The structure was confirmed by syntheses of (9) by BAKER *et al.* (*13, 13 a*).

After HILL and EDWARDS (*85*) had assigned the *S*-configuration to C-2′ of febrifugine as a result of degradation experiments, BARRINGER *et al.* (*14, 15*) showed that the substituents on the piperidine ring were *trans*-orientated by investigating the thermal equilibration of *cis*-(3-methoxy-2-piperidyl)-2-propanone and some analogous *cis*-(3-substituted-2-piperidyl)-2-propanones with their *trans* isomers. Assignment of the *trans* configuration to the more stable isomer was based on the nmr spectra, chemical data, and conformational free energy calculations. *Cis*- and *trans*-(9) were synthesized from the corresponding piperidine derivatives, with the *trans* isomer and the natural product having identical ir spectra. The nmr spectra of the corresponding O-acetates were also identical, and exhibited the H-2′-signal as a quartet at δ 3.98 (J = 7 Hz) which collapsed to a doublet

(9) (32) (33) (34) (35)

KMnO₄

semicarbazide

IO₄⁻

zinc dust

upon irradiation of the side chain methylene signal. A multiplet at δ 5.0 (J = 4—5.7 and 9—10 Hz) could be attributed to H-3′. The observed couplings are compatible with a *trans-* diaxial arrangement and the structure of febrifugine is accordingly (2′ S, 3′ R)-3-[3-(3-hydroxy-2-piperidyl)acetonyl]-quinazolin-4-one. The absolute configuration of the C-3′-OH group in (9) is the same as that in S-hydroxylysine, an attractive biogenetic precursor.

BAKER *et al.* (*13*) have suggested that isofebrifugine might be a hemiketal formed by interaction of the 3′-hydroxyl with the side chain ketone group. Stabilization of the hemiketal ring by hydrogen bonding to the carbonyl of the quinazolinone would reduce the ketonic reactivity, but probably not enough to prevent formation of ketone derivatives under all of the conditions tried. Furthermore, two diastereomeric forms of the hemiketal should exist, but only one form was found. Therefore, some doubts remain concerning the details of the structure of isofebrifugine (*6*). A comprehensive treatment of the properties of this alkaloid is beyond the scope of this review; the reader is referred to reviews by ARMAREGO (*6*) and others (*25, 94, 147, 169*). Syntheses of a number of febrifugine analogues and their biological activities are discussed in Chapter IX.

III. The Pyrroloquinazolines

The numbering of the pyrroloquinazoline ring system is a source of some confusion. SPÄTH (*198*), YUNUSOV *et al.* (*208*), ARNDT *et al.* (*9*), and other workers (e. g. *58*) have numbered this ring system in different ways. The nomenclature used in this chapter is that used in *Chemical Abstracts;* compounds are listed in Chart 2.

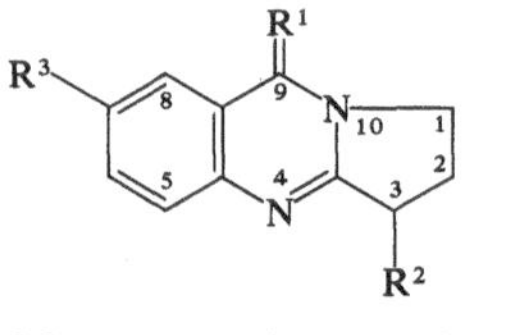

(36) R^1=H$_2$, R^2=OH, R^3=H
(37) R^1=O, R^2=OH, R^3=H
(38) R^1=H$_2$, R^2+R^3=OH
(39) R^1=O, R^2+R^3=OH
(42) R^1=O, R^2+R^3=H

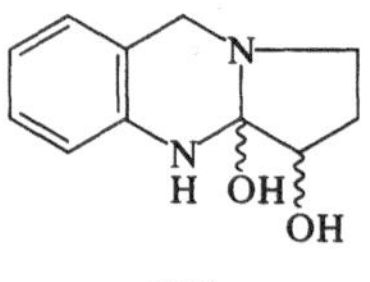

(40)

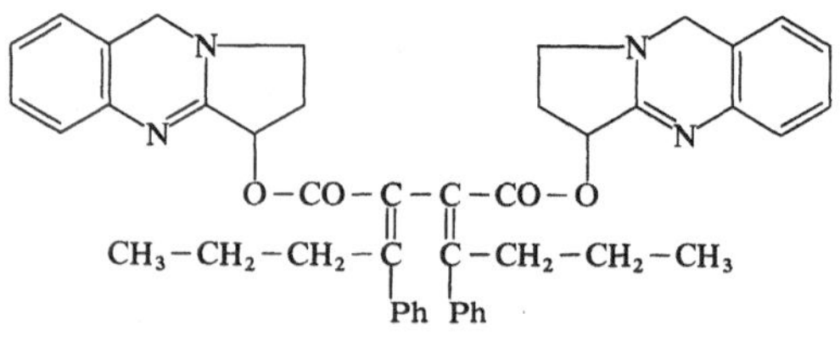

(41)

Chart 2. Pyrroloquinazolines

Chart 2 (continued)

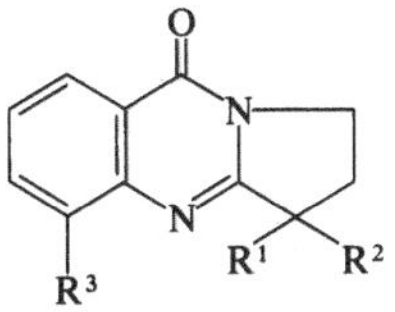

(43) R^1, R^3=H, R^2 =

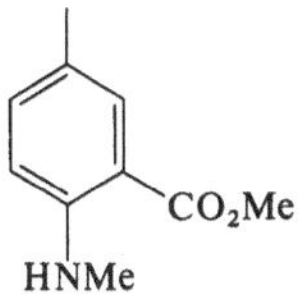

(48) R^1=H, R^3=OMe, R^2 =

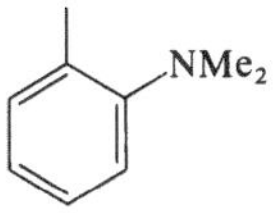

(45) R^1=OH, R^3=H, R^2 =

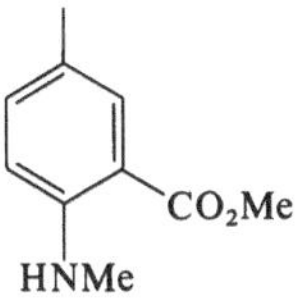

(49) R^1=H, R^3=OMe, R^2 =

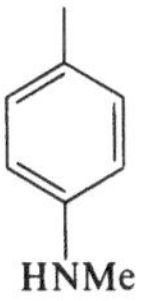

(46) R^1, R^3=H, R^2 =

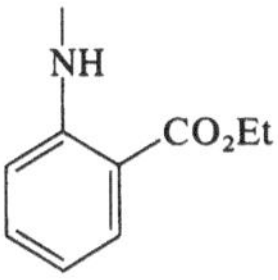

(47) R^1=OH, R^3=OMe, R^2 =

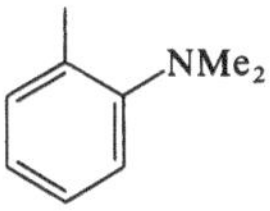

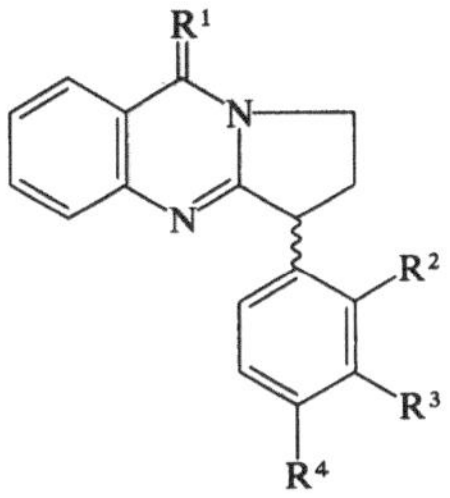

(51) R^1=H$_2$, R^2=NMe$_2$, R^3+R^4=H
(52) R^1=O, R^2=NMe$_2$, R^3+R^4=H
(53) R^1=H$_2$, R^2=H, R^3=CO$_2$Me, R^4=NHMe

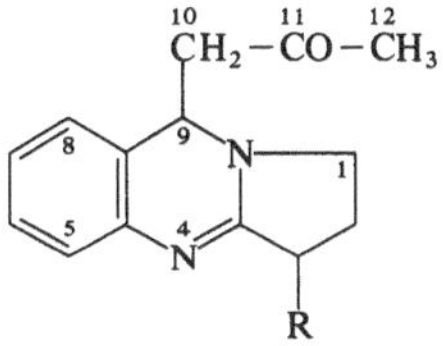

(54) R=OH
(55) R=H

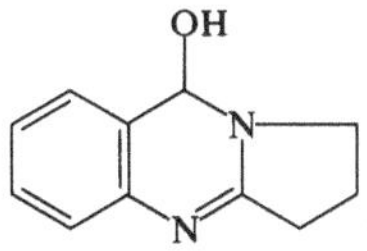

(56)

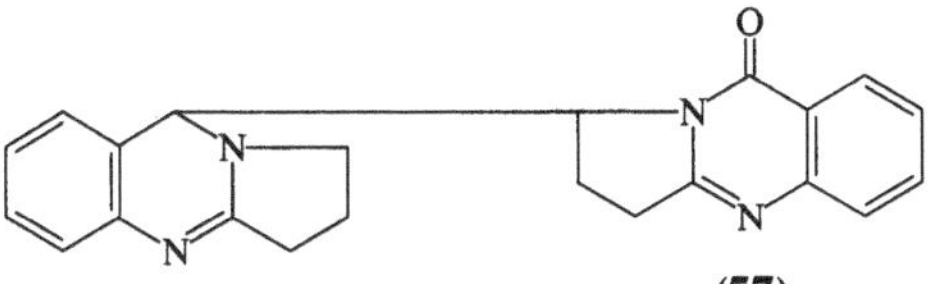

(57)

1. Vasicine (Peganine), Vasicinone, 7-Hydroxypeganine, Vasicinolone, Vasicol, Nordine and Deoxyvasicinone

Vasicine (Peganine) (**36**) is the oldest and best known quinazoline alkaloid. It was discovered in *Adhatoda vasica* Nees (Acanthaceae), and later found in *Peganum harmala* (Zygophyllaceae) and a number of other species (see Table 1). The names vasicine and peganine are both widely used in the literature, and a similar situation exists for the derivatives of this alkaloid — deoxyvasicine is identical with deoxypeganine, 7-hydroxy-vasicine to 7-hydroxypeganine, etc.

Structure elucidation of this alkaloid was accomplished by SPÄTH *et al.*, who made perhaps the most important contributions, by Indian workers, and by MORRIS, HANFORD and ADAMS. The subject has been reviewed in detail (*6, 25, 94, 147, 169, 198*) and only some spectroscopic properties and syntheses will be briefly mentioned. The ir spectrum has bands at 1038, 1061, 1079, 1113, 1132, 1158, 1178, 1193, 1233, 1290, 1308, 1338, 1488, 1508, 1577, 1603, 1637 cm^{-1} (*86*) and the uv spectrum has a maximum at 293 nm (log ε 3.84). The ^{1}H-nmr in CDCl$_3$ exhibits signals for four aromatic protons at δ 7.3—6.8, a one-proton triplet at δ 4.80 (J = 7 Hz), two multiplets, each representing two protons, centered at δ 3.5 and 2.8, assigned to the C-1 and C-2 protons, respectively, and a two-proton singlet at δ 4.62, assigned to the C-9 protons (*9*, cf. also *95*).

The first synthesis of vasicine was performed by SPÄTH *et al.* (*200*). Condensation of methyl 4-amino-2-hydroxybutyrate with o-nitrobenzyl chloride gave 1-o-nitrobenzyl-3-hydroxypyrrolid-2-one, which, after reduction, cyclized readily to dl-vasicine. SPÄTH and PLATZER (*201*) subsequently described a second, simpler synthesis: γ-butyrolactone is brominated, and the product hydrolyzed to α-hydroxybutyrolactone, which is then condensed with o-aminobenzylamine at 200° to give dl-vasicine. SOUTHWICK and CASANOVA (*197*) prepared (**36**) from o-nitrotoluene by a reaction sequence in which o-nitrobenzylamine hydro-chloride, ethyl β-(o-nitrobenzylamino)-propionate hydrochloride, 1-(o-nitrobenzyl)-4-carbethoxy-2,3-dioxopyrrolidine, 1-(o-nitrobenzyl)-2,3-di-oxopyrrolidine, and 1-(o-nitrobenzyl)-3-hydroxy-2-oxopyrrolidine were isolated as intermediates. A parallel series of reactions, with only slight modification of the procedural details, allowed preparation of the analogous 7-methoxyvasicine from 3-methyl-4-nitroanisole. In 1960, LEONHARD and MARTELL (*126*) reported the condensation of the long-sought γ-amino-α-hydroxybutyraldehyde with o-aminobenzaldehyde to form (**36**) in 39% yield. Finally, a facile synthesis of (**36**) was reported in 1970, in the course of which condensation of o-nitrobenzyl chloride with 3-hydroxypyrrolidine, reduction of the nitro function followed by treat-ment with mercuric acetate/EDTA afforded (**36**) exclusively (*138*).

The crystal and molecular structures of (−)-vasicine hydrochloride have been determined by X-ray crystallography. The absolute configuration was deduced using the anomalous scattering of the chlorine atom. (−)-Vasicine possesses the R-configuration at C-3, and the pyrrolidine ring is in the envelope conformation (205). ^{13}C-nmr spectra of vasicine, vasicinone and related compounds have been reported (96), and the mass spectral fragmentations of vasicine, deoxyvasicine, deoxyvasicinone and vasicinone have been discussed (20, 171). Fragmentation proceeds via cleavage of the C-2/C-3 bond, and subsequent cleavage of the methyleneimido and C-3/C-1 chain radicals with migration of H to neutral or charged fragments. The mass spectral behavior of 2,3-polymethylene-1,2,3,4-tetrahydroquinazolin-4-ones was compared (166) and the mass spectra of 2,3-polymethylene-3,4-dihydroquinazolin-4-ones substituted at C-3 have been studied (165).

The seasonal distribution of vasicine in different parts of Adhatoda vasica (76, 89), Sida cordifolia (79), Linaria genistifolia and L. vulgaris (129), Peganum harmala (115, 163) and Galega officinalis (178, 179) has been determined.

Vasicinone (37) is an oxidation product of vasicine. MEHTA et al. (136) have found that the crude total alkaloids from Adhatoda vasica contain predominantly vasicine, but that gradual conversion of vasicine to vasicinone takes place during countercurrent distribution and partition chromatography, probably as a result of autooxidation. Pure vasicine similarly undergoes autooxidation to vasicinone. (37) has also been isolated from the seeds (121, 122, 136) and other parts (163, cf. 115) of Peganum harmala as well as Linaria transiliensis (164). The physical properties are as follows: uv, λ_{max} 227, 272, 302 and 315 nm; ir, the short wavenumber part of the spectrum contains absorptions at 3115, 2924, 1658, 1454, 1385, 1331, 1291, 1209, 1179, 1104, 1080, 1028, 987, 970, 896, 883, 867, 854, 774, 749, 720, 694 cm^{-1}; ^{1}H-nmr (CF$_3$COOH), δ 7.3—7.8, aromatic protons; 8.05, C8-H; 5.56, C3-H; 2.64 + 2.21, C2-H$_2$; 3.92 + 4.32, C1-H$_2$.

Vasicinone can be obtained by oxidation of vasicine with 30% hydrogen peroxide (136, 141). ONAKA (146) reported a biomimetic-type synthesis of (37) starting from anthranilic acid and O-methylbutyrolactim in 17% overall yield. Treatment of (37) with sulfuryl chloride yields chloropegenone which can be reduced to deoxyvasicine. Further details are presented in (94).

7-Hydroxypeganine (38) was isolated by SPÄTH and KESZTLER-GANDINI (199) from the leaves of Adhatoda vasica and detected in the seeds of the same species by GRÖGER and JOHNE (76). Structure elucidation and synthesis were achieved by KUFFNER et al. (123). "Vasicinol", isolated by RAJAGOPALAN et al. from the roots of A. vasica (170), is identical with 7-hydroxypeganine (19).

Vasicinolone (39), isolated from the roots of *Adhatoda vasica (90)*, appears to be an oxidation product of 7-hydroxypeganine and possesses the following properties: uv, λ_{max} 225, 276 and 325 nm; ir, 3400, 1670, 1624, 1600 and 1490 cm^{-1}; ^{1}H-nmr, δ 2.4, C2-H$_2$; 4.2, C1-H$_2$; 5.6, C3-H; 7.6, C5-H, C6-H and C8-H. On acetylation, the alkaloid yields a diacetyl derivative.

Vasicol (40), isolated from the roots of *A. vasica,* was characterized as 2,3,4,9-tetrahydropyrrolo-[2,1-b]-quinazolin-3,3a(1H)-diol by chemical and spectroscopic methods *(58)*. It has uv absorption at λ_{max} 238 and 291 nm and prominent absorptions in the infrared at 1493 and 1603 (aromatic) and 3330—3360 cm^{-1} (NH and OH). The ^{1}H-nmr spectrum (CDCl$_3$) contained two multiplets (each 2H), centered at δ 2.1 and 3.16, attributed to C2-H$_2$ and C1-H$_2$ respectively. Deuterium oxide exchange simplified a complex multiplet centered at δ 4.2 (6H) into broad singlets at δ 4.33 and 4.36 (each 1H) attributed to C9-H$_a$ and C9-H$_b$, respectively, and δ 4.5 (1H, t, partially obscured by the C9-H$_2$ signals) attributed to the C-3 methine proton. Two multiplets, centered at δ 6.63 and 7.03 and integrating for a total of four protons, were due to the aromatic protons. The appearance of the benzylic protons, C9-H$_2$, as two broad singlets rather than as the single multiplet observed in vasicine can be attributed to a somewhat less symmetric environment resulting from the presence of a hydroxy at C-3a. The mass spectrum of (40) shows the molecular ion, m/z 206, a base peak at m/z 106, and other significant fragments at m/z 162 (14%), 161 (30%), and 133 (16%).

Acetylation of (40) afforded the 3,4-diacetyl derivative, and attempts to further acetylate this product failed to yield a triacetylated product. Passing hydrogen chloride gas through a dry methanolic solution of (40) resulted in the formation of vasicine hydrochloride, indicating dehydration of the tertiary alcohol at C-3a. Reaction of phosphorus oxychloride with (40) in pyridine yielded chlorodeoxyvasicine. Partial synthesis of (40) was accomplished by heating vasicine and water in a sealed tube at 140—150° for 16 hours.

Nordine (41) is a constituent of the water-insoluble red pigment of *Daemonorops draco* (?) (Palmae) called "Dragon's Blood" *(160)*. Acid hydrolysis of this symmetric compound yields two moles of vasicine and one mole of 4,7-diphenyl-deca-4,6-dien-5,6-dicarboxylic acid.

Deoxyvasicinone (42) has the following spectral properties: uv (EtOH), λ_{max} 224, 267, 272, 302 and 314 nm (log ε 4.2, 3.63, 3.59, 3.4, 3.33); ir, no NH absorption, but strong bands at 1681 and 1626 cm^{-1} due to the amide carbonyl and imino moieties; ^{1}H-nmr (CF$_3$COOH), δ 7.3—7.8, aromatic protons; 8.00, C8-H; 3.43, C3-H$_2$; 2.29, C2-H$_2$; 4.2, C1-H$_2$ *(9, 40, 208)*. SIDDIQUI *(195)* has isolated a crystalline compound, C$_{11}$H$_{12}$N$_2$, possibly deoxyvasicine, from the seeds of *Peganum harmala,* but unequivocal

identification has not been accomplished. The distribution of deoxyvasicinone (42) in *Peganum harmala* has been determined (*115, 163*).

Many syntheses of (42) have been reported, but only three will be
mentioned here (cf. *94*). CHATTERJEE and GANGULY (*40*) reported the
condensation of anthranilic acid and γ-aminobutyric acid with phosphorus
pentoxide in boiling xylene to give (42) in 50% yield, while refluxing
anthranilic acid with O-methylbutyrolactim in benzene afforded (42) in
82% yield (*146, 161*). KAMETANI et al. (*108*) synthesized (42) regiospecifically
in good yield by treatment of the sulfinamide anhydride of anthranilic acid
with O-methylbutyrolactim. Synthesis of (42) is also possible by oxidation
of deoxyvasicine with hydrogen peroxide (cf. *94*).

SAKAMOTO and SEMEJIMA (*175, 176*) reported the conversion of Δ^1-
pyrroline to deoxyvasicinone, a reaction which could be applied to the
immunological determination of Δ^1-pyrroline, since a specific antibody for
(42) had been prepared by these authors.

In a continuing investigation of the synthesis of quinazoline derivatives,
Soviet workers have condensed deoxyvasicinone with a number of aromatic
aldehydes (*191*). The same group (*190*) has synthesized 7-nitro-, amino- and
3,3-dibromodeoxyvasicinone   (*188*)   and   methylenebis-(7,7'-
deoxyvasicinone) and its homologs by cyclocondensation of 5,5'-
methylenedianthranilic acid with the corresponding lactams (*190*).
Azepinoquinazolinones were obtained by cyclocondensation of anthranilic
acid with lactams. SHARMA and JAIN (*193*) have also reported the
condensation of deoxyvasicinone with aromatic aldehydes. 9-Thio analogues of deoxyvasicinone, its derivatives and homologs were obtained by
boiling the corresponding oxygen compounds with phosphorus pentasul-
fide in m-xylene. The desulfurated derivatives were obtained by heating
with zinc in 10% HCl (*189*). ORIPOV et al. (*148*) have reported some
reactions of 3-hydroxy- and 3-dimethylaminoformylidene-pyrrolo[2,1-b]
quinazoline derivatives. Deoxyvasicine and its derivatives have been
synthesized by reaction of anthranilic acids with 2-pyrrolidone, followed by
reduction with zinc and hydrochloric acid (*187*). KARIMOV et al. (*112*) have
reported the synthesis of methoxy- and hydroxy-substituted deoxyvasicinones and deoxyvasicines.

2. Anisotine, Anisessine, Aniflorine, Deoxyaniflorine
and Sessiflorine

These five optically inactive alkaloids were isolated in low yields from
Anisotes sessiliflorus C. B. Cl. (*9*). Formulas are listed in Chart 2. Anisotine
has also been isolated from *Adhatoda vasica* (*95*). The main alkaloid in both
species is vasicine and the minor alkaloids are closely related. Their

structures have been determined by a combination of spectroscopic techniques and a few appropriate chemical reactions.

Anisotine (43). The uv spectrum of this alkaloid is very similar to that of quinazolin-4-one. Infrared absorptions indicate the presence of an NH and two carbonyl functions which belong to an amide and an aromatic carbomethoxy group. In the ^{1}H-nmr spectrum, the methoxy resonance appears at δ 3.80. The doublet at lowest field, δ 8.27 (J = 9 Hz), is characteristic of a proton at C-8, situated *peri* to the amide carbonyl of a quinazolin-4-one derivative. That the ester group is situated ortho to the amino substituent on the aromatic ring is evident from the facile mass spectral loss of methanol, the (M-32) or (M-1-32) fragments being attributed to the "ortho effect". This interpretation was supported by the loss of methanol-d (M-33) from the N-methyl(d) compound obtained by treatment of (43) with deuterium oxide. Chemical evidence for the position of attachment was obtained from the inability to form the benzal derivative of (43), a derivative which readily forms in the case of deoxyvasicinone yielding (44). Potassium permanganate oxidation of (43) affords the product (45).

(44)

Anisessine (46). Infrared absorptions indicate the presence of an NH group along with amide and aromatic ester carbonyls. In the ^{1}H-nmr spectrum, the presence of an ethyl ester was evident from the characteristic ethyl A_2X_3 pattern, a triplet at δ 1.37 (J = 7 Hz) and a quartet at δ 4.38 (J = 7 Hz). The presence of eight aromatic protons suggested that the phenyl substituent on the deoxyvasicinone skeleton is disubstituted. H-8 is found at low field, δ 8.27 (d, J = 4 Hz), and while another signal at δ 8.00 is a doublet of doublets (J = 4 and 1 Hz) the shift being typical of an aromatic proton which is adjacent to an ester function and coupled to ortho and meta protons. The mass spectrum is in accord with the proposed structure. Oxidation of (46) with potassium permanganate yielded ethyl anthranilate. A synthesis of (46) is outlined in Scheme 5.

Aniflorine (47). The similarity of the uv spectrum of aniflorine to those of (43) and (46) indicates a close relationship to those alkaloids. In the ^{1}H-nmr spectrum, the presence of a dimethylamino group is evident from a 6-

proton singlet at δ 2.80, while an aromatic methoxyl gives rise to a singlet at δ 3.88. That this methoxyl is attached to the quinazolinone aromatic ring can be concluded from the mass spectrum. The para-position can be ruled out for the dimethylamino group due to the absence of the expected A_2B_2 pattern in the nmr spectrum.

Exposure of (47) to Raney nickel in ethanol in a hydrogen atmosphere yields **deoxyaniflorine** (48), identical in all respects with the naturally-occurring alkaloid.

Sessiflorine (49). The ^{1}H-nmr spectrum revealed the presence of an aromatic methoxyl and a tertiary methylamino group by signals at δ 3.94 and at δ 2.95. A doublet of doublets at δ 7.89 (J = 8 and 1 Hz) which could be attributed to H-8, together with the mass spectral data allowed placement of the methoxyl at C-5. Structure (50), originally proposed for sessiflorine by ARNDT et al. (9), was revised to (49) by ONAKA (146), who compared the thin-layer chromatographic behavior and spectral data for natural sessiflorine and a synthetic product which had been prepared by the reaction outlined in Scheme 5. Anisessine and vasicinone were synthesized in an analogous manner. The synthetic compound (50) exhibits a one proton triplet (J = 8.2 Hz) at δ 5.47 and no NH absorption in the ir spectrum. Presence of the A_2 portion of an A_2B_2 spin system at δ 7.0—6.7 (2H) and the deshielded nature of the N-methyl signal (δ 2.95) favors structure (49) for sessiflorine.

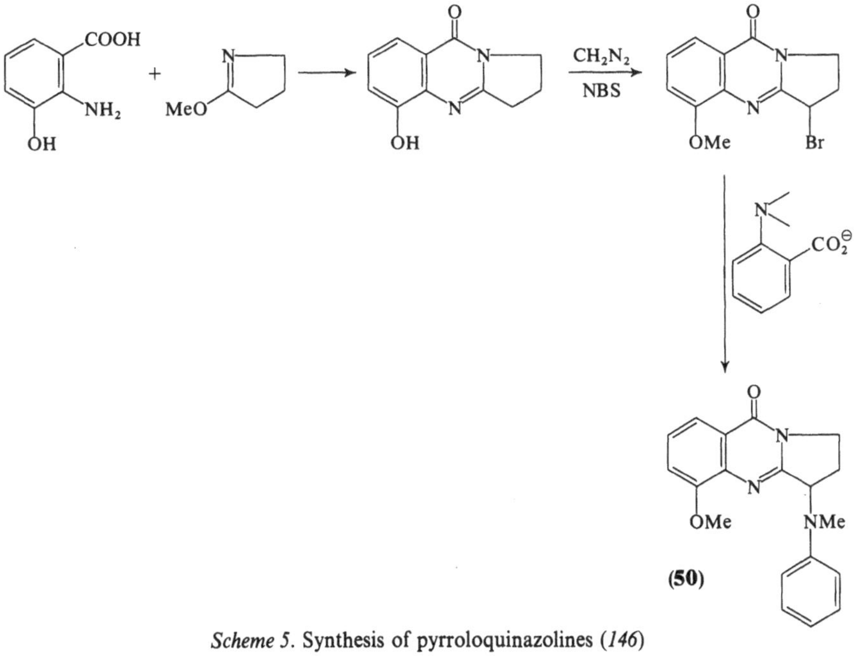

Scheme 5. Synthesis of pyrroloquinazolines (146)

3. Vasicoline, Vasicolinone and Adhatodine

The alkaloids, anisotine, vasicoline (**51**), vasicolinone (**52**), and adhatodine (**53**) were found (*95*) in young plants of *Adhatoda vasica* (Acanthaceae) in addition to vasicine, vasicinone, and 7-hydroxypeganine. Structures which are listed in Chart 2 were assigned primarily on the basis of spectral properties (uv, ir, nmr and ms). Vasicoline and adhatodine are oxidized at C-9 by atmospheric oxygen to vasicolinone and anisotine. It is, therefore, remarkable that the parent bases are present as such in the plant material. Spectral properties are listed below.

Vasicoline (51). The ^{1}H-nmr spectrum shows, in addition to the aromatic multiplet at δ 6.8—7.3 (8 H), a singlet at δ 4.60 (2H), assigned to the C-9 methylene protons, overlapping a triplet, δ 4.50 (1H), due to the C-3 methine proton. The remaining methylene protons, at C-2 and C-1, appear as multiplets at δ 3.1—3.4 (2H), 1.6—2.1 (1H) and 2.25—3.8 (1H). A singlet at δ 2.67 (6H) is readily assigned to the dimethylamino moiety. The mass spectrum of (**51**) shows interesting major and minor fragmentation pathways, (**58**) → (**59**) and (**58**) → (**60**), respectively.

(59) (m/z 247)   **(58)** (m/z 291)   **(60)** (m/z 276)

Vasicolinone (52). Uv: λ_{max} 227, 268, 305 and 317 nm (log ε 4.33, 3.91, 3.52 and 3.42). Ir: 1678 (aromatic ester), 1664 (aromatic amide) and 1618 cm^{-1} (imine). The proton nmr spectrum shows the presence of two N-methyl groups (δ 2.64, s, 6H), C8-H (δ 8.31, d, J = 7 Hz), seven additional aromatic protons (multiplets between δ 7.05 and 7.70), and C3-H (δ 5.04, t, J = 9 Hz). The proton at C-3 is coupled to the two protons attached to C-2 (δ 1.9—2.9, multiplet), which are, in turn, coupled to the two protons at C-1 (δ 4.00—4.55, multiplet). Chemical shifts of the protons at C-1, C-2, C-3 and C-8 are all very similar to those of deoxyaniflorine.

Adhatodine (53). Uv: λ_{max} 225, 262, 300 and 361 nm (log ε 4.51, 4.03, 3.86 and 3.70). Ir: 3378 (NH) and 1678 cm^{-1} (aromatic methoxycarbonyl). Acetylation yields a monoacetyl derivative.

4. Peganidine, Isopeganidine, Deoxypeganidine, Dipegine and Peganol

From different parts of *Peganum harmala,* Soviet workers have isolated the quinazolines peganidine (**54**) (*116*), isopeganidine (*229*), deoxypeganidine (**55**) (*230*), dipegine (**57**) (*229*), and peganol (**56**) (*207*). Formulas are listed in Chart 2. Dipegine was the first known dimeric quinazoline alkaloid. Isopeganidine was described as a "racemic diastereomer of peganidine". It is indeed possible that both alkaloids are artifacts formed during the isolation procedure. Deoxypeganidine can be converted to vasicinone by oxidation.

Nmr spectroscopy has been very helpful in the structure elucidation of these alkaloids (Table 1). The methylene protons at C-9 of vasicine and deoxyvasicine appear as a singlet around δ 4.5 (*9, 95*). In isopeganidine and deoxypeganidine, the methine proton at C-9 appears between δ 5.02—5.14 as a triplet (*230*). Further details can be found in the original papers. The alkaloids of *Peganum harmala* have been comprehensively reviewed by TELEZHENETSKAYA and YUNUSOV (*208*).

Table 1. *Nmr Spectra of Some Quinazoline Alkaloids (δ, PPM, CF₃COOH*)*

Alkaloid	Aromatic protons	C-1	C-2	C-3	C-9	C-10	C-12
Peganidine (**54**)	6.6—7.1 (m)	3.73 (m) 3.35 (m)	2.02 (m) 2.38 (m)	5.02 (t)	5.02 (t)	3.07 (d)	1.86 (s)
Isopeganidine	6.6—7.0 (m)	3.47 (m)	2.05 (m) 2.45 (m)	5.04 (m)	5.04 (m)	2.95 (d)	1.81 (s)
Deoxy-peganidine (**55**)	6.9—7.2 (m)	3.42 (m)	2.00 (m) 2.60 (m)	3.42 (m)	5.14 (t)	2.76 (m)	1.95 (s)
Peganol (**56**)	7.5—8.0	4.71 (t)	2.40 (m)	3.40 (t)			

* CDCl₃ in the case of deoxypeganidine.

IV. The Pyridoquinazolines

FITZGERALD, JOHNS *et al.* (*67, 99*) have isolated two alkaloids (**61**) and (**62**) from the leaves and stems of *Mackinlaya subulata* and *M. macrosciadia.* Both were readily identified by a combination of physical and chemical methods, and the structures were confirmed by direct comparison with synthetic samples.

The first of these is 6,7,8,9-tetrahydropyrido[2,1-b]quinazolin-11-one (**61**). Of the signals in the nmr spectrum due to the aromatic protons of (**61**), a low field multiplet, δ 8.12—8.33 which exhibits the expected ortho and meta couplings has been attributed to H-1. The downfield shift of approximately 0.7 PPM relative to the protons on C-2, C-3 and C-4 is due to deshielding by the lactam carbonyl, C-11. A broad, two-proton triplet, δ 3.90—4.20, has been assigned to the C-9 methylene protons. A second broad, two-proton triplet, δ 2.80—3.15, can then be assigned to the methylene protons of C-6. Reduction of (61) with sodium borohydride gave the corresponding 5,5a-dihydro derivative.

(61) R=O
(62) R=H₂

    The second alkaloid was the 11-deoxy derivative, 6,7,8,9-tetrahydro-11H-pyrido[2,1-b]quinazoline (62). Its ¹H-nmr spectrum is similar to that of (61), if the absence of the lactam carbonyl is taken into account. The chemical shift of C1-H of (62) is similar to that of the protons at C-2, C-3 and C-4, its signal appearing as part of the multiplet between δ 6.67—7.13. The C-11 methylene protons give rise to a sharp singlet at δ 4.33. No definitive assignments have been made for two broad, two-proton multiplets, δ 2.80—3.17 and 2.28—2.73, however it has been suggested that the multiplet which is further downfield may arise from the two protons on C-9. If this were so, presence of the lactam carbonyl in (61) would have produced a paramagnetic shift of this multiplet by nearly 1 PPM. If the multiplet at higher field were assigned to the protons at C-9, this paramagnetic shift would increase to approximately 1.5 PPM. As the paramagnetic shift of such protons caused by the presence of a lactam group in other quinazoline alkaloids, is approximately 1.0 PPM, the assignment of the multiplet at δ 2.80—3.17 to C9-H₂ in (62) seems reasonably sense. Reduction of (62), either catalytically or with sodium borohydride, affords a 5,5a-dihydro derivative.

    The mass spectra of both alkaloids show that the tetrahydropyrido-quinazoline ring system is particularly stable. The molecular ion, m/z 200, is the base peak in the spectrum of (61) and the only major fragment (85% of the base peak) is at m/z 199, due to loss of a hydrogen radical and formation of a stable ion. In the mass spectrum of (62), the M-1 ion, m/z 185, is the base peak, and the molecular ion, m/z 186 (85% of the base peak), is the only other major peak found.

    Reduction of (61) with zinc and acetic acid, as described by SPÄTH and PLATZER (202), affords (62). (61) can be prepared by the method of STEPHEN and STEPHEN (203) or that of SCHÖPF et al. (184). A number of methods exist for the synthesis of (62) (cf. 144).

V. The Indolopyridoquinazolines

1. Rutaecarpine and Related Alkaloids

Rutaecarpine (**63**), whose formula is listed in Chart 3, was isolated from *Evodia rutaecarpa* Hook. f. & Thoms. in 1915 by ASAHINA *et al.* Significant infrared absorption occurs at 1655 and 3325 cm^{-1}, uv absorption at λ_{max} 276, 288, 330, 344, and 360 nm, and non-aromatic proton signals in the ^{1}H-nmr spectrum at δ 3.23 and 4.60, each two protons (t, $J = 7$ Hz).

The action of a solution of potash in amyl alcohol on rutaecarpine produces anthranilic acid and a second acid which, when boiled with hydrochloric acid, is readily decarboxylated to tryptamine (*6, 25, 94, 134, 135, 169*). A close relationship between rutaecarpine and evodiamine was demonstrated by fusion of isoevodiamine hydrochloride. Rutaecarpine was formed with liberation of chloromethane. A number of syntheses of (**63**) have been reported (*101, 135, 161*), including some under so-called physiological conditions. Some of the more recent examples will be mentioned. KAMETANI *et al.* (*102, 108*) obtained (**63**) in 80% yield through a regiospecific Π^4s + Π^2s cycloaddition of a keteneimine (generated *in situ* by extrusion of sulfur dioxide from the sulfinamide anhydride of anthranilic acid) with 3,4-dihydro-β-carboline or with 1,2,3,4-tetrahydro-1-keto-β-carboline (*109*) (also called 1,2,3,4-tetrahydronorharman-1-one or, as in *Chemical Abstracts*, 2,3,4,9-tetrahydro-1H-pyrido[3,4-b]indol-1-one) ac-

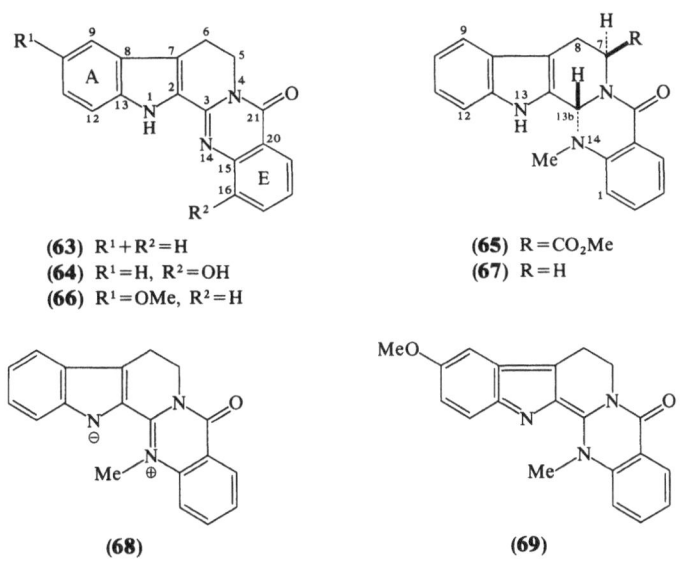

(**63**) R^1+R^2=H
(**64**) R^1=H, R^2=OH
(**66**) R^1=OMe, R^2=H

(**65**) R=CO$_2$Me
(**67**) R=H

(**68**)   (**69**)

Chart 3. Indolopyridoquinazolines

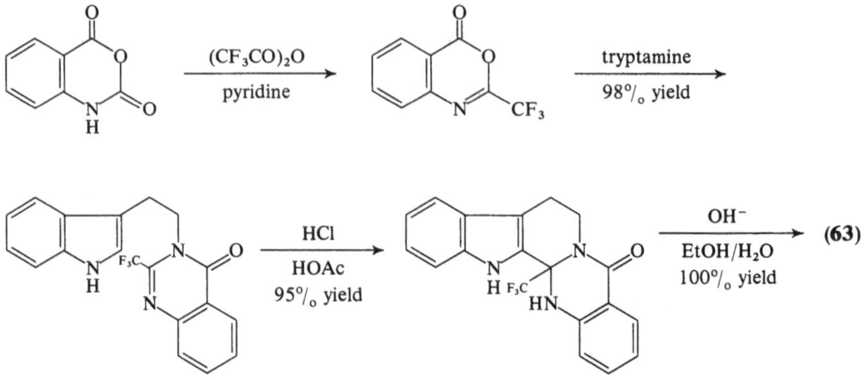

Scheme 6. Synthesis of rutaecarpine according to BERGMAN and BERGMAN (17)

cording to Scheme 3. Reaction of 1,2,3,4-tetrahydro-1-keto-β-carboline with isatoic anhydride or treatment of 3,4-dihydro-β-carboline with isatoic anhydride also led to (63) (109). Condensation of N-formyltryptamine with the sulfinamide anhydride derived from anthranilic acid gave 3-indolyl-ethylquinazolin-4-one in 63% yield. Heating this product with HCl in acetic acid afforded (63) in 45% yield (104). BERGMAN and BERGMAN (17) synthesized (63) in high yield according to Scheme 6. Further aspects of the synthesis and chemistry of the indolopyridoquinazoline alkaloids are discussed in (56a, 56b). DANIELI and PALMISANO (54) have described a new synthesis of evodiamine (see below) which can then be regiospecifically oxidized to (63) by manganese dioxide. MÖHRLE et al. (139) obtained (63) by treatment of 1,2,3,4-tetrahydro-β-carboline with 2-nitrobenzyl chloride to yield the tertiary amine (70) which, on reduction of the nitro group and dehydrogenating cyclization, yielded the pentacyclic amidine (71), oxidation of which produced (63) in high yield.

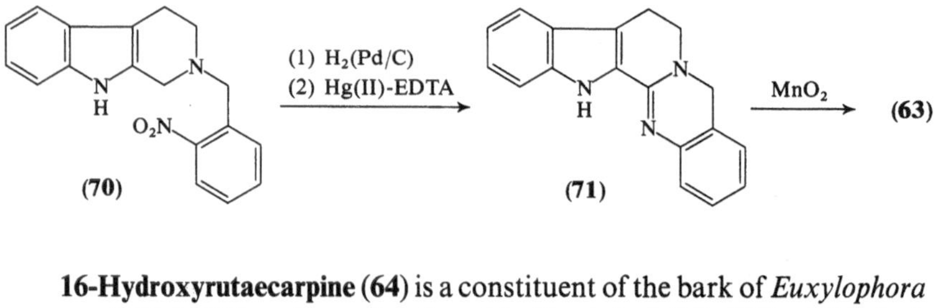

16-Hydroxyrutaecarpine (64) is a constituent of the bark of *Euxylophora paraensis* Hub. (55) and was also found in *Vepris louisii* G. Gilbert (12a). A bathochromic shift in the presence of base evidenced the phenolic nature of the hydroxyl group. The mass spectrum (m/z 169, 168, 167, 155, 115)

indicates that the carboline system was unsubstituted and that the hydroxyl group was located on ring E. The exact location of this group was determined by synthesis of the four possible hydroxyrutaecarpines. **(64)** was obtained from 3-hydroxyanthranilic acid and 1,2,3,4-tetrahydro-1-keto-β-carboline.

5,6-Dehydro-16-hydroxyrutaecarpine. Evidence for the location at a double bond between C-5 and C-6 (C-7 and C-8 in the alternative ring numbering system) was obtained from the compound's ^{1}H-nmr spectrum

which lacked signals for the $=\overset{|}{C}-CH_2-CH_2-\overset{|}{N}$ moiety present in **(64)**

but showed an AX pattern at δ 7.85 and 8.55 ($J_{5,6} = 7.5$ Hz). The structure was confirmed by DDQ dehydrogenation of **(64)** (*12a*).

3,14-Dihydrorutaecarpine and **14-formyl-dihydrorutaecarpine** were isolated from ripe fruits of *Evodia rutaecarpa* by KAMIKADO *et al.* (*111*). A signal at δ 6.06 (1H, s) in the ^{1}H-nmr spectrum was ascribed to a proton on C-3. The infrared spectrum contains two NH bands, 3330 and 3280 cm^{-1}, and a carbonyl absorption at 1633 – 1609 cm^{-1}. The presence of fragment ions m/z 170 and 119 from a retro Diels-Alder type cleavage in the mass spectrum supports the structure proposed for the former compound. Oxidation with manganese dioxide affords rutaecarpine. KAMIKADO *et al.* have synthesized this alkaloid by treatment of isatoic anhydride with 3,4-dihydro-β-carboline (71.5% yield). The product prepared by HORVATH-DORA and CLAUDER (*88*) by heating **(72)** with triethyl orthoformate is not 3,14-dihydrorutaecarpine but is, in fact, identical with the noncyclized compound **(73)** (*17*).

(72) (73)

The presence of a formyl group in 14-formyl-dihydrorutaecarpine was suggested by a carbonyl absorption at 1695 cm^{-1} in the infrared spectrum, as well as by a signal at δ 9.10 in the nmr spectrum. Upon heating with 5% ethanolic KOH, the compound loses formaldehyde to afford rutaecarpine.

(7S,13bS)-7-Carboxy-8,13,13b,14-tetrahydro-14-methylindolo[2′,3′- :3,4]pyrido[2,1-b]quinazolin-5(7H)-one (65) has been isolated from the fruits of *Evodia rutaecarpa* (Juss.) Benth. & Hook. by DANIELI *et al.* (*49*). After methylation with CH_2N_2 the presence of a carbomethoxy group was suggested by infrared absorption at $1745 \, \text{cm}^{-1}$ and was confirmed by a singlet at δ 3.59 in the ^{1}H-nmr spectrum. The mass spectrum shows, *inter alia*, peaks at m/z 302 ($M^+ - \text{COOMe}$), and 300 ($M^+ - \text{COOMe} - 2H$), reminiscent of a tetrahydro-β-carboline substituted by a carbomethoxy group at the position alpha to N_b. Localization of the carbomethoxy group at C-7, and absolute configuration of this center, were established by synthesis of (65) from (3S)-3-carbomethoxy-4,9-dihydro-3H-pyrido[3,4-b]indole hydrochloride and N-methylanthranilic acid in the presence of $Ph_3P - CBr_4$. The authors expected to obtain two C-13b epimeric compounds, but only the naturally-occurring stereoisomer was formed (in 25% yield) under these mild conditions.

2. Evodiamine and Dehydroevodiamine

Evodiamine (67) was isolated from the dried, unripe fruits of *Evodia rutaecarpa* and is one of the few optically active indole alkaloids possessing only a single chiral center. The racemate of evodiamine, (*152*), has also been isolated from *Xanthoxylum rhetsa* DC., and was apparently given the name rhetsine (*39*). Ir: 1640 and $3475 \, \text{cm}^{-1}$. Uv: λ_{max} 268, 273, 283 and 292 nm. ^{1}H-nmr: δ 2.53 (3H, s, NMe) and 5.90 (1H, s, C3-H). The structure of (67) was partially revealed by alkaline hydrolysis which yields N-methyl-anthranilic acid and 3,4-dihydro-β-carboline. Boiling alcoholic potassium hydroxide degrades (67) to N-methylanthranilic acid, carbon dioxide and tryptamine. When boiled with alcoholic hydrochloric acid (67) yields optically inactive isoevodiamine [evodiamine hydrate (74)] which can be recyclized with acetic anhydride or oxalic acid to optically inactive evodiamine. (67) can be converted to rutaecarpine *via* the dry hydrochloride of (74), which, on heating, yields (63). The chemistry of evodiamine has been reviewed by ARMAREGO (*6*); see also (*25, 94, 134, 169*).

(74)

(75)   (67)

DANIELI et al. (51) have recently determined the absolute configuration of (+)-evodiamine as (7S,13bS), by correlation with (S)-tryptophan via naturally-occurring (7S,13bS)-carboxyevodiamine (65) (cf. 49).

Reaction of tryptamine and N-methylisatoic anhydride gives the intermediate o-methylaminobenzoyltryptamine (75), which could be cyclized to evodiamine by boiling with ethyl orthoformate (11). KAMETANI et al. (102, 108) have reported a "retro mass spectral" synthesis of (67). This synthetic approach is based upon a fragmentation process observed in the mass spectrum. Mass spectral cleavage of (67) involves a retrograde Diels-Alder type fragmentation to form two characteristic ions, the 3,4-dihydro-β-carboline ion (76) and the iminoketene ion (77). Heating N-methylanthranilic acid with thionyl chloride in dry benzene gives, according to Scheme 3, an unstable sulfinamide anhydride which, on treatment with 3,4-dihydro-β-carboline at room temperature affords (67) regiospecifically in 65% yield with evolution of sulfur dioxide. DANIELI and PALMISANO (54) have reported the oxidation of the lactam (78), obtained through condensation of tryptamine with N-methylanthranilic acid followed by intramolecular amidomethylation, with mercuric acetate to (67) in 92% yield. Formation of (67) takes place via acid-catalyzed interaction of the electrophilic carbon of the intermediate 3,4-dihydro-4-oxoquinazolinium salt (79) with the nucleophilic carbon-2 of the indole moiety. In a synthesis reported by KAMIKADO et al. (111), 3,4-dihydro-β-carboline, on treatment with N-methylisatoic anhydride in pyridine and dry benzene, furnishes (67).

"Rhetsinine", isolated from Xanthoxylum rhetsa DC. (59, 70, 152) and a few other species (54a, 65a, 209, 217a), was recognized as being identical with a substance obtained by mild permanganate oxidation of (67) called hydroxyevodiamine. PACHTER and SULD (152) have demonstrated that hydroxyevodiamine is, in fact, the dicarbonyl compound (80).

(80)

Hydroxyevodiamine has been isolated from the leaves (*145*) and fruits (*209, 226*) of *Evodia rutaecarpa*. Recrystallization of this compound from benzene affords dehydroevodiamine, which is the anhydronium base **(68)** (Chart 3). NAKASATO *et al.* have found that this compound is the main alkaloid of the leaves of *E. rutaecarpa*. Heating hydroxyevodiamine in 5% ethanolic potassium hydroxide gives 2,3,4,9-tetrahydro-1H-pyrido[3,4-b]indol-1-one, while catalytic reduction in acetic acid or treatment with sodium borohydride gives dl-evodiamine (rhetsine). Heating the hydrochloride of dehydroevodiamine *in vacuo* at 260 – 80° gives rutaecarpine. Hydroxyevodiamine (rhetsinine) can also be obtained by photo-oxidation of evodiamine (*226*).

PACHTER and SULD have prepared **(80)** in 68% yield by POCl₃-promoted condensation of 2,3,4,9-tetrahydro-1H-pyrido[3,4-b]indol-1-one with methyl N-methylanthranilate. Catalytic reduction of **(80)** yields **(67)**. DANIELI and PALMISANO (*54*) have found that **(67)** can be converted into **(68)** by a variety of oxidants. In particular, Tl(OAc)₃ and DDQ give **(68)** in 85 and 67% yield, respectively. The yield of **(68)** from alkaline permanganate oxidation of **(67)** (in acetone at 0°: 65% yield) was improved markedly to 91%, by use of dicyclohexyl-18-crown-6 (methylene chloride at room temperature).

3. Hortiamine and Hortiacine

These alkaloids **(69)** and **(66)** were isolated from the bark of *Hortia arborea* Engl. by PACHTER *et al.* Hortiamine **(69)** was also found in extracts of *Hortia braziliana* Vel. Degradation and synthetic studies have allowed elucidation of the structures of hortiamine **(69)** and hortiacine **(66)** (*149—151*). Hortiamine, a red compound, yields 6-methoxy-1,2,3,4-tetrahydropyrido[3,4-b]indol-1-one and N-methylanthranilic acid when heated with ethanolic potassium hydroxide. The alkaloid was regenerated by condensation of these degradation products and has been synthesized in quantitative yield by treatment of the accessible amide **(81)** with polyphosphoric acid. Hydrolysis of **(69)** gives rise to 2-(o-methylaminobenzoyl)-6-methoxy-1,2,3,4-tetrahydropyrido[3,4-b]indol-1-one (6-methoxyrhetsinine) and pyrolysis of the hydrochloride of **(69)** yields 10-methoxyrutaecarpine (= hortiacine, **66**), the second alkaloid, which was found in this plant in 0.001% yield. The synthesis of hortiacine can be accomplished by reaction of 6-methoxy-1,2,3,4-tetrahydropyrido[3,4-b]indol-1-one with methyl anthranilate in the presence of phosphorus oxychloride. (**66**) undergoes O-demethylation when heated with 48% hydrobromic acid, producing 10-hydroxyrutaecarpine.

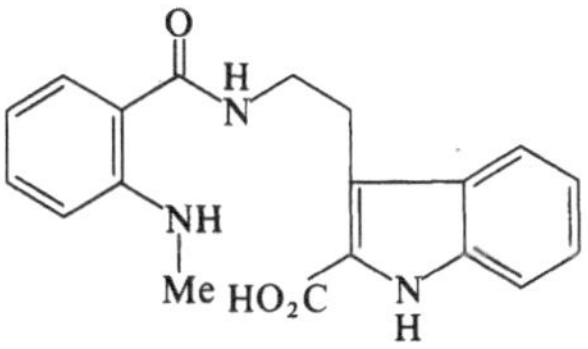

(**81**)

4. Euxylophorine A—D and Euxylophoricine A—F

Euxylophora paraensis Hub. is a Brazilian plant, locally called "Pao Amarello". CANONICA, DANIELI *et al.* (*32, 48, 52, 56*) have isolated the following alkaloids whose structures are listed in Chart 4 from the bark of this plant.

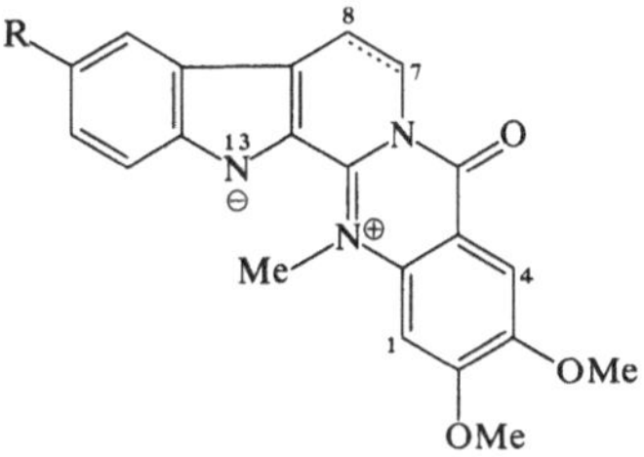

(**82**) R=H
(**83**) R=H double bond at 7,8
(**84**) R=OMe
(**85**) R=OMe double bond at 7,8

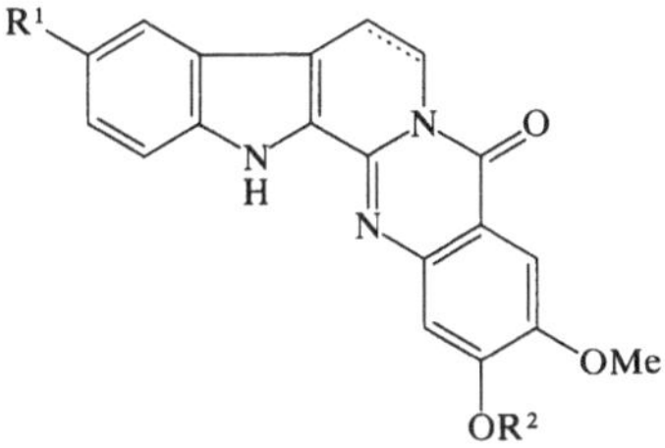

(**86**) R^1=H, R^2=Me
(**87**) R^1=H, R^2=Me double bond at 7,8
(**88**) R^1=OMe, R^2=Me
(**89**) R^1=OMe, R^2=Me double bond at 7,8
(**90**) R^1+R^2=H

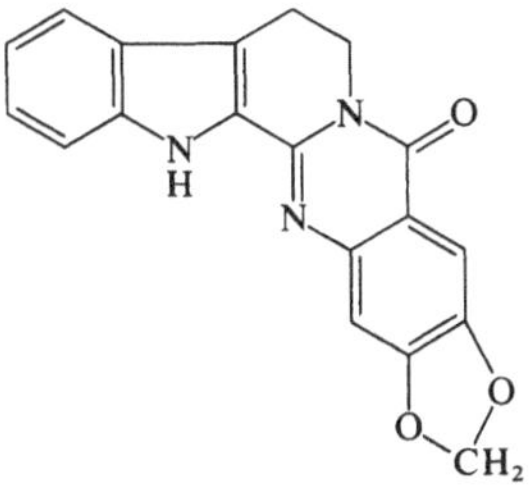

(**91**)

Chart 4. Euxylophorines and euxylophoricines

Euxylophorine A (82) crystallized as orange-red needles from benzene. The ^{1}H-nmr spectrum showed two symmetrical triplets, each two protons,

at δ 3.30 and 4.75 ($=\overset{|}{C}-(CH_2)_2-\overset{|}{N}$), singlets at δ 3.90 and 3.96

(2 × OMe), a singlet at δ 5.23 (NMe), and a complex multiplet between δ 7.2 and 8.2 corresponding to six aromatic protons. Treatment with potassium hydroxide in refluxing amyl alcohol afforded 1,2,3,4-tetrahydro-1-keto-β-carboline and 6-methylaminoveratric acid. The structure of euxylophorine A was confirmed by condensation of 1,2,3,4-tetrahydro-1-keto-β-carboline with methyl 6-methylaminoveratrate (*32*).

DANIELI *et al.* (*50*) have investigated the regioselectivity of (**82**) toward nucleophilic reagents, as well as its reduction and catalytic hydrogenation. Spectral data have disclosed the existence of a dynamic equilibrium between the orange-red anhydronium base and the yellow "open-chain" form (**94**) of euxylophorine A in solution. Yellow (**94**), obtained by crystallization from wet benzene, reverts to red (**82**) when heated in a suitable anhydrous solvent (*e.g.* benzene) or in the crystalline state. The transformation of (**82**) into (**94**) can be envisaged as a four-way equilibrium, taking place through the quaternary ammonium salt (**92**), the neutral covalent hydrate (carbinolamine, pseudobase, **93**) and fission of ring-D.

Different behavior is observed in polar protic (hydrogen-bonding) solvents (*e.g.* ethanol). Both (**82**) and (**94**) give the same uv spectrum, with maxima at 329 and 390 nm which indicates that the abovementioned equilibrium is displaced toward the anhydronium base and/or ammonium salt forms with increasing solvent polarity. There is no evidence for the intermediacy of the carbinolamine (93) or its equivalent.

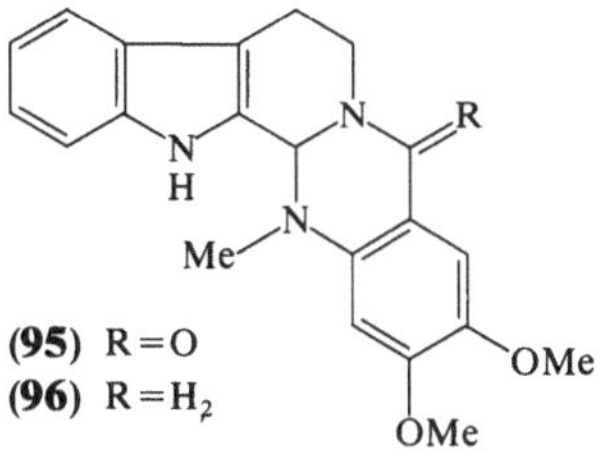

(95) R = O
(96) R = H₂

(82) gives a colorless dihydroderivative (95) in quantitative yield on treatment with sodium borohydride in methanol. (82) is smoothly reduced to the deoxo-derivative (96) in 75% yield by lithium aluminium hydride in refluxing tetrahydrofuran. Hydrogenation of (82) over platinum oxide results in exclusive formation of 9,10,11,12-tetrahydro-(82).

(83) ⟷   (97)   ⟷   (98)

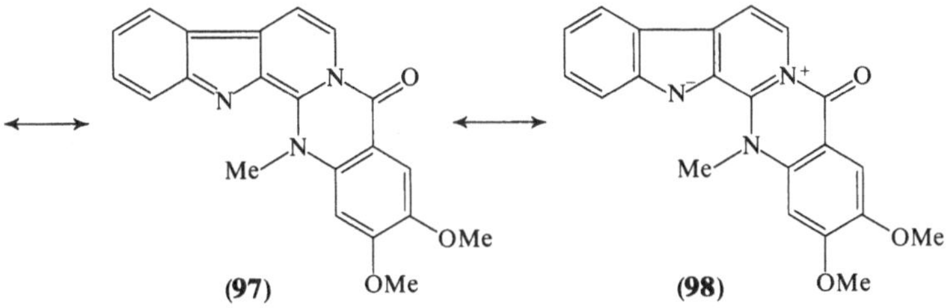

Euxylophorine B (83) is an example of an anhydronium base whose structure can be represented by resonance formulas (83), (97), and (98). The alkaloid shows uv maxima (in acetonitrile) at 278 and 352 nm (log ε 4.47 and 4.70, respectively). The ¹H-nmr spectrum contains singlets at δ 4.30 and 4.22 for two methoxyl groups, a slightly broadened singlet at δ 4.81 for an N-methyl group, singlets at δ 8.05 and 7.37 for the two protons on the aromatic nucleus bearing the two methoxyls, a multiplet of four aromatic protons, δ 7.4—8.3, and an AB spin system, δ 9.34 and 8.32 (J = 9 Hz), attributed to the =C—CH=CH—N moiety. A marked color change is observed when the yellow-orange methanolic solution of (83) is allowed to stand at ambient temperature. The color fades and a single compound (99) can then be isolated in quantitative yield. (83) dissolved in 1N NaOH, undergoing a facile cleavage to give the colorless carboxylate (100) which reverts to yellow (83) on acid treatment.

(99) R=CO₂Me
(100) R=COO⁻Na⁺
(101) R=CH₂OH

Attempted reduction of (83) with excess sodium borohydride in methanol is unsuccessful; heating with lithium aluminium hydride in THF or with sodium borohydride in methanol cleaves the $C_5 - N_6$ bond yielding the alcohol (101). Hydrogenation of (83) over platinum oxide affords 9,10,11,12-tetrahydro-(82) (50). Vacuum pyrolysis of the hydrochloride of (83) causes elimination of methyl chloride and formation of euxylophoricine B (87). Synthesis of (83) was achieved via dehydrogenation of (82) with 2,3-dichloro-5,6-dicyano-1,4-benzoquinone.

Euxylophorine C (84) exists as red needles (from benzene). The infrared spectrum of this alkaloid exhibits a carbonyl absorption at 1665 cm⁻¹ and unsaturation bands at 1620 and 1555 cm⁻¹. The uv spectrum shows a maximum at 425 nm (log ε 4.51) in chloroform, which shifts to 408 nm in acetonitrile and to 389 nm in absolute ethanol. Reduction with sodium borohydride gives a dihydroderivative. The influence of moisture on a solution of (84) in boiling benzene, or addition of water to a solution of (84) in pyridine causes separation of a pale yellow solid. This compound turns red on heating and has the same melting point as (84). Thus, the yellow compound is a hydrated form of (84), 2-(4,5-dimethoxy-2-methylaminobenzoyl)-1,2,3,4-tetrahydro-1-keto-6-methoxy-β-carboline [the 6-methoxy derivative of (94)]. Conclusive proof for the structure of euxylophorine C (84) was obtained through synthesis from 1,2,3,4-tetrahydro-1-keto-6-methoxy-β-carboline and methyl-4,5-dimethoxy-2-methylaminobenzoate in the presence of POCl₃ in toluene (56).

Euxylophorine D (85). On the basis of analogous spectroscopic behavior and co-occurrence with (84), structure (85) was proposed for this alkaloid. This proposal was confirmed by partial synthesis of (85) via dehydrogenation of (84) with 2,3-dichloro-5,6-dicyano-1,4-benzoquinone.

Euxylophoricine A (86). From its nmr spectrum, (86) contains a

$$=C-(CH_2)_2-\overset{\diagup}{\underset{\diagdown}{N}}$$ system (symmetrical triplets at δ 3.10 and 4.60), two

methoxyl groups (singlets at δ 3.85 and 3.80), a proton on nitrogen and six aromatic protons (multiplet at δ 7.2—7.9). Hydrolysis of (86) with potassium hydroxide in amyl alcohol gives 1,2,3,4-tetrahydro-1-keto-β-carboline and 6-aminoveratric acid. Selenium dehydrogenation of (86) furnishes (87). Synthesis of (86) was accomplished by treatment of 1,2,3,4-tetrahydro-1-keto-β-carboline with methyl 6-aminoveratrate (32). KAMETANI et al. (103) have realized the synthesis of (86) by converting 2-amino-4,5-dimethoxybenzoic acid into the corresponding sulphinamide anhydride by heating with thionyl chloride, then condensing that product with 3,4-dihydro-β-carboline at ambient temperature.

Euxylophoricine B (87). Evidence for the presence of a $C_7 - C_8$ double bond in (87) was obtained from the ^{1}H-nmr spectrum, which lacks signals

for the $=C-(CH_2)_2-N$ system present in (86) and contains an AM spin

system at δ 8.20 and 9.13 (J = 7 Hz).

Euxylophoricine C (91), an optically inactive alkaloid, was present in very small amounts. The structure was elucidated on the basis of its spectroscopic and chemical properties and confirmed by synthesis. The ^{1}H-nmr spectrum contained two symmetrical triplets at δ 3.58 and 4.32,

corresponding to the $=C-(CH_2)_2-N$ moiety, singlets at δ 7.28 and 7.72,

each due to one aromatic proton, a complex multiplet, δ 7.2—7.8, for four additional aromatic protons, a two-proton singlet at δ 6.28 indicating a methylenedioxy group, and an indolic NH group at δ 10.60. Synthesis of (91) was carried out by refluxing 1,2,3,4-tetrahydro-1-keto-β-carboline with an excess of phosphorus oxychloride followed by addition of 6-amino-3,4-methylenedioxybenzoate and heating to give the alkaloid. Euxylophoricine C can also be synthesized by condensation of 3,4-dihydro-β-carboline with the sulphinamide anhydride derived from 2-amino-4,5-methylenedioxybenzoic acid (103).

Euxylophoricine D (88) is a white solid. Pyrolysis of the trifluoroacetate of (91) gives (88), identical in all respects with the natural product.

Euxylophoricine E (89) is a yellow compound which gives intensely fluorescent solutions. Its structure was deduced from its spectroscopic properties and confirmed by dehydrogenation of (88) with selenium at 300°.

Euxylophoricine F (90) exhibits the same uv spectrum as the other euxylophoricines. Methylation with methyl iodide and potassium carbonate yields N_{13}-methyleuxylophoricine A, indicating that euxylophoricine F is a 2,3-disubstituted rutaecarpine derivative. The structure was confirmed by synthesis. Condensation of 4-benzyloxy-5-methoxyanthranilic acid methyl ester with 1,2,3,4-tetrahydro-1-keto-β-carboline in the presence of phosphorus oxychloride and subsequent hydrogenolysis gave (90).

5. Paraensine

Paraensine (102) was isolated from the bark of *Euxylophora paraensis* Hub. by DANIELI *et al.* (*53*). The ir spectrum shows the presence of an NH group (3310 cm^{-1}), an amide carbonyl (1650 cm^{-1}), double bond absorption and an aromatic system (1640, 1600 and 1550 cm^{-1}). Ultraviolet absorption maxima were found at 342, 358 and 376 nm (log ε 4.32, 4.42 and 4.29); the ^{1}H-nmr spectrum (CDCl$_3$) had signals at δ 1.48, s, 6H ($-$C(CH$_3$)$_2$); δ 3.12, t (J = 7 Hz), 2 H and 4.51, t (J = 7 Hz), 2 H

$(=\overset{|}{C}-(CH_2)_2-N=)$; δ 3.89, s, 3 H ($-OCH_3$); δ 5.63, d (J = 10 Hz), 1H (olefinic proton); and δ 7.2—7.7, multiplet, 7H (aromatic, benzylic methine and NH) including a singlet at δ 7.53 for C4-H. The structure (102) of this alkaloid was confirmed by synthesis. Condensation of 5-amino-6-carbomethoxy-2,2-dimethyl-8-methoxychroman with 1,2,3,4-tetrahydro-1-keto-β-carboline gives dihydroparaensine which is identical with the hydrogenation product of (102). Paraensine is the first known indolopyridoquinazoline alkaloid containing an isoprenoid moiety.

(102)

VI. Quinazolines Produced by Microorganisms

1. Tryptanthrin

Tryptanthrin (103) is an antibiotic produced by the yeast *Candida lipolytica* and by other organisms only in the presence of large amounts of L-tryptophan (*183*). Structure (103) was deduced from nmr, uv and mass spectra and X-ray analysis by BRUFANI *et al.* (*27*). The carbonyl groups absorb at 1688 and 1725 cm^{-1} in the infrared spectrum. Proton nmr spectra at 90 and 100 MHz were not very informative. At this frequency, spin-spin decoupling experiments allowed identification of two sets of four adjacent aromatic protons (δ 8.7 to 7.4), but assignment of these signal groups to the relevant rings or to the individual protons could not be accomplished. Since no NH resonance appears in the spectrum, the two nitrogen atoms are thought to be parts of an extended aromatic molecule. In contrast, the nmr spectrum at 300 MHz was more useful, allowing, through comparison with nmr spectra of some monosubstituted tryptanthrines, the tentative assignment of signals to the protons on C-7, C-8, C-9 and C-10 (*92*). X-ray crystallographic analysis (*64*) has established that tryptanthrin is indolo-[2,1-b]-quinazolin-6,12-dione, having an approximately planar arrangement of nearly all atoms. (**103**) is a compound with a long history (*130*), and has been synthesized by BIRD (*24*) and by FRIEDLÄNDER and ROSCHDESTWENSKY (*68*).

(**103**)

chemical shifts of (**103**) in τ

2. Quinazolines of Pseudomonas Species

Several 4-methylquinazolines have been found in the culture broths of various strains of *Pseudomonas aeruginosa, Ps. putida, Ps. fluorescens, Sarcina lutea* and *Bacillus mesentericus* when tryptophan is supplied as the nitrogen source (*131, 132*). Simple compounds, including 4-methyl-quinazoline, 2,4-dimethylquinazoline, 2-ethyl-4-methylquinazoline, 2-hydroxymethyl-4-methylquinazoline, 4-methylquinazoline-2-carbonamide and 2-carboxy-4-methylquinazoline have been detected. Identification of these substances was accomplished by comparison with synthetic compounds using paper chromatography and micro-analytical methods. The highly reactive nature of the protons of the C-4 methyl group is responsible for a number of color reactions characteristic of this type of quinazoline derivative, such as the blue color reaction with Ehrlich's reagent. Interestingly, these quinazolines are accompanied by 2-aminoaceto-phenone. MANN has postulated the existence of a so-called "Quinazoline Pathway" in this connection (see Section VIII).

3. Quinazolines of Aspergillus Species

Tryptoquivaline (**104**) and **Nortryptoquivalone**. A strain of the fungus, *Aspergillus clavatus,* isolated from mold-damaged rice, produces these two, highly toxic, tremorigenic metabolites (*46*). The quinazoline structures of these alkaloids were deduced from spectral data, X-ray crystallographic analysis of the p-bromophenylurethane derivative of a transformation product of tryptoquivaline, and from chemical studies. (104) exhibits a positive triphenyltetrazolium chloride (TTC) test suggesting the presence of a hydroxylamine moiety. Methanolysis gives a desacetyl derivative with an unchanged uv spectrum, characterized as the p-bromobenzoate and p-bromophenylurethane, both derivatives giving negative TTC tests. Alkaline hydrolysis of tryptoquivaline affords an uncharacterized carboxylate which, on acidification, is transformed to a hydroxy-γ-lactone, apparently 19-epitryptoquivaline. This behavior was verified by formation of a deuterolactone. Tryptoquivaline C (FTC), isolated by YAMAZAKI et al. (222), is identical with (104) (226a). Nortryptoquivalone gives positive TTC and 2,4-DNP tests. The configurations of (104) and nortryptoquivalone are identical.

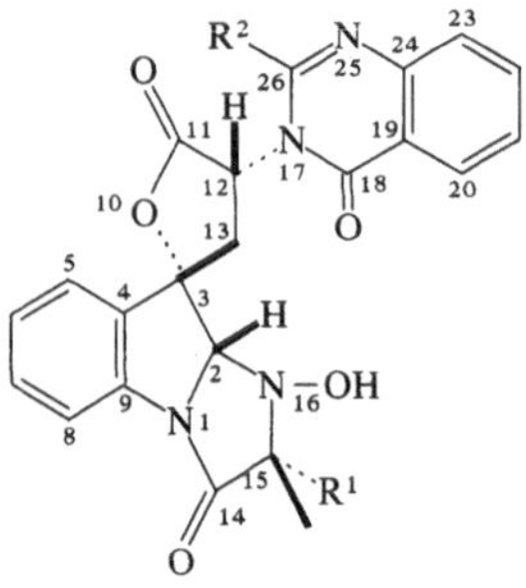

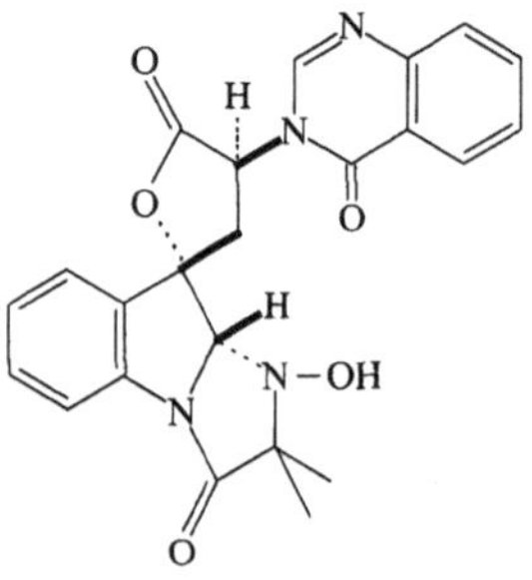

(107)

(104) R¹=Me, R²= ...

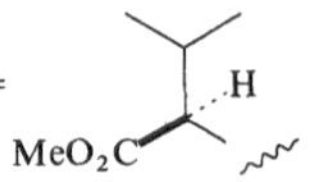

(105) R¹=H, R²= ...

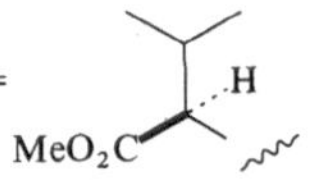

(106) R¹=Me, R²=H

Chart 5. Quinazoline alkaloids from *Aspergillus* spec.

BÜCHI et al. (30) have isolated four new mycotoxins related to (104) which are produced by the same species.

Nortryptoquivaline (105). The ¹H-nmr spectrum of (105) is identical with that of tryptoquivaline, except that signals associated with the geminal dimethyl group in the latter have been replaced by those of a secondary methyl group. The CD curve of (105) indicates the relative stereochemistry shown, and an X-ray crystallographic analysis by SPRINGER (211) has established the absolute configuration as (2*S*, 3*S*, 12*R*, 15*S*, 27*S*). Tryptoquivaline D (FTD) isolated by YAMAZAKI *et al.* (*222*) is identical with (**105**) (*226 a*).

Deoxytryptoquivaline gives a negative TTC test for hydroxylamines. It is only weakly basic and is not extracted from organic solvents by 1N aqueous HCl. The alkaloid differs from tryptoquivaline in the absence of an oxygen atom. The proposed structure was confirmed by oxidation with m-chloroperbenzoic acid to (**104**).

Deoxynortryptoquivalone can be oxidized to nortryptoquivalone by m-chloroperbenzoic acid. Tryptoquivaline N (FTN), isolated from *Aspergillus fumigatus* by YAMAZAKI *et al.* (*226 a*) is identical with deoxynortryptoquivalone.

Deoxynortryptoquivaline is also a secondary amine, oxidation of which leads to (**105**).

All these fungal metabolites are toxic.

YAMAZAKI *et al.* have isolated and identified some additional metabolites related to tryptoquivaline from *Aspergillus fumigatus* (*221—223*).

Tryptoquivaline E (FTE) and the five tryptoquivalines F—J have structures which are closely related to (**104**) and (**105**). FTE is dextrorotatory and exhibits a positive Cotton effect in the ord spectrum like (**104**) and (**105**). In the nmr spectrum of FTE, an ABX spin system assigned to the protons of the five-membered spirolactone ring closely resembles similar systems in the spectra of (**104**) and (**105**). These observations strongly suggest that all three metabolites have the same stereochemistry.

Tryptoquivaline F (FTF) gives a negative TTC test and differs from FTE by the lack of one oxygen atom. This suggests that FTF is a secondary amine rather than a hydroxylamine. Structure elucidation has established that FTF is an epimer of tryptoquivaline J, and may indeed be an artifact formed from FTJ.

Tryptoquivaline G (FTG) (106). The uv, ir and nmr spectra of this compound closely resemble those of FTE. Analysis of the nmr data excludes the possibility of the presence of an isobutyl side chain. Signals due to the secondary methyl group are replaced by two singlets arising from a gem-dimethyl group at C-15. (**106**) is strongly dextrorotatory. Epimerization of (**106**) to FTL, isolated by YAMAZAKI *et al.* (*226 a*), is possible with 0.1% KOH/MeOH. The first total synthesis of (**106**) was achieved by BÜCHI *et al.* (*29 a*), who have also established the absolute configuration. A formal synthesis of (**106**) was reported by BAN *et al.* (*145 c*). NAKAGAWA *et al.* (*110*) have reported a facile biogenetic-type total synthesis of (+)- and (−)-(**106**) by a different approach, which utilizes a newly devised oxidative double cyclization of an N-acyltryptophan precursor, and allows efficient formation of the unique ring system of (**106**) in one step.

Tryptoquivaline H (FTH) exhibits uv and ir spectra very similar to those of FTE. The molecular formula and mass spectral fragmentation are identical with those of FTE. FTH is an epimer of FTE and can be obtained by treatment of FTE with dilute alkali. The facile epimerization suggests that FTH may well be an artifact.

Tryptoquivaline I (FTI). The uv spectrum of FTI differs characteristically from those of the other tryptoquivalines, but resembles that of nortryptoquivalone. The presence of a geminal dimethyl group at C-15 is indicated by a sharp six-proton singlet at δ 1.49 in the ^{1}H-nmr spectrum, as in the case of (**104**). The presence of an isobutyryl group in FTI is also suggested by the absence of a signal due to C27-H, and the downfield position of the multiplet due to C28-H relative to the position of the analogous signal in the nmr spectrum of (**104**). Strong dextrorotation indicates that FTI belongs to the same stereochemical series as (**104**).

Tryptoquivaline J (FTJ) is an epimer of FTF, and can readily be converted to that compound by treatment with dilute alkali.

Tryptoquivaline L (FTL) (107) is an epimer of FTG. In the ^{1}H-nmr spectrum, a singlet due to the aromatic proton C26-H appears at δ 8.53, while two singlets due to the gem-dimethyl group at C-15 appear at δ 1.26 and 1.36, and a hydroxyl singlet, removed by addition of D$_2$O, is found at δ 8.74.

Tryptoquivaline M (FTM). The spectral data for FTM are all quite similar to those of nortryptoquivaline. The optical rotation of FTM is opposite in sign to that of nortryptoquivaline; thus, FTM is thought to be an epimer of nortryptoquivaline.

In summary, tryptoquivaline, nortryptoquivaline, nortryptoquivalone, tryptoquivalines E, G, I, and J are dextrorotatory, while tryptoquivalines F, H, L, and M are levorotatory. FTF, H, and L are epimers of FTJ, E, and G, respectively, and each of the former can be derived from the corresponding dextrorotatory isomer by treatment with dilute alkali.

4. Quinazolines of Streptomyces

An echinomycin-producing strain of *Streptomyces* (X-53), grown in the presence of quinazolin-4-one-3-acetic acid, has been found to produce an echinomycin analogue which has been designated quinazomycin. In this interesting compound, which has uv maxima at λ_{max} 225, 244, 290 and 328 nm, one of the two quinoxaline-2-carboxyl residues of echinomycin is replaced by a quinazolin-4-one-3-acetyl moiety (**108**) (*113*). The same group reported a cell-free synthesis of an echinomycin analogue (uv: λ_{max} 225 and 290 nm), tentatively designated biquinazomycin in which both quinoxaline-

2-carboxyl residues are replaced by quinazolin-4-one-3-acetyl residues (4, 59).

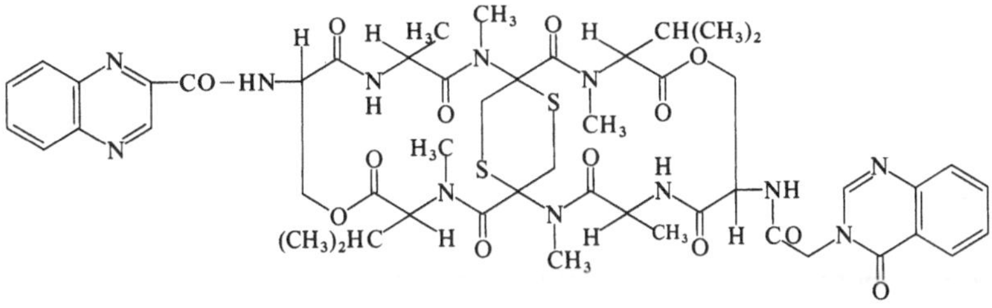

(108)

5. Chrysogine

Chrysogine (109) was isolated from the culture broth of *Penicillium chrysogenum* and exhibited the following properties: uv, λ_{max} 226, 230, 238, 265, 273, 292, 305 and 316 nm (log ε 4.34, 4.32, 4.11, 3.86, 3.82, 3.46, 3.59 and 3.51); ir, 770, 1471, 1610, 1632, 1680, 3050, 3190 cm^{-1}; ^{1}H-nmr, δ 1.45 (d, 3H, J = 6 Hz, CH₃), 4.60 (1H, C9-H), 5.59 (d, 1H, J = 6 Hz, OH), 7.35— 7.85 (m, 3H, C6-H, C7-H and C8-H), and 8.08 (dd, 1H, J = 2, 8 Hz, C5-H). Fermentation with anthranilic acid plus lactic acid, or with anthranilic acid alone, increased the yield of chrysogine while fermentation with lactic acid alone decreased the yield of this compound. Fermentation with anthranilic acid plus lactic acid analogues such as propionic acid, glycolic acid, α-hydroxybutyric acid, pyruvic acid and malic acid, afforded no congeners but increased the yield of chrysogine (84). Chrysogine has been synthesized by KAMETANI *et al.* (103) according to Scheme 3. Reaction of the N-unsubstituted sulphinamide anhydride with an O-benzyllactamide, followed by debenzylation of the resulting 2-(1-benzyloxyethyl)-quinazolin-4-one yields (109).

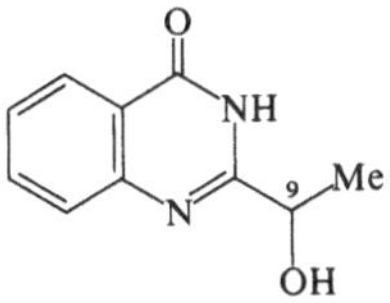

(109)

VII. Quinazolines Found in Animals

1. Glomerin and Homoglomerin

Glomerin and homoglomerin, two crystalline alkaloids, were isolated from the defensive secretions of the glomerid millipede, Glomeris marginata (137, 182). This millipede discharges its secretion in response to pinching, tapping, or, on occasion, even prodding. The liquid oozes as discreet droplets from eight pores, evenly spaced in a row along the dorsal midline of the animal. The structure of glomerin was elucidated independently by MEINWALD et al. (137) and SCHILDKNECHT and WENNEIS (181). ^{1}H-nmr spectra of both alkaloids show the presence of four aromatic protons, three as a complex multiplet at δ 7.5 and one as a doublet (J = 9 Hz) of closely spaced doublets at δ 8.2. Methyl groups appear as singlets, δ 3.75, in both compounds. Glomerin contains an additional methyl group, seen as a singlet at δ 2.65, while homoglomerin contains an ethyl group, the characteristic pattern of which is seen at δ 2.89 (quartet) and 1.39 (triplet). From these data and high resolution mass spectra, the compounds were identified as 1-methyl-2-ethylquinazolin-4-one (homoglomerin) and 1,2-dimethylquinazolin-4-one (glomerin, 18) (Scheme 2). Syntheses of both alkaloids were described by CHAKRAVARTI et al. (34) and ZIEGLER et al. (231) (cf. Scheme 2). KAMETANI et al. (107) have synthesized (18) according to Scheme 3, from the sulfinamide anhydride of N-methylanthranilic acid and thioacetamide. Another synthesis by KAMETANI et al. is described in (103).

2. Tetrodotoxin

Tetrodotoxin (110) has a long history (142). Originally isolated from certain varieties of the Japanese puffer fish (tora fugu, Sphoeroides rubripes; ma fugu, Sphoeroides phyreus), tetrodotoxin is also found in the embryos of the California newt (salamander, Taricha torosa), a goby (Gobius criniger), Central American frogs of the genus Atetopus (117), the blue-ringed octopus, Hapalochlaena maculosa (194) and in shells (145a, 145b). Tetrodotoxin was first isolated in crystalline form by YOKOO in 1950. It is one of the most potent non-protein neurotoxins, and its structure was elucidated by three groups led by HIRATA and GOTO and by TSUDA in Japan and by WOODWARD in the United States. The structure elucidation and synthesis of tetrodotoxin are highlights in natural products chemistry. A great number of investigations have been reported, and a comprehensive review is beyond the scope of this chapter. The reader is referred to the works of TSUDA (212, 213) and HIRATA (71) and their collaborators for appropriate references. The work of the Harvard group has been summarized by WOODWARD (219). The reader is further referred to the review by ARMAREGO (6, cf. also 94).

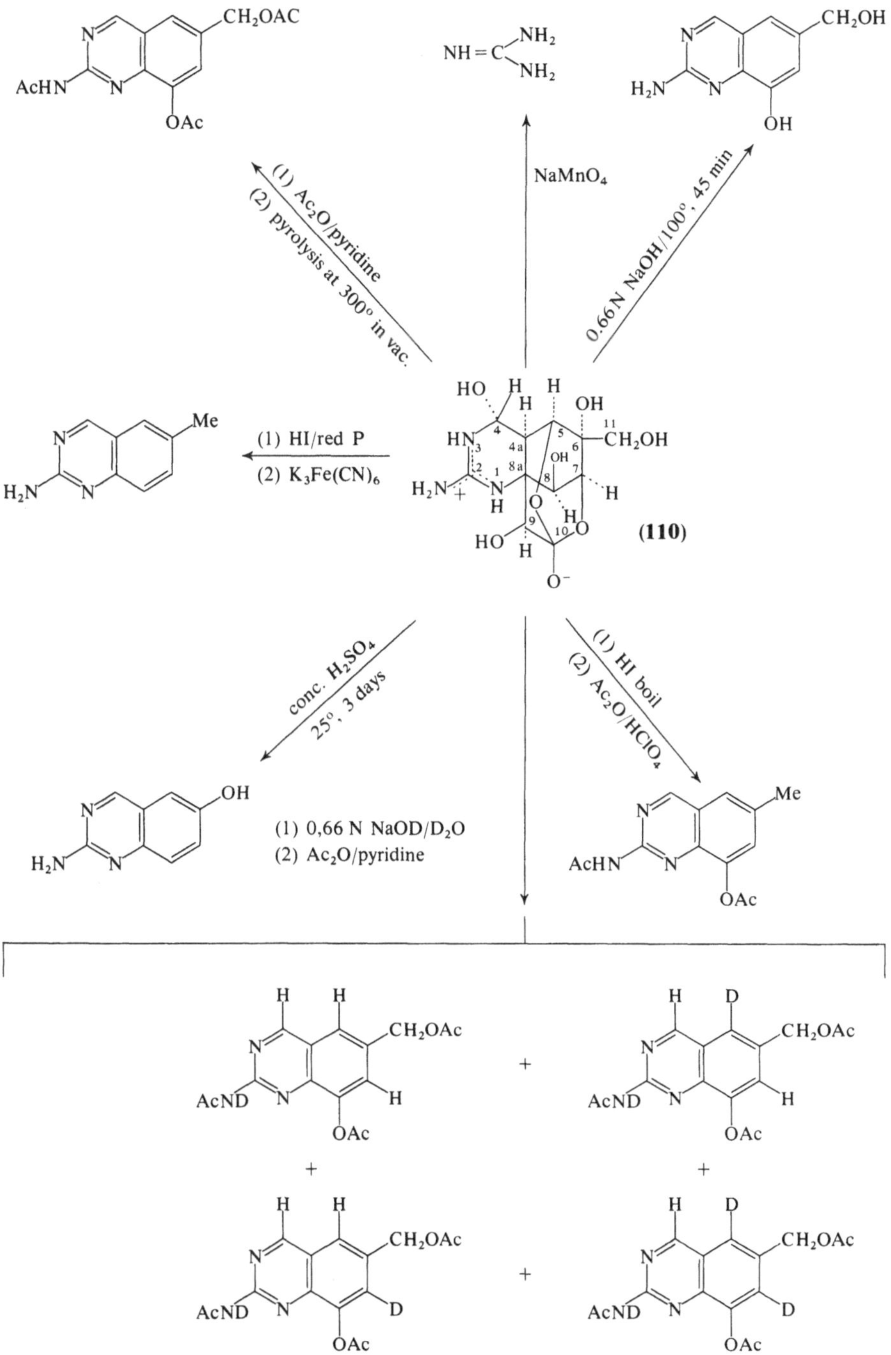

Scheme 7. Some transformation reactions of tetrodotoxin

Scheme 7 (continued)

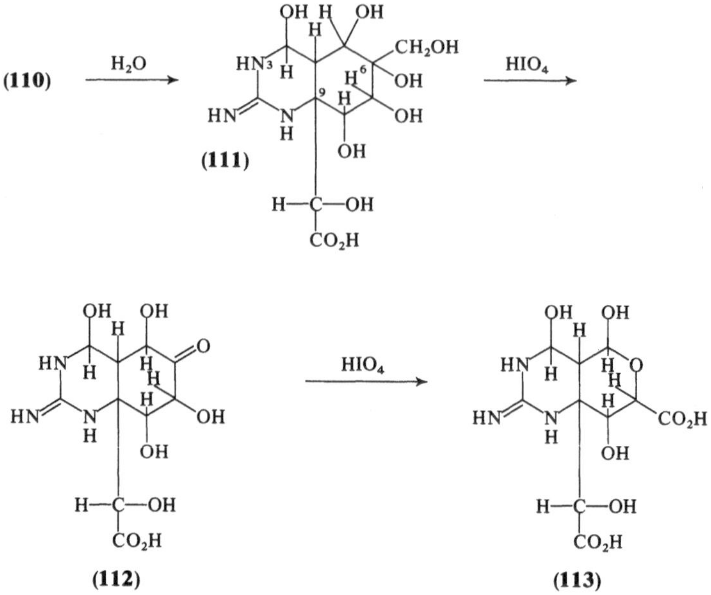

(110) → (111) → (112) → (113)

Tetrodotoxin (110) is a zwitterionic polyhydroxy-perhydro-2-imino-quinazoline with a hemilactal structure. Its chemical lability is indeed considerable. The three nitrogen atoms are part of a guanidine system, and permanganate oxidation of (110) gives guanidine. The presence of a quinazoline skeleton is indicated by transformation of (110) to the many quinazoline derivatives shown in Scheme 7 (after ARMAREGO, 6). Heating (110) with water in a sealed tube gives tetrodoic acid (111), which consumes one mole of periodic acid to afford formaldehyde and nortetrodoic acid (112). The latter compound consumes an additional mole of periodic acid to yield seconortetrodoic acid (113). Reaction of (112) with HCl gives 2-amino-5,6-dihydroxyquinazoline. Formation of 2-amino-8-hydroxy-6-hydroxymethylquinazoline and oxalic acid by degradation of (110) or tetrodoic acid indicates the presence of a glyoxylic acid group at the bridgehead carbon, C-9. The complete structure of (110), including the absolute configuration, was elucidated by detailed, three-dimensional X-ray crystallographic analyses of tetrodotoxin and some of its derivatives (cf. 220).

An elegant total synthesis of DL-(110) was accomplished by KISHI *et al.* (118). Several reactions were developed expressly for this purpose and should prove useful in other areas of organic synthesis. The pyrimidine ring is formed only in the final stages of this multi-step synthesis. The key intermediate dihydrofuranacetamide (114) can be converted to the diacetylguanidinediol (**115**) and then to (**110**). A more direct route involves conversion of (**114**) to (**110**) *via* the dihydrofuran-diacetylguanidine (**116**). The maximum yield of (**110**) from (**115**) was 15%. The second route was superior, both in terms of overall yield [from (**116**) to (**110**), 25%] and reproducibility.

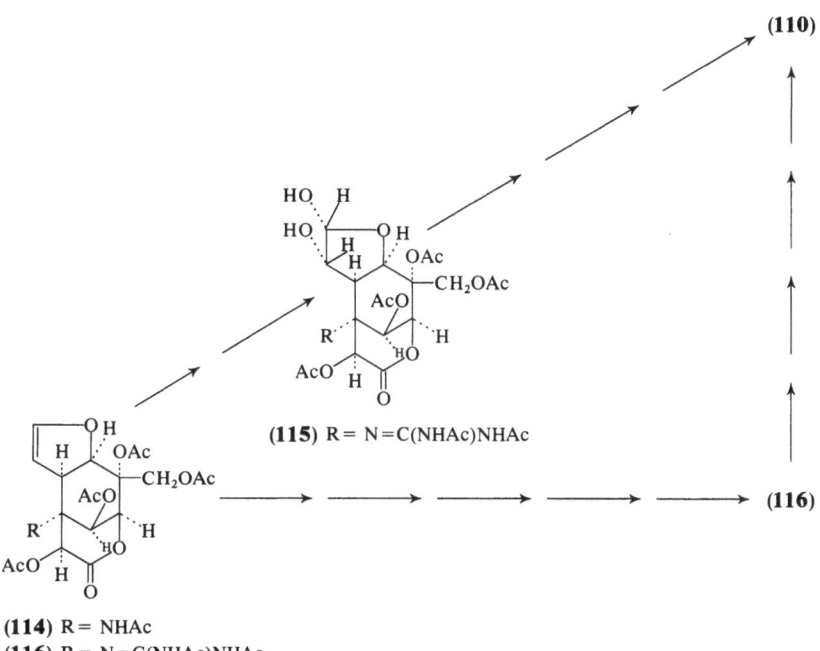

(**110**)

HO H
HO   O H
       H   OAc
    H   CH₂OAc
    AcO
    R   'H
           HO
AcO H
      O

(**115**) R = N=C(NHAc)NHAc

O H
H   OAc
    CH₂OAc
AcO
R   'H
       HO
AcO H
      O

(**116**)

(**114**) R = NHAc
(**116**) R = N=C(NHAc)NHAc

Scheme 8. Key intermediates of the total synthesis of tetrodotoxin according to KISHI *et al.*

VIII. Biosynthesis

Almost all theories on the biosynthesis of quinazolines consider anthranilic acid, or an equivalent compound, to be one of the essential building blocks (*72—74, 125, 173*). Thus, vasicine may be derived from anthranilic acid and proline or closely related metabolites. N-Methylanthranilic acid, ammonia and phenylacetic acid can be considered precursors of arborine, while febrifugine could be formed from anthranilic acid, formic acid, ammonia, a C_3-unit and lysine, or equivalent compounds. SAKAR and CHAKRABORTY (*177, 177a*) have suggested that the isolation of glycophymoline, glycosminine, arborine and glycomide (phenylacetanilide) from *Glycosmis pentaphylla* provides circumstantial evidence for the formation of such simple substituted quinazoline alkaloids from the precursor phenylacetanilide carboxylic acid. It has been postulated that the biosynthesis of 6,7,8,9-tetrahydropyrido-[2,1-b]-quinazolin-11-one and related alkaloids starts from anthranilic acid and lysine (67). Several quinazolines have been synthesized under so-called physiological conditions.

1. Pyrroloquinazolines

As early as 1960, it could be demonstrated that aseptically grown plants of *Peganum harmala* are capable of incorporating [^{14}COOH] anthranilic acid into vasicine (**36**), cf. Scheme 9. It could be shown that the incorporation of anthranilic acid (**117**) involves retention of the carboxyl group in *Adhatoda vasica* plants.

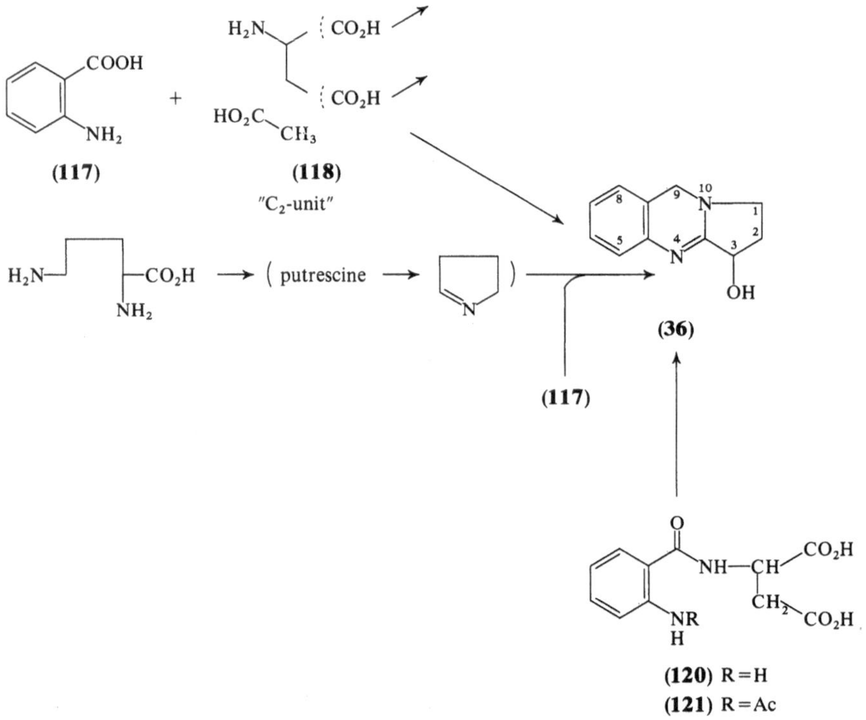

Scheme 9. Possible biosynthetic pathways of vasicine

The origin of the "non-anthranilic acid moiety" is still controversial. It was found that, in *Adhatoda vasica*, C-1, C-2 and N-10 of vasicine are derived from aspartic acid (**118**) and C-3 and C-3a from a C$_2$-unit (97, 216).

This was supported experimentally by the incorporation of [3-^{14}C]-aspartic acid, [3-^{14}C]-malic acid and [2,3-^{14}C]-succinic acid, with localization of the label at C-1 and C-2 of vasicine. Incorporation of succinic acid could proceed *via* aspartic acid, as succinic acid is readily converted to oxalacetic acid in the citric acid cycle and oxalacetic acid can be easily converted to aspartic acid. After feeding [3-^{14}C]-malic acid, 72% of the radioactivity of vasicine is found at carbons 1 (26%) and 2 (46%) without equilibration. The labelling pattern following administration of [3-^{14}C]-aspartic acid or [3-^{14}C]-malic acid can be explained by assuming conversion of aspartic acid to fumaric acid (by aspartase) or oxalacetic acid, which, in turn, can be converted to fumarate *via* malate. As these processes are reversible, the malate and aspartate resynthesized from fumarate will contain equal amounts of label at carbon atoms 2 and 3, and together with the precursor originally fed, would result in radioactivity at C-1 and preferentially at C-2, as found. The radioactivity of [4-^{14}C]-aspartic acid is only non-specifically incorporated and does not label C-3 of vasicine. The incorporations of [2-^{14}C]-acetate, [2-^{14}C]-glycine, [3-^{14}C]-pyruvic acid and the specific incorporation of [2'-^{14}C,^{15}N]-N-acetylanthranilic acid and especially of anthranoyl-[3-^{14}C]-aspartic acid (120) are in accord with the other results. After administration of the latter compound, 77% of the radioactive label is found at C-3. This derivative has been suspected of being an intermediate in the biosynthesis of vasicine. Thus, it is possible that acetylanthranoyl-aspartic acid (121) may be the most direct precursor of vasicine. Feeding experiments using anthranoylaspartic acid and [^{15}N]-aspartic acid indicate that N-10 of vasicine may be derived from aspartic acid. [^{15}N]-Anthranilic acid and [CO-^{15}NH$_2$]-glutamine are converted into vasicine with a high degree of specific incorporation. These compounds apparently provide N-4 of the alkaloid. [CO-^{15}NH$_2$]-Anthranilic acid amide is poorly incorporated. Members of the α-ketoglutaric acid family (glutamic acid, 4-hydroxyglutamic acid, 4-amino-2-hydroxybutyric acid, ornithine, proline, putrescine) as well as [^{14}C]-oxalic acid, N-[^{14}C]-oxalylanthranilic acid, [1-^{14}C]-glycine and [1-^{14}C]-glyoxylate are only incorporated nonspecifically. [^{14}CHO]-N-Formylanthranilic acid and [^{14}CH$_3$]-N-methylanthranilic acid are not incorporated. Likewise, it has been shown that direct hydroxylation of deoxyvasicine does not occur, and [U-^{14}C]-vasicinone is poorly incorporated into vasicine.

In *Peganum harmala*, [2-^{14}C]-ornithine (119) and labeled proline, putrescine, and related compounds are more or less specifically incorporated into the pyrrolidine ring system of vasicine (*127, 128*). This result suggests that ornithine is decarboxylated affording putrescine, a symmetrical molecule, which is then incorporated, a process which results in equal labeling at C-1 and C-3a. The results can be rationalized by postulating that a symmetrical intermediate such as putrescine is involved in the pathway from ornithine to vasicine, or by invoking [2-^{14}C]-ornithine as a source of [1-^{14}C]-acetate (*via* the citric acid cycle) which can then be incorporated by way of N-acetylanthranilic acid. If both pathways to vasicine are valid, and this remains to be established, then these results are unique since alkaloid biosynthesis in plants has, up to now, not been found to be species-specific (with the well-known exception of the radically different pathways to coniine and N-methylpelletierine). The reported investigations have not established which of the alkaloids is formed first as the result of the condensation yielding the pyrroloquinazoline system.

The paths of the interconversions which occur among the alkaloids of *Peganum harmala* have also not been elucidated so far. Such studies have been hindered by the rapid metabolism of these compounds. After feeding [^{3}H]-vasicine, the label was detected in vasicinone and deoxyvasicinone. As only 65% of the administered radioactivity was recovered, it is possible that vasicine may have been degraded in the plant. Pegamine (**6**) is a potential intermediate for deoxyvasicinone (**42**).

Adhatodine, vasicoline, vasicolinone, anisotine and sessiflorine may arise from pyrroloquinazolines and derivatives of anthranilic acid.

2. Arborine and Glomerin

In *Glycosmis arborea*, arborine (**2**) is elaborated from phenylalanine and anthranilic acid (*77*). After administration of [U-^{14}C]-phenylalanine, 92% of the label is found in the phenylacetic part of arborine. [^{3}H]-Anthranilic acid is specifically incorporated into the N-methylanthranilic acid moiety of arborine. Feeding of [^{14}CH$_3$]-methionine results in specific labelling of the N-methyl group of arborine. The same result is obtained on administration of [^{14}CH$_3$]-N-methylanthranilic acid. These results seem to indicate that anthranilic acid is methylated prior to condensation with phenylalanine leading to arborine. In arborine, N-1 is derived from the nitrogen of anthranilic acid. Experiments with [3-^{14}C,^{15}N]-phenylalanine clearly show that, except for the carbonyl group, the aliphatic side-chain of phenyl-alanine is incorporated into arborine. Thus, phenylacetic acid can be excluded as an immediate precursor of (**2**). [U-^{3}H]-Anthranoyl-phenylalanine is a very poor precursor of (**2**). N-Methyl-N-phenyl-[^{14}CO$-$CH$_2$]-acetylanthranilamide is efficiently transformed into arborine in *Glycosmis arborea*. The role of N-methyl-N-phenylacetylanthranilamide as a true precursor in arborine biosynthesis remains doubtful, as, surpris-ingly, it could be shown that transformation of this compound to arborine also occurs in pea plants. Ring closure of N-methyl-N-phenylacetyl-anthranilamide is apparently catalyzed by an enzyme or enzyme system which is nonspecific and widely distributed (*98*). In contrast to these results, another research group has postulated that the incorporation of phenyla-lanine proceeds *via* phenylacetic acid and N-methyl-N-phenylacetyl-anthranilic acid (*61*).

It has also been demonstrated that [^{14}COOH]-anthranilic acid is a precursor of the animal alkaloids glomerin (**18**) and homoglomerin (*180*).

3. Indolopyridoquinazolines

Evodiamine (**67**) and rutaecarpine (**63**) are derived, in *Evodia rutae-carpa*, from tryptophan (**122**), anthranilic acid (**117**) and formic acid (*224*) (see Scheme 10). After administration of [^{14}CH$_3$]-methionine, 99% of the label was found at C-13b of rutaecarpine, while in evodiamine, the label is distributed between C-13b and the N-methyl group. Surprisingly, the level of specific incorporation of label in (**67**) is lower than that in (**63**); therefore,

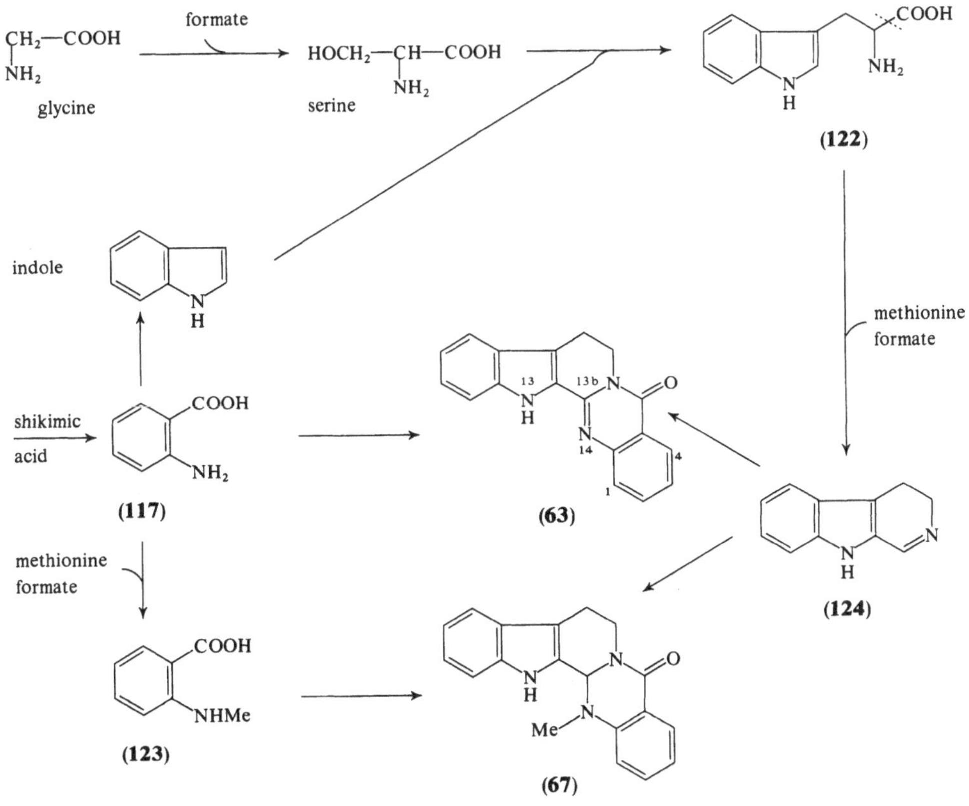

Scheme 10. Biosynthesis of evodiamine and rutaecarpine

rutaecarpine cannot be regarded as the precursor of evodiamine. It has been postulated (225) that the C_1-unit is introduced at an earlier stage of the biosynthesis. Possibly, N-methylanthranilic acid (123) is formed first and then condenses with dihydronorharman (124) to give evodiamine. In such a case, the radioactivity of the administered C_1-labeled compound would be diluted by the non-labeled N-methylanthranilic acid already present, explaining the lower specific incorporation in evodiamine. Experimental evidence for the possible involvement of a β-carboline in the biosynthesis of *Evodia* alkaloids has not yet been reported.

4. Fungal Metabolites

A small number of simple substituted quinazolines can be formed in the course of tryptophan degradation by microorganisms (*131, 132*). In *Pseudomonas* species, three pathways of tryptophan degradation are known: the aromatic pathway in *Ps. fluorescens*, the quinoline pathway in *Ps. acidovorans*, and the quinazoline pathway in *Ps. aeruginosa*. Investigations with [β-^{14}C]-tryptophan have provided evidence for a new pathway from tryptophan through the intermediates, formylkynurenine and N-formylaminoacetophenone, forming 4-methylquinazoline with ammonia and free 2-aminoacetophenone. Reacylation of the product and cyclization with ammonia produces other derivatives of 4-methylquinazoline.

YAMAZAKI et al. (*221*) have postulated that the tryptoquivalines may be biogenetically derived from four amino acids; tryptophan, anthranilic acid, valine and alanine. Deoxynortryptoquivalone is thought to be the first compound formed in the biosynthesis of the tryptoquivaline series. Oxidation of the secondary amine in this compound to a hydroxylamine would result in nortryptoquivalone. Oxidative loss of the side-chain would lead to FTE or FTJ. Alternatively, reduction of the side-chain carbonyl group would afford nortryptoquivaline. The geminal dimethyl group at C-15 may result from incorporation of a C_1-unit into deoxynortryptoquivalone or from the direct participation of methylalanine rather than alanine in the initial step of the biosynthesis.

The antibiotic tryptanthrin (103) has been biosynthetically prepared from one mole of tryptophan and one mole of anthranilic acid. Feeding tryptophan plus substituted anthranilic acids, or substituted tryptophans plus anthranilic acid has resulted in the generation of the expected tryptanthrin derivatives. No substrate specificity (except for bromotryptophan) was observed in the enzymes involved in tryptanthrin biosynthesis. The anthranilic acid moiety of these compounds is the result of tryptophan degradation (*65, 183*).

IX. Biological Activity of Natural and Synthetic Quinazolines

Several quinazoline alkaloids are known to elicit a variety of biological responses. This has spurred the preparation and pharmacological evaluation of a great number of quinazoline derivatives, and intensive research in the quinazoline field is still in progress. At present, approximately fifty quinazoline derivatives with a variety of biological activities are available for clinical use. This topic has been reviewed by JOHNE (93). Only a few aspects of the biological activity of the quinazoline alkaloids will be discussed in this chapter.

Adhatoda vasica Nees (vasaka) is used in Indian indigenous medicine, where use of the leaves and roots is said to have beneficial effects in bronchitis, asthma, diarrhea and dysentery. These parts are also claimed to possess antiseptic, antiperiodic and anthelminthic properties.

Vasicine is also of interest for its action on bronchial muscles. CHOPRA *et al.* have observed that vasicine produces slight but persistent bronchodilation, causes mild hypotension (due in part to direct cardiac depression) and in high concentrations inhibits peristalsis in the isolated gut. AMIN *et al.*, however, have found vasicine to be a bronchoconstrictor and to produce a negative inotropic effect with reduced coronary flow in the isolated heart. CAMBRIDGE *et al.* have observed slight relaxation of the tracheal chain at low vasicine concentrations, but contraction at higher concentrations. Vasicine-induced bronchoconstriction has been related to its acetylcholine potentiating properties. The same alkaloid causes psychomotor excitation in dogs and rats.

AMIN *et al.* have observed that vasicinone, an auto-oxidation product of vasicine from extracts of *A. vasica* leaves, causes definite bronchodilation in the presence of histamine-induced bronchospasm, slight hypotension and a positive cardiac inotropic action with increased coronary flow. The investigators attribute the bronchodilator effect of vasaka leaves to the small amount of vasicinone produced by auto-oxidation of vasicine (cited after *124*). MEHTA *et al.* (*136*) have isolated 1-vasicinone from *A. vasica* and demonstrated that it is not an artifact as it has bronchodilator activity in contrast to the bronchoconstrictor activity of 1-vasicine.

The bronchodilator activities of vasicine and quinazolin-4-one have been studied in detail. The activity of these compounds is in no way comparable to other known bronchodilator drugs (*31*). The same authors, however, failed to obtain bronchodilation with vasicinone *in vivo* in the guinea pig although it was observed in the isolated guinea pig tracheal chain. This difference in activity was attributed to rapid metabolism of the drug. Vasicinone was found to be less powerful than vasicine in antagonizing histamine-induced bronchospasms (cited after *124*, cf. *3*).

Vasicine has been found to possess slight anticholinesterase activity (*124*), and recently it has come into prominence as an uterine stimulant and uterotonic abortifacient, comparable to oxytocin and methergine (*80, 232*). Some discrepancies in the results of the pharmacological tests of vasicine and related alkaloids and their interpretations need further clarification. The text by ATAL (*12*) is a leading reference on the pharmacology of vasicine.

Vasicinone has been used as a model for the development of the expectorant, 2-amino-3,5-dibromo-N-cyclohexyl-N-methylbenzene-methanamine. PLUGAR et al. have investigated the metabolism of the quinazolin-4-one family of alkaloids in animals. Deoxyvasicinone, injected intraperitoneally into rats at a level of 50 mg/kg, was metabolized primarily to vasicinone which was then excreted in the urine. Vasicinone administered i. p. was excreted unchanged (*162*, cf. *167*).

The biological activities of vasicine have prompted the synthesis and pharmacological evaluation of vasicine and vasicinone analogues (*57*). 5,6-Methylenedioxyvasicine possesses uterine stimulant and bronchodilator activities comparable to those of vasicine, but with more marked hypotensive and cardiac depressant activities (*192*).

LAHIRI and PRADHAN (*124*) have investigated the biological activities of 7-hydroxypeganine. This alkaloid produces transient hypertension in cats, contraction of the isolated guinea pig intestine, and depression of the isolated guinea pig heart. All of the activities can be blocked by atropine. 7-Hydroxypeganine produces slight contraction of the isolated guinea pig tracheal chain, but antagonizes histamine-induced bronchospasm in the guinea pig *in vivo*. Mild anticholinesterase activity is also observed.

Febrifugine is about 100 times as active as quinine against *Plasmodium lophurae* in ducks and against the trophozoites of *Pl. cynomolgi* in monkeys. It is also very active against *Pl. gallinaceum* in chicks. Unfortunately, a high degree of toxicity renders it unsuitable for clinical use in malaria (*33*). At present, toxicity and drug resistance are significant problems.

After side-chain modification, as in the synthetic 3-[β-keto-γ-(3-hydroxy-2-pyridyl)-propyl]-quinazolin-4-one (*66*), was found to be fruitless, attempts at structural modification of febrifugine were focused on the synthesis of some methylenedioxy analogues by CHIEN and CHENG (*44*). The 5,6-, 6,7- and 7,8-methylenedioxy analogues were found to be active against *Pl. berghei*. Toxicity of these compounds in mice is much lower than that of febrifugine and their therapeutic indices are comparable to that of the parent compound.

Febrifugine is also a powerful emetic (with a low therapeutic index) and is one of the most effective coccidiostatics known. 6-Chloro-7-bromofebrifugine is active prophylactically against *Eimeria tenella* when added to fodder at a level of 0.0003% (*217, 227*). BUDĚŠINSKY et al. (*28*) have prepared a number of 3-(3-amino-2-hydroxypropyl)- and 3-(3-amino-acetonyl)-quinazolin-4-ones in which the 3-piperidinol part of febrifugine has been replaced by readily available bases (aniline, pyrrolidine, piperidine, morpholine). The coccidiostatic efficacy of these compounds was evaluated in chicks infected with *Eimeria tenella* using the battery test. Replacement of the 3-piperidinol moiety with other amines appears to result in a decrease of coccidiostatic activity. Some compounds show marked activity against the test organism and are also active against *Nippostrongylus brasiliensis* in rats and against the liver fluke, *Fasciola hepatica,* in animals. Febrifugine and isofebrifugine also possess anti-moth activity (*228*). Further references may be found in (*93*).

Tetrodotoxin is a highly potent, fast-acting substance, and one of the most toxic of low molecular weight compounds. Poisoning due to this compound has long been a serious problem in Japan, as puffer fish is a delicacy highly prized at the table (*212*). Tetrodotoxin acts by specifically obstructing the sodium channels in neuronal membranes without significantly affecting their permeability for potassium ions. The action potential of desheathed nerves is blocked, producing muscular weakness, paralysis, a fall in blood pressure and respiratory failure. The mechanism of action resembles that of local anesthetics. The threshold effective dose of tetrodotoxin is only 1/160,000 that of cocaine. This alkaloidal toxin is presently used in neurophysiological research.

The combination of unique structural features, extensive functionalization and extreme biological activity found in tetrodotoxin have presented a formidable challenge to the synthetic chemist. During the last few years, a number of research groups have reported a variety of synthetic approaches to tetrodotoxin analogues, but the biological activities of these derivatives have been rather disappointing (cf. *7, 8, 204, 210*). The results of biological evaluation indicate that the "cage" structure does not enhance the activity of the hydroquinazoline in these tests, and that most of the observed activities are due to the cyclic guanidino or ureido moiety present in these molecules (*8*).

When the hydroxymethyl side-chain of the natural toxin is removed by oxidation, the 11-nor-tetrodotoxin formed (with C-6 − OH and CH_2OH replaced by a carbonyl group on C-6) is several hundred times less active than the parent toxin. Some biological activity is regained, however, when a methoxyimino group is inserted at C-6 in 11-nor-tetrodotoxin. 11-Succinylanhydrotetrodotoxin is a synthetic derivative containing a free carboxy group, potentially useful for attachment to an affinity chromatography column or to a large molecule. The latter application may prove useful for development of an immunologic antidote for the toxin (cited after *5*).

Glycosmis arborea is a shrub found along roadsides throughout West

Bengal. In the Ayurvedic system of medicine, it is said to be useful in flatulence, cold, rheumatism, intestinal worms, anemia, fever and jaundice. The outstanding effect of arborine, isolated from this plant, is inhibition of the peripheral action of acetylcholine on rat uterus, on guinea pig ileum and on skeletal muscle. It also markedly inhibits the action of pituitrin on the rat uterus. The arborine-induced fall in blood pressure in the intact cat is central in origin, being absent in the spinal cat, but present in the vagotomized animal (37, 43).

Rutaecarpine, dihydrorutaecarpine and its 14-formyl derivative, and evodiamine inhibit the growth of silkworm larvae when administered orally (111). The synthesis and structure of hydrogenated rutaecarpine derivatives is discussed in (208a). Evodiamine is useful as a diuretic and diaphoretic (110). It must be mentioned once again that glomerin and homoglomerin are the active principles of the defensive secretion of Glomeris marginata. MUKHERJEE and DEY (143) have investigated the changes in amino acid metabolism in rat brain following administration of arborine.

The minimum inhibitory concentration of tryptanthrin against 17 fungal strains, including Trichophyton mentagrophytes and T. rubrum have been investigated (206). The antibiotic activity of this compound depends on the composition of the agar medium [cf. (183) for further details]. Quinazomycin is active against Staphylococcus aureus (113).

From the preceding remarks it should be obvious that investigations over the last few years have demonstrated that the natural quinazoline alkaloids and their synthetic derivatives exhibit a wide variety of pharmacological activities. In the continuing search for compounds producing interesting biological activities, the quinazoline alkaloids should provide an excellent starting point for further investigations.

Table 2. Naturally Occurring Quinazoline Alkaloids

Alkaloid	Mol. formula	m.p. °C	$[\alpha]_D$	Source (Ref.)
Quinazolin-4-one**	$C_8H_6N_2O$	212		Dichroa febrifuga Lour. (45)
4-Methylquinazoline*	$C_9H_8N_2$			Pseudomonas spec. (131, 132)
Glycorine (3)	$C_9H_8N_2O$	145—47		Glycosmis arborea (Roxb.) DC. (153, 156)
				G. bilocularis Thw. (26)
Glycosmicine (4)	$C_9H_8N_2O_2$	270—71		Glycosmis arborea (Roxb.) DC. (153, 156)
2-Carboxy-4-methylquinazoline*	$C_{10}H_8N_2O_2$			Pseudomonas spec. (131, 132)
4-Methylquinazoline-2-carbonamide*	$C_{10}H_9N_3O$			Pseudomonas spec. (131, 132)
2,4-Dimethylquinazoline*	$C_{10}H_{10}N_2$			Pseudomonas spec. (131, 132)
Glomerin (18)	$C_{10}H_{10}N_2O$	204		Glomeris marginata Vill. (137, 182)
2-Hydroxymethyl-4-methylquinazoline*	$C_{10}H_{10}N_2O$			Pseudomonas spec. (131, 132)
Chrysogine (109)	$C_{10}H_{10}N_2O_2$	189—90	$-26\pm4°$	Penicillium chrysogenum (84)
Deoxyvasicinone (42)	$C_{11}H_{10}N_2O$	110—11		Linaria spec. (164)
				Mackinlaya macrosciadia (F. Muell.) F. Muell. (82)
				Peganum nigellastrum (16)
				Adhatoda vasica Nees (2)
Vasicinone (37)	$C_{11}H_{10}N_2O_2$	203—04	$-129°$	Linaria spec. (164)
				Peganum harmala L. (121, 122)
				P. nigellastrum (16)
				Sida spec. (69, 168)
Vasicinolone (39)	$C_{11}H_{10}N_2O_3$	279		Adhatoda vasica Nees (90)
2-Ethyl-4-methylquinazoline*	$C_{11}H_{12}N_2$			Pseudomonas spec. (131, 132)
Vasicine (peganine) (36)	$C_{11}H_{12}N_2O$	DL: 209—10	$-254°**$	** Adhatoda vasica Nees (87)
		L: 211—12	$+162,5°*$	Anisotes sessiliflorus C.B. Cl. (9)
		D: 170—73		* Galega officinalis L. (185)
				G. orientalis Lam. (119)
				Linaria spec. (see 75)
				** Peganum harmala L. (174)
				P. nigellastrum (16)
				Sida spec. (69, 168)
Homoglomerin	$C_{11}H_{12}N_2O$	149		Glomeris marginata Vill. (137, 182)
Peganol (56)	$C_{11}H_{12}N_2O$	178—80		Peganum harmala L. (207)

* The alkaloids were identified by TLC.
** This compound is may be an artefact, compare (169).

Table 2 (continued)

Alkaloid	Mol. formula	m.p. °C	[α]$_D$	Source (Ref.)
7-Hydroxypeganine (38)	C$_{11}$H$_{12}$N$_2$O$_2$	272—73	+45,8°	Adhatoda vasica Nees (199)
				Peganum harmala L. (121, 122)
				Sida spec. (69, 168)
Pegamine (6)	C$_{11}$H$_{12}$N$_2$O$_2$	160—61		Peganum harmala L. (114)
Vasicol (40)	C$_{11}$H$_{14}$N$_2$O$_2$	204—06 (HCl)	−17,34°	Adhatoda vasica Nees (58)
Tetrodotoxin (110)	C$_{11}$H$_{17}$N$_3$O$_8$ · H$_2$O		−8,1°	Atelopus spec. (cf. 117)
				Charonia sauliae (145a)
				Babylonia japonica (145b)
				Gobius criniger
				Hapalochlaena maculosa (194)
				Spoeroides rubripes
				Taricha torosa
				Tritus ensicanda (see 212)
(62)	C$_{12}$H$_{14}$N$_2$	85—87		Mackinlaya macrosciadia (F. Muell.) F. Muell.
				M. subulata Philipson (67, 99)
(61)	C$_{12}$H$_{12}$N$_2$O	99		Mackinlaya macrosciadia (F. Muell.) F. Muell.
				M. subulata Philipson (67, 99)
Deoxypeganidine (55)	C$_{14}$H$_{16}$N$_2$O	76—79		Peganum harmala L. (230)
Peganidine (54)	C$_{14}$H$_{16}$N$_2$O$_2$	189—90		Peganum harmala L. (116)
Isopeganidine	C$_{14}$H$_{16}$N$_2$O$_2$	169—70		Peganum harmala L. (229)
Tryptanthrin (103)	C$_{15}$H$_8$N$_2$O$_2$	268 (261)		Candida lipolytica (183)
				Couroupita guaianensis Aubl. (18, 186)
				Leucopaxillus cereralis var. piceina (Peck) (92)
				Polygonum tinctorum (206)
				Strolanthes cusia (206)
Glycosminine (5)	C$_{15}$H$_{12}$N$_2$O	249		Glycosmis arborea (Roxb.) DC. (41, 153)
Arborine (2)	C$_{16}$H$_{14}$N$_2$O	155—56		Ruta spec. (214)
				Glycosmis arborea (Roxb.) DC. (36, 38, 42)
				G. bilocularis Thw. (26)
Glycophymoline	C$_{16}$H$_{14}$N$_2$O	165		Glycosmis pentaphylla (177a)
Febrifugine (9)	C$_{16}$H$_{19}$N$_3$O$_3$	139—40	+28°	Dichroa febrifuga Lour. (45, 91, 120)
		154—56		Hydrangea umbellata Rheder (1)
Isofebrifugine	C$_{16}$H$_{19}$N$_3$O$_3$	129—31	+120°	Dichroa febrifuga Lour. (45, 91, 120)
				Hydrangea umbellata Rheder (1)

Compound	Formula	m.p. (°C)	[α]	Occurrence
(7)	$C_{17}H_{16}N_2O_2$	100—02		Zanthoxylum arborescens (Rose) (62)
5,6-Dehydro-16-hydroxyrutaecarpine	$C_{18}H_{11}N_3O_2$	>340		Vepris louisii G. Gilbert (12a)
Rutaecarpine (63)	$C_{18}H_{13}N_3O$	258		Evodia rutaecarpa Hook. f. & Thoms. (10, 111, 209); Hortia arborea Engl. (151); H. badinii (56c)
16-Hydroxyrutaecarpine (64)	$C_{18}H_{13}N_3O_2$	318—20		Zanthoxylum pluviatile Hartl. (217a); Euxylophora paraensis Hub. (55); Vepris louisii G. Gilbert (12a)
3,14-Dihydrorutaecarpine (8)	$C_{18}H_{15}N_3O$	214—16	−564°	Evodia rutaecarpa Hook. f. & Thoms. (111); Zanthoxylum arborescens (Rose) (62)
Euxylophoricine C (91)	$C_{18}H_{18}N_2O_3$	133—34		Euxylophora paraensis Hub. (52)
Dehydroevodiamine (68)	$C_{19}H_{13}N_3O_3$	310—12		Evodia rutaecarpa Hook. f. & Thoms. (145)
Hortiacine (66)	$C_{19}H_{15}N_3O$	189—90		Hortia arborea Engl. (150); H. badinii (56c), H. longifolia (56d)
14-Formyl-dihydrorutaecarpine	$C_{19}H_{15}N_3O_2$	252		Evodia rutaecarpa Hook. f. & Thoms. (111)
Euxylophoricine F (90)	$C_{19}H_{15}N_3O_2$	280—81	+260°	Euxylophora paraensis Hub. (48)
Evodiamine (67)	$C_{19}H_{15}N_3O_3$	226	+352°	Araliopsis tabouensis Aubrev. et Pellegr. (65a); Evodia rutaecarpa Hook. f. & Thoms. (10, 111); Zanthoxylum rhetsa DC. (39, 39a, 70, 226)
Vasicolinone (52)	$C_{19}H_{17}N_3O$	278		Adhatoda vasica Nees (95)
Sessiflorine (49)	$C_{19}H_{19}N_3O$	152		Anisotes sessiliflorus C. B. Cl. (9)
Vasicoline (51)	$C_{19}H_{19}N_3O_2$	195—97		Adhatoda vasica Nees (95)
Euxylophoricine B (87)	$C_{19}H_{21}N_3$	135		Euxylophora paraensis Hub. (32)
Hortiamine (69)	$C_{20}H_{15}N_3O_3$	310—12		Hortia arborea Engl. (150); H. braziliana Vel. (149)
Euxylophoricine A (86)	$C_{20}H_{17}N_3O_2$	208		Euxylophora paraensis Hub. (32)
(65)	$C_{20}H_{17}N_3O_3$	295—98		Evodia rutaecarpa (Juss.) Benth. & Hook. (49)
Anisotine (43)	$C_{20}H_{17}N_3O_3$	189—90		Adhatoda vasica Nees (95)
Anisessine (46)	$C_{20}H_{19}N_3O_3$	170		Anisotes sessiliflorus C. B. Cl. (9)
Adhatodine (53)	$C_{20}H_{19}N_3O_3$	183		Anisotes sessiliflorus C. B. Cl. (9)
Deoxyaniflorine (48)	$C_{20}H_{21}N_3O_2$	168—72		Adhatoda vasica Nees (95)
Aniflorine (47)	$C_{20}H_{21}N_3O_2$	197		Anisotes sessiliflorus C. B. Cl. (9)
Euxylophorine B (83)	$C_{21}H_{17}N_3O_3$	268—71		Euxylophora paraensis Hub. (52)
Euxylophoricine E (89)	$C_{21}H_{17}N_3O_4$	290		Euxylophora paraensis Hub. (56)
Euxylophorine A (82)	$C_{21}H_{19}N_3O_3$	227—30		Euxylophora paraensis Hub. (32)
Euxylophoricine D (88)	$C_{21}H_{19}N_3O_4$	293—95		Euxylophora paraensis Hub. (56)
Tryptoquivaline F (FTF)	$C_{22}H_{18}N_4O_4$	277	−109°	Aspergillus fumigatus (221)

Table 2 *(continued)*

Alkaloid	Mol. formula	m.p. °C	$[\alpha]_D$	Source (Ref.)
Tryptoquivaline J (FTJ)	$C_{22}H_{18}N_4O_4$	254—58	$+135°$	*Aspergillus fumigatus (221)*
Tryptoquivaline E (FTE)	$C_{22}H_{18}N_4O_5$	257	$+257°$	*Aspergillus fumigatus (221)*
Tryptoquivaline H (FTH)	$C_{22}H_{18}N_4O_5$	274	$-155°$	*Aspergillus fumigatus (221)*
Euxylophorine D (85)	$C_{22}H_{19}N_3O_4$	256—60		*Euxylophora paraensis* Hub. (56)
Dipegine (57)	$C_{22}H_{20}N_4O$	221—23		*Peganum harmala* L. (229)
Euxylophorine C (84)	$C_{22}H_{21}N_3O_4$	207—09		*Euxylophora paraensis* Hub. (56)
Tryptoquivaline G (FTG) (106)	$C_{23}H_{20}N_4O_5$	240—41.5	$+215°$	*Aspergillus fumigatus (221)*
Tryptoquivaline L (FTL) (107)	$C_{23}H_{20}N_4O_5$	265—68	$-229°$	*Aspergillus fumigatus (226a)*
Paraensine (102)	$C_{24}H_{21}N_3O_3$	281—82		*Euxylophora paraensis* Hub. (53)
Deoxynortryptoquivalone	$C_{26}H_{24}N_4O_5$	192—93	$+171°$	*Aspergillus clavatus (30)*
				A. fumigatus (226a)
Nortryptoquivalone	$C_{26}H_{24}N_4O_6$	208—09	$+255°$	*Aspergillus clavatus (30, 46)*
Tryptoquivaline I (FTI)	$C_{27}H_{26}N_4O_6$	232—35.5	$+239°$	*Aspergillus fumigatus (221)*
Deoxynortryptoquivaline	$C_{28}H_{28}N_4O_6$	158—60	$+69.5°$	*Aspergillus clavatus (30)*
Nortryptoquivaline (105)	$C_{28}H_{28}N_4O_7$	256—58	$+170°$	*Aspergillus clavatus (30)*
				A. fumigatus (226a)
Tryptoquivaline M (FTM)	$C_{28}H_{28}N_4O_7$	157—64	$-154°$	*Aspergillus fumigatus (226a)*
Deoxytryptoquivaline	$C_{29}H_{30}N_4O_6$	150—52	$+56.8°$	*Aspergillus clavatus (30)*
Tryptoquivaline (104)	$C_{29}H_{30}N_4O_7$	155—57	$+130°$	*Aspergillus clavatus (30, 46)*
				A. fumigatus (226a)
Nordine (41)	$C_{46}H_{46}N_4O_4$	292		*Daemonorops draco (160)*
Quinazomycin (108)	$C_{49}H_{62}N_{12}O_{13}S_2$			*Streptomyces (113)*
Biquinazomycin	$C_{50}H_{64}N_{12}O_{14}S_2$	>300		*Streptomyces (4, 59)*

Acknowledgement. I wish to thank Dr. ROBERT P. BORRIS, Merck Sharp & Dohme Research Laboratories, Rahway, New Jersey, U.S.A., for his help in the preparation of this manuscript.

References

1. ABLONDI, F., S. GORDON, J. MORTON II, and J. H. WILLIAMS: An Antimalarial Alkaloid from *Hydrangea.* II. Isolation. J. Org. Chem. **17,** 14 (1952).
2. AMIN, A. H., and D. R. MEHTA: A Bronchodilator Alkaloid (Vasicinone) from *Adhatoda vasica* Nees. Nature **184,** 1317 (1959).
3. AMIN, A. H., D. R. MEHTA, and S. S. SAMARTH: Biological Activity in the Quinazolone Series. Progr. Drug Res. **14,** 218 (1970).
4. ARIF, A. J., CH. SINGH, A. P. BHADURI, C. M. GUPTA, A. W. KHAN, and M. M. DHAR: Actinomycetes Studies: Part II — Cell-free Synthesis of Echinomycin & an Echinomycin Analogue. Indian J. Biochem. **7,** 193 (1970).
5. ARMAREGO, W. L. F.: Quinazolines. Adv. Heterocyclic Chem. **24,** 1 (1979).
6. — In: The Chemistry of Heterocyclic Compounds Vol. 24/I: Quinazolines. New York-London-Sydney: Interscience Publishers Ltd. 1967.
7. ARMAREGO, W. L. F., and PH. A. REECE: Quinazolines. Part XXI. Synthesis of *cis*-2-Amino-8a-carboxymethyl-3,4,4a,5,6,7,8,8a-octahydroquinazoline and Related Compounds. Conversions of Perhydroquinazolin-2-ones into 2-Amino-3,4,4a,5,6,7,8,8a-octahydroquinazolines. J. Chem. Soc. Perkin Trans. I **1975,** 1470.
8. ARMAREGO, W. L. F., and P. G. TUCKER: Quinazolines. XXIV. The Synthesis of 2,4-Diazatetracyclo[7,3,1,1^{7,11},O^{1,6}]tetradecanes (Adamantano[1,2-d]pyrimidines). Aust. J. Chem. **32,** 1805 (1979), and earlier papers.
9. ARNDT, R. R., S. H. EGGERS, and A. JORDAAN: The Alkaloids of *Anisotes sessiliflorus* C. B. Cl. (Acanthaceae) — Five New 4-Quinazolone Alkaloids. Tetrahedron **23,** 3521 (1967).
10. ASAHINA, Y., and K. KASHIWAKI: Chemical Constituents of the Fruits of *Evodia rutaecarpa.* J. Pharm. Soc. Japan No. 405, 1293 (1915).
11. ASAHINA, Y., and T. OHTA: Eine Synthese des Evodiamins. Ber. dtsch. Chem. Ges. **61,** 319; 869 (1928).
12. ATAL, C. K.: Chemistry and Pharmacology of Vasicine — A New Oxytocic and Abortifacient. New Delhi: Ray Bandhu Industrial Co. 1980.
12a. AYAFOR, J. F., B. L. SONDENGAM, and B. T. NGADJUI: Quinoline and Indolopyridoquinazoline Alkaloids from *Vepris louisii.* Phytochemistry **21,** 2733 (1982).
13. BAKER, B. R., F. J. MC EVOY, R. E. SCHAUB, J. P. JOSEPH, and J. H. WILLIAMS: An Antimalarial Alkaloid from *Hydrangea.* XXI. Synthesis and Structure of Febrifugine and Isofebrifugine. J. Org. Chem. **18,** 178 (1953).
13a. BAKER, B. R., and F. J. MC EVOY: An Antimalarial Alkaloid from *Hydrangea.* XXIII. Synthesis by the Pyridine Approach. II. J. Org. Chem. **20,** 136 (1955).
14. BARRINGER JR., D. F., G. BERKELHAMMER, S. D. CARTER, L. GOLDMAN, and A. E. LANZILOTTI: The Stereochemistry of Febrifugine. I. The Equilibrium between *cis*- and *trans*-(3-Substituted 2-piperidyl)-2-propanones. J. Org. Chem. **38,** 1933 (1973).
15. BARRINGER JR., D. F., G. BERKELHAMMER, and R. S. WAYNE: The Stereochemistry of Febrifugine. II. Evidence for the Trans Configuration in the Piperidine Ring. J. Org. Chem. **38,** 1937 (1973).
16. BATSUREN, D., M. V. TELEZHENETSKAYA, S. YU. YUNUSOV, and T. BALDAN: Alkaloids of *Peganum nigellastrum.* Khim. Prir. Soedin. **1978,** 418 (Russ).
16a. BERGMAN, J.: The Quinazolinocarboline Alkaloids. In: (*133*); Vol. XXI, p. 30 (1983).
17. BERGMAN, J., and S. BERGMAN: Synthesis of Rutaecarpine and Related Indole Alkaloids. Heterocycles **16,** 347 (1981).

18. BERGMAN, J., B. EGESTAD, and J.-O. LINDSTRÖM: The Structure of Some Indolic Constituents in *Couroupita guaianensis* Aubl. Tetrahedron Lett. **1977**, 2625.
19. BHATNAGAR, A. K., S. BHATTACHARJI, and S. P. POPLI: On the Identity of Vasicinol. Indian J. Chem. **3**, 525 (1965).
20. BHATNAGAR, A. K., and S. P. POPLI: Mass Fragmentation of the Alkaloids of *Adhatoda vasica* Nees. Indian J. Chem. **4**, 291 (1966).
21. BHATTACHARYYA, J., and S. C. PAKRASHI: Carbon-13 NMR Analysis of Some 4-Quinazolinone Alkaloids and Related Compounds. Heterocycles **14**, 1469 (1980).
22. — — The Identity of Glycophymine and Glycosminine: an Alkaloid of *Glycosmis arborea* (Roxb.) DC. Heterocycles **12**, 929 (1979).
23. BHATTACHARYYA, P., M. SAKAR, T. ROYCHOWDHURY, and D. P. CHAKRABORTY: New Synthesis of the Quinazolone Alkaloids Arborine and Glycorine. Chem. and Ind. **1978**, 532.
24. BIRD, C. W.: The Structure of Methylisatoid. Tetrahedron **19**, 901 (1963).
25. BOIT, H. G.: Chinazolin-Alkaloide. In: Fortschritte der Alkaloid-Chemie seit 1933, p. 331, Berlin: Akademie-Verlag 1950; Ergebnisse der Alkaloid-Chemie bis 1960, p. 741, Berlin: Akademie-Verlag 1961.
26. BOWEN, I. H., K. P. W. C. PERERA, and J. R. LEWIS: Alkaloids of the Leaves of *Glycosmis bilocularis*. Phytochemistry **17**, 2125 (1978).
27. BRUFANI, M., W. FEDELI, F. MAZZA, A. GERHARD, and W. KELLER-SCHIERLEIN: The Structure of Tryptanthrin. Experientia **27**, 1249 (1971).
28. BUDĚSINSKY, Z., P. LEDERER, and J. DANEK: 3-(3-Amino-2-hydroxypropyl)- and 3-(3-Aminoacetoxy)-4(3H)-quinazolines. Collect. Czech. Chem. Comm. **42**, 3473 (1977).
29. BUDZIKIEWICZ, H., C. DJERASSI, and D. H. WILLIAMS: Structure Elucidation of Natural Products by Mass Spectrometry, Vol. I, p. 212. San Francisco: Holden-Day, Inc. 1964.
29a. BÜCHI, G., P. R. DE SHONG, S. KATSUMURA, and Y. SUGIMURA: Total Synthesis of Tryptoquivaline G. J. Amer. Chem. Soc. **101**, 5084 (1979).
30. BÜCHI, G., K. CH. LUK, B. KOLBE, and J. M. TOWNSEND: Four New Mycotoxins of *Aspergillus clavatus* Related to Tryptoquivaline. J. Org. Chem. **42**, 244 (1977).
31. CAMBRIDGE, G. W., A. B. A. JANSEN, and D. A. JARMAN: Bronchodilating Action of Vasicinone and Related Compounds. Nature **196**, 1217 (1962).
32. CANONICA, L., B. DANIELI, P. MANITTO, and G. RUSSO: New Quinazolinocarboline Alkaloids from *Euxylophora paraensis* Hub. Tetrahedron Lett. **1968**, 4865.
33. CHAKRAVARTI, D., and R. N. CHAKRAVARTI: Quinazolone Alkaloids. J. Proc. Inst. Chemists (India) **39** (Part III), 131 (1967).
34. CHAKRAVARTI, D., R. N. CHAKRAVARTI, L. A. COHEN, B. DAS GUPTA, S. DUTTA, and H. K. MILLER: Alkaloids of *Glycosmis arborea* — II. Structure of Arborine. Tetrahedron **16**, 224 (1961).
35. CHAKRAVARTI, D., R. N. CHAKRAVARTI, and S. C. CHAKRAVARTI: Arborine and Glycosine. Science and Culture (India) **18**, 553 (1953).
36. — — — Alkaloids of *Glycosmis arborea*. Part I. Isolation of Arborine and Arborinine: The Structure of Arborine. J. Chem. Soc. **1953**, 3337.
37. CHAKRAVARTI, R. N.: Chemistry of Arborine. Bull. Calcutta School Trop. Med. **11**, 37 (1963).
38. CHAKRAVARTI, R. N., and S. C. CHAKRAVARTI: Chemical Investigation of *Glycosmis arborea* Correa. J. Proc. Inst. Chemists (India) **24**, 96 (1952).
39. CHATTERJEE, A., S. BOSE, and C. GHOSH: Rhetsine and Rhetsinine: The Quinazoline Alkaloids of *Xanthoxylum rhetsa*. Tetrahedron **7**, 257 (1959).
39a. CHATTERJEE, A., and J. MITRA: Chemistry of Rhetine and Synthesis of Rhetsine. The Alkaloids of *Zanthoxylum rhetsa*. Science and Culture (India) **25**, 493 (1960).
40. CHATTERJEE, A., and M. GANGULY: Alkaloidal Constituents of *Peganum harmala* and Synthesis of the Minor Alkaloid Deoxyvasicinone. Phytochemistry **7**, 307 (1968).

The Quinazoline Alkaloids 221

41. CHATTERJEE, A., and S. GHOSH MAJUMDAR: Alkaloids of *Glycosmis pentaphylla* (Retz.) DC. Part I. J. Amer. Chem. Soc. **76**, 2459 (1954).

42. — — Constitution and Synthesis of Glycosin, the New Alkaloid of *Glycosmis pentaphylla* Retz. DC. J. Amer. Chem. Soc. **75**, 4365 (1953).

43. CHATTERJEE, M. L., and M. S. DE: Pharmacological Studies with Arborine. Bull. Calcutta School Trop. Med. **8**, 102 (1960).

44. CHIEN, P.-L., and C. C. CHENG: Structural Modification of Febrifugine. Some Methylenedioxy Analogs. J. Med. Chem. **13**, 867 (1970).

45. CHOU, T. Q., F. Y. FU, and Y. S. KAO: Antimalarial Constituents of Chinese Drug, Ch'ang Shan, *Dichroa febrifuga* Lour. J. Amer. Chem. Soc. **70**, 1765 (1948).

46. CLARDY, I., J. P. SPRINGER, G. BÜCHI, K. MATSUO, and R. WIGHTMAN: Tryptoquivaline and Tryptoquivalone, Two Tremorgenic Metabolites of *Aspergillus clavatus*. J. Amer. Chem. Soc. **97**, 663 (1975).

47. CULBERTSON, H., J. C. DECIUS, and B. E. CHRISTENSEN: Quinazolines. XIII. A Study of the Infrared Spectra of Certain Quinazoline Derivatives. J. Amer. Chem. Soc. **74**, 4834 (1952).

48. DANIELI, B., C. FARACHI, and G. PALMISANO: A New Indolopyridoquinazoline in the Bark of *Euxylophora paraensis*. Phytochemistry **15**, 1095 (1976).

49. DANIELI, B., G. LESMA, and G. PALMISANO: A New Tryptophan Derived Alkaloid from *Evodia rutaecarpa* (Juss.) Benth. et Hook. Experientia **35**, 156 (1979).

50. — — — Quinazolinocarboline Alkaloids Chemistry: Reactivity of Euxylophorines — Part I. Heterocycles **12**, 353 (1979).

51. — — — The Configuration of (+)-Evodiamine: a Long-standing Problem in the Chemistry of Indole Alkaloids. J. Chem. Soc. Chem. Commun. **1982**, 1092.

52. DANIELI, B., P. MANITTO, F. RONCHETTI, and G. RUSSO: New Indolopyridoquinazoline Alkaloids from *Euxylophora paraensis*. Phytochemistry **11**, 1833 (1972).

53. DANIELI, B., P. MANITTO, F. RONCHETTI, G. RUSSO, and G. FERRARI: Paraensine, a New Indolopyridoquinazoline Alkaloid from *Euxylophora paraensis* Hub. Experientia **28**, 249 (1972).

54. DANIELI, B., and G. PALMISANO: A New Approach to Quinazolinocarboline Alkaloids: Synthesis of (±)-Evodiamine, Rutaecarpine and Dehydroevodiamine. Heterocycles **9**, 803 (1978).

54a. — — Unusually Simple Methylation at N-13 of Indolo-pyridoquinazoline Alkaloids. Gazz. Chim. Ital. **105**, 45 (1975).

55. DANIELI, B., G. PALMISANO, G. RAINOLDI, and G. RUSSO: 1-Hydroxyrutaecarpine from *Euxylophora paraensis*. Phytochemistry **13**, 1603 (1974).

56. DANIELI, B., G. PALMISANO, G. RUSSO, and G. FERRARI: Minor Indolopyridoquinazoline Alkaloids from *Euxylophora paraensis*. Phytochemistry **12**, 2521 (1973).

56a. DANIELI, B., and G. PALMISANO: Ind-N-Alkylation of Rutecarpine and Synthesis of Two Novel Related Heterocyclic Ring Systems: Indolo[1',2':3,4]pyrazo[1,2-a]quinazoline and Indolo[1',2':3,4][1,4]diazepino[1,2-a]quinazoline. J. Heterocycl. Chem. **14**, 839 (1977).

56b. DANIELI, B., G. LESMA, and G. PALMISANO: Quinazolinocarboline Alkaloids Chemistry: Thermal Rearrangement of 14-Alkylindolo[2',3':3,4]pyrido[2,1-b]quinazolin-5-one. Part II. Heterocycles **12**, 1433 (1979).

56c. DE CORRÊA, D. B., O. R. GOTTLIEB, and A. P. DE PADUA: Chemistry of Brazilian Rutaceae. I. Dihydrocinnamic Acids from *Hortia badinii*. Phytochemistry **14**, 2059 (1975).

56d. DE CORRÊA, D. B., O. R. GOTTLIEB, A. P. DE PADUA, and A. I. DA ROCHA: The Chemistry of Brazilian Rutaceae. II. Constituents of *Hortia longifolia*. Rev. Latinoamer. Quim. **7** (1), 43 (1976) [Chem. Abstr. **84**, 161790g (1976)].

57. DEVI, G., R. S. KAPIL, and S. P. POPLI: Potential CNS & CVS Agents: Syntheses Based on Vasicinone. Indian J. Chem. **14B**, 354 (1976).

222   S. JOHNE:

58. DHAR, K. L., M. P. JAIN, S. K. KOUL, and C. K. ATAL: Vasicol, a New Alkaloid from *Adhatoda vasica*. Phytochemistry **20**, 319 (1981).
59. DHAR, M. M., CH. SINGH, A. W. KHAN, A. J. ARIF, C. M. GUPTA, and A. P. BHADURI: Cell-free Synthesis of Echinomycin and an Echinomycin Analog. Pure Appl. Chem. **28**, 469 (1971).
60. DÖPKE, W.: Ergebnisse der Alkaloid-Chemie 1960—1968. 2 Bände. Berlin: Akademie-Verlag. 1976/1978.
61. DONOVAN, D. G. O., and H. HORAN: The Biosynthesis of Arborine. J. Chem. Soc. (C) **1970**, 2466.
62. DREYER, D. L., and R. C. BRENNER: Alkaloids of Some Mexican *Zanthoxylum* Species. Phytochemistry **19**, 935 (1980).
63. ELGUERO, J., C. MARZIN, A. R. KATRITZKY, and P. LINDA: In: The Tautomerism of Heterocycles. Advances in Heterocyclic Chemistry (A. R. KATRITZKY and A. J. BOULTON, eds.), Suppl. 1, p. 129. New York-San Francisco-London: Academic Press. 1976.
64. FEDELI, W., and F. MAZZA: Crystal Structure of Tryptanthrin (Indolo[2,1-b]quinazoline-6,12-dione). J. Chem. Soc. Perkin Trans. II **1974**, 1621.
65. FIEDLER, E., H. P. FIEDLER, A. GERHARD, W. KELLER-SCHIERLEIN, W. A. KÖNIG, and H. ZÄHNER: Stoffwechselprodukte von Mikroorganismen. 156. Mitteilung. Synthese und Biosynthese substituierter Tryptanthrine. Arch. Mikrobiol. **107**, 249 (1976).
65a. FISH, F., I. A. MESHAL, and P. G. WATERMAN: Minor Alkaloids of *Araliopsis tabouensis*. Planta Med. **29**, 310 (1976).
66. FISHMAN, M., and PH. A. CRUICKSHANK: Febrifugine Antimalarial Agents. I. Pyridine Analogs of Febrifugine. J. Med. Chem. **13**, 155 (1970).
67. FITZGERALD, J. S., S. R. JOHNS, J. A. LAMBERTON, and A. H. REDCLIFFE: 6,7,8,9-Tetrahydropyridoquinazolines, a New Class of Alkaloids from *Mackinlaya* Species (Araliaceae). Aust. J. Chem. **19**, 151 (1966).
68. FRIEDLÄNDER, P., and W. ROSCHDESTWENSKY: Über ein Oxydationsprodukt des Indigblaues. Ber. dtsch. Chem. Ges. **48**, 1841 (1915).
69. GHOSAL, S., R. B. P. S. CHAUHAN, and R. MEHTA: Alkaloids of *Sida cordifolia*. Phytochemistry **14**, 830 (1975).
70. GOPINATH, K. W., T. R. GOVINDACHARI, and U. RAMADAS RAO: The Alkaloids of *Zanthoxylum rhetsa* DC. Tetrahedron **8**, 293 (1960).
71. GOTO, T., Y. KISHI, S. TAKAHASHI, and Y. HIRATA: Tetrodotoxin. Tetrahedron **21**, 2059 (1965), and references cited therein.
72. GRÖGER, D.: Alkaloids Derived from Tryptophan and Anthranilic Acid. In: Encyclopedia of Plant Physiology, New Series, Vol. 8 (BELL, E. A., V. CHARLWOOD, eds.), p. 128. Berlin-Heidelberg: Springer-Verlag. 1980.
73. — Anthranilic Acid as Precursor of Alkaloids. Lloydia **32**, 221 (1969).
74. — Chinazolinalkaloide. In: Biosynthese der Alkaloide (MOTHES, K., H. R. SCHÜTTE, eds.), p. 551. Berlin: VEB Deutscher Verlag der Wissenschaften. 1969.
75. JOHNE, S., and D. GRÖGER: Alkaloide in *Linaria*-Arten. Pharmazie **23**, 35 (1968).
76. GRÖGER, D., and S. JOHNE: Zur Analytik und Biochemie der Alkaloide von *Adhatoda vasica* Nees. In: Festschrift K. Mothes, p. 205. Jena: G. Fischer Verlag. 1965.
77. — — Zur Biosynthese einiger Alkaloide von *Glycosmis arborea* (Rutaceae). Z. Naturforsch. **23b**, 1072 (1968).
78. GRUNDON, M. F.: Quinazoline Alkaloids. Alkaloids **6**, 108 (1976); **8**, 83 (1978); **9**, 85 (1979); **10**, 80 (1980); **12**, 88 (1982) and in other chapters of this periodical.
79. GUNATILAKA, A. A. L., S. SOTHEESWARAN, S. BALASUBRAMANIAM, A. I. CHANDRASEKARA, and H. T. B. SRIYANI: Studies on Medicinal Plants of Sri Lanka. III: Pharmacologically Important Alkaloids of Some *Sida* Species. Planta Med. **39**, 66 (1980).
80. GUPTA, O. P., K. K. ANAND, B. J. R. GHATAK, and C. K. ATAL: Vasicine, Alkaloid of

Adhatoda vasica, a Promising Uterotonic Abortifacient. Indian J. Exp. Biol. **16,** 1075 (1978).

81. HAGIWARA, Y., M. KURIHARA, and N. YODA: Intramolecular Rearrangement — IV. Intramolecular Alkyl Rearrangements and Tautomerism of Quinazolinone Derivatives. Tetrahedron **25,** 783 (1969).

82. HART, N. K., S. R. JOHNS, and J. A. LAMBERTON: The Identification of a Minor Alkaloid of *Mackinlaya macrosciadia* (Araliaceae) as Deoxyvasicinone. Aust. J. Chem. **24,** 223 (1971).

83. HEARN, J. M., R. A. MORTON, and J. C. E. SIMPSON: Ultra-violet Absorption Spectra of Some Derivatives of Quinoline, Quinazoline, and Cinnoline. J. Chem. Soc. **1951,** 3318.

84. HIKINO, H., S. NABETANI, and T. TAKEMOTO: Structure and Biosynthesis of Chrysogine, a Metabolite of *Penicillium chrysogenum.* Yakugaku Zasshi **93,** 619 (1973). (Japan) [Chem. Abstr. **79,** 40 922q (1973)].

85. HILL, R. K., and A. G. EDWARDS: The Absolute Configuration of Febrifugine. Chem. and Ind. **1962,** 858.

86. HOLUBEK, J., and O. STROUF (eds.): Spectral Data and Physical Constants of Alkaloids, Vol. I. Prag: Publishing House of the Czechoslovak Academy of Sciences. 1965.

87. HOOPER, D.: Blätter von *Adhatoda Vasica,* Nees. Phar. J. **18,** 841 (1888).

88. HORVATH-DORA, K., and O. CLAUDER: Alkaloids Containing the Indolo[2,3-c]quinazolino[3,2-a]pyridine Skeleton. III. 3,14-Dihydrorutaecarpine. Acta Chim. Acad. Hung. **84,** 93 (1975).

89. HUQ, E. M., M. IKRAM, and S. A. WARSI: Chemical Composition of *Adhatoda vasica.* II. Pakistan J. Sci. Ind. Res. **10,** 224 (1967) [Chem. Abstr. **69,** 16 807d (1968)].

90. JAIN, M. P., and V. K. SHARMA: Phytochemical Investigation of Roots of *Adhatoda vasica.* Planta Med. **46,** 250 (1982).

91. JANG, C. S., F. Y. FU, C. Y. WANG, K. C. HUANG, G. LU, and T. C. CHOU: Ch'ang Shan, a Chinese Antimalarial Herb. Science **103,** 59 (1946).

92. JARRAH, M. Y., and V. THALLER: 300 MHz ^{1}H N.m.r. Spectra of Indolo[2,1-b]quinazoline-6,12-dione, Tryptanthrine, and its 2- and 8-Chloro-derivatives. J. Chem. Res. (S) **1980,** 186; J. Chem. Res. (M) **1980,** 2601.

93. JOHNE, S.: Search for Pharmaceutically Interesting Quinazoline Derivatives: Efforts and Results (1969—1980). Progr. Drug Res. **26,** 259 (1982).

94. JOHNE, S., and D. GRÖGER: Natürlich vorkommende Chinazolin-Derivate. Pharmazie **25,** 22 (1970).

95. JOHNE, S., D. GRÖGER, and M. HESSE: Neue Alkaloide aus *Adhatoda vasica* Nees. Helv. Chim. Acta **54,** 826 (1971).

96. JOHNE, S., B. JUNG, D. GRÖGER, and R. RADEGLIA: Synthese und ^{13}C-NMR-Spektroskopie einiger Pyrrolo[2,1-b]chinazoline. J. Prakt. Chem. **319,** 919 (1977).

97. JOHNE, S., K. WAIBLINGER, and D. GRÖGER: Untersuchungen zur Biosynthese des Chinazolinalkaloids Peganin in *Adhatoda vasica* Nees. Pharmazie **28,** 403 (1973).

98. — — — Zur Biosynthese des Chinazolin-Alkaloids Arborin. Eur. J. Biochem. **15,** 415 (1970).

99. JOHNS, S. R., and J. A. LAMBERTON: Alkaloids of *Mackinlaya* Species (Family Araliaceae). J. Chem. Soc. Chem. Commun. **1965,** 267.

100. KAMETANI, T.: Evodiamine. Japan. Kokai 77 77,098 [Chem. Abstr. **87,** 201 858f (1977)].

101. — Rutecarpine. Japan. Kokai 78 77,100 [Chem. Abstr. **89,** 163 836d (1978)].

102. KAMETANI, T., T. HIGA, K. FUKUMOTO, and M. KOIZUMI: One-Step Synthesis of Evodiamine and Rutecarpine. Heterocycles **4,** 23 (1976).

103. KAMETANI, T., C. V. LOC, T. HIGA, M. IHARA, and K. FUKUMOTO: Studies on the Syntheses of Heterocyclic Compounds. Part 724. Total Syntheses of the Quinazolinone Alkaloids Glycorine, Glomerine, Homoglomerine, Crysogine, and Euxylophoricines A and C. J. Chem. Soc. Perkin Trans. I **1977,** 2347.

104. KAMETANI, T., C. V. LOC, T. HIGA, M. KOIZUMI, M. IHARA, and K. FUKUMOTO: Iminoketene Cycloaddition. 2. Total Syntheses of Arborine, Glycosminine, and Rutecarpine by Condensation of Iminoketene with Amides. J. Amer. Chem. Soc. **99**, 2306 (1977).

105. — — — — — — Simple Synthesis of Quinazolone Alkaloids, Arborine and Rutecarpine Through Iminoketene. Heterocycles **4**, 1487 (1976).

106. KAMETANI, T., C. V. LOC, T. HIGA, T. OHSAWA, M. KOIZUMI, M. IHARA, and K. FUKUMOTO: The Synthesis of Quinazolinone Alkaloids Through Potential Iminoketene Intermediate. Heterocycles **12**, 208 (1979).

107. KAMETANI, T., C. V. LOC, M. IHARA, and K. FUKUMOTO: Modified Total Synthesis of Arborine, Glycosminine, and Glomerine by Condensation of Sulfinamide Anhydride with Thioamides. Heterocycles **9**, 1585 (1978).

108. KAMETANI, T., T. HIGA, C. V. LOC, M. IHARA, M. KOIZUMI, and K. FUKUMOTO: Iminoketene Cycloaddition. 1. A Facile Synthesis of Quinazolone System by Condensation of Iminoketene with Imines — A Total Synthesis of Evodiamine and Rutaecarpine by Retro Mass-Spectral Synthesis. J. Amer. Chem. Soc. **98**, 6186 (1976).

109. KAMETANI, T., T. OHSAWA, M. IHARA, and K. FUKUMOTO: Studies on the Syntheses of Heterocyclic Compounds. DCCLV. Iminoketene Cycloaddition. (4). Alternative Syntheses of 5,6,7,8-Tetrahydro-2,3-dimethoxy-8-oxoisoquinolo[1,2-b]quinazoline and Rutecarpine. Chem. Pharm. Bull. **26**, 1922 (1978).

110. NAKAGAWA, M., M. TANIGUCHI, M. SODEOKA, M. ITO, K. YAMAGUCHI, and T. HINO: Total Synthesis of (+)- and (−)-Tryptoquivaline G by Biomimetic Double Cyclization. J. Amer. Chem. Soc. **105**, 3709 (1983).

111. KAMIKADO, T., S. MURAKOSHI, and S. TAMURA: Structure Elucidation and Synthesis of Alkaloids Isolated from Fruits of *Evodia rutaecarpa*. Agric. Biol. Chem. **42**, 1515 (1978).

112. KARIMOV, A., M. V. TELEZHENETSKAYA, and S. YU. YUNUSOV: The Synthetic Analogs of the *Peganum* Alkaloids. I. Synthesis of the Methoxy- and Oxysubstituted Deoxyvasicinones and Deoxypeganines. Khim. Prir. Soedin. **1982**, 498 (Russ).

113. KHAN, A. W., A. P. BHADURI, C. M. GUPTA, and M. M. DHAR: Actinomycetes Studies: Part I — Microbiological Synthesis of Quinazomycin, an Echinomycin Analogue Containing One Quinazol-4-one-3-acetyl Residue. Indian J. Biochem. **6**, 220 (1969).

114. KHASHIMOV, KH. N., M. V. TELEZHENETSKAYA, YA. V. RASHKES, and S. YU. YUNUSOV: Pegamine, a New Alkaloid from *Peganum harmala*. Khim. Prir. Soedin. **1970**, 453 (Russ).

115. KHASHIMOV, KH. N., M. V. TELEZHENETSKAYA, N. N. SHARAKHIMOV, and S. YU. YUNUSOV: Dynamics of the Accumulation of Alkaloids in *Peganum harmala*. Khim. Prir. Soedin. **1971**, 382 (Russ).

116. KHASHIMOV, KH. N., M. V. TELEZHENETSKAYA, and S. YU. YUNUSOV: Peganidine, a New Base from *Peganum harmala*. Khim. Prir. Soedin. **1969**, 599 (Russ).

117. KIM, Y. H., G. B. BROWN, H. S. MOSHER, and F. A. FUHRMAN: Tetrodotoxin: Occurrence in Atelopid Frogs of Costa Rica. Science **189**, 151 (1975) and references cited therein.

118. KISHI, Y., T. FUKUYAMA, M. ARATANI, F. NAKATSUBO, T. GOTO, S. INOUE, H. TANINO, S. SUGIURA, and H. KAKOI: Synthetic Studies on Tetrodotoxin and Related Compounds. IV. Stereospecific Total Syntheses of DL-Tetrodotoxin. J. Amer. Chem. Soc. **94**, 9219 (1972) and earlier papers.

119. KÖHLER, H.: Die Prüfung von *Galega*-Arten auf ihren Gehalt an Giftstoffen mit Hilfe biologischer Methoden. I. Die Giftigkeit der Geißraute (*Galega officinalis* L.) für Warmblüter. Biol. Zentralbl. **88**, 165 (1969).

120. KOEPFLI, J. B., J. F. MEAD, and J. A. BROCKMAN: Alkaloids of *Dichroa febrifuga*. I. Isolation and Degradative Studies. J. Amer. Chem. Soc. **71**, 1048 (1949).

121. KORETSKAYA, N. I.: Alkaloids of *Peganum harmala*. I. Isolation of Two New Alkaloids. Zhur. Obshchei Khim. **27**, 3361 (1957) (Russ).

122. KORETSKAYA, N. I., and L. M. UTKIN: Alkaloids of *Peganum harmala.* II. Structure of Two New Alkaloids. Zhur. Obshchei. Khim. **28,** 1087 (1958) (Russ).

123. KUFFNER, F., G. LENNEIS, and H. BAUER: Über die Konstitution eines Nebenalkaloids aus *Adhatoda vasica* Nees. Monatsh. Chem. **91,** 1152 (1960).

124. LAHIRI, P. K., and S. N. PRADHAN: Pharmacological Investigation of Vasicinol — An Alkaloid from *Adhatoda vasica* Nees. Indian J. Exp. Biol. **2,** 219 (1964) and references cited therein.

125. LEETE, E.: Alkaloid Biogenesis. In: Biogenesis of Natural Compounds (P. BERNFELD, ed.), 2nd. edn., p. 953. Oxford etc.: Pergamon Press. 1967.

126. LEONHARD, N. J., and M. J. MARTELL: Laboratory Realization of the Schöpf-Oechler Scheme of Vasicine Synthesis. Tetrahedron Lett. **1960,** 44.

127. LILJEGREN, D. J.: Biosynthesis of Quinazoline Alkaloids of *Peganum harmala.* Phytochemistry **10,** 2661 (1971).

128. — The Biosynthesis of Quinazoline Alkaloids of *Peganum harmala* L. Phytochemistry **7,** 1299 (1968).

129. LUPU, K. G.: Dynamics of the Alkaloid Levels in *Linaria vulgaris* and *L. genistifolia* Grown in Moldavia. Rast. Resur. **9,** 206 (1973) (Russ) [Chem. Abstr. **79,** 15 906 e (1973)].

130. MACHEMER, H.: Die Autoxydation der Metallkomplexe des Indigos. Ber. dtsch. Chem. Ges. **63,** 1341 (1930) and references cited therein.

131. MANN, S.: Chinazolinderivate bei Pseudomonaden. Arch. Mikrobiol. **56,** 324 (1967).

132. — Besonderheiten im Tryptophanstoffwechsel von *Pseudomonas aeruginosa.* Arch. Hygiene Bakteriol. **151,** 474 (1967).

133. MANSKE, R. H. F. (ed.): The Alkaloids. New York: Academic Press.

134. — The Quinazolinocarbolines. In: (*133*); Vol. VIII, p. 55 (1965).

135. MARION, L.: In: (*133*); Vol. II, p. 369 (1952).

136. MEHTA, D. R., J. S. NARAVANE, and R. M. DESAI: Vasicinone. A Bronchodilator Principle from *Adhatoda vasica* Nees. (N. O. Acanthaceae). J. Org. Chem. **28,** 445 (1963).

137. MEINWALD, Y. C., J. MEINWALD, and TH. EISNER: 1,2-Dialkyl-4(3H)-quinazolinones in the Defensive Secretion of a Millipede *(Glomeris marginata).* Science **154,** 390 (1966).

138. MÖHRLE, H., and P. GUNDLACH: Eine neue Synthese für DL-Vasicin. Tetrahedron Lett. **1970,** 3249.

139. MÖHRLE, H., CHR. KAMPER, and R. SCHMID: Eine neue Synthese von Rutaecarpin. Arch. Pharm. **313,** 990 (1980).

140. MÖHRLE, H., and C.-M. SEIDEL: Eine neue Synthese für Arborin und strukturanaloge 4(1H)-Chinazolinone, 1. u. 2. Mitt. Arch. Pharm. **309,** 503; 572 (1976).

141. MORRIS, R. C., W. E. HANFORD, and R. ADAMS: Structure of Vasicine. III. Position of the Hydroxyl Group. J. Amer. Chem. Soc. **57,** 951 (1935).

142. MOSHER, H. S., F. A. FUHRMAN, H. D. BUCHWALD, and H. G. FISCHER: Tarichatoxin — Tetrodotoxin: A Potent Neurotoxin. Science **144,** 1100 (1964).

143. MUKHERJEE, A., and P. K. DEY: Changes in Amino Acid Metabolism in Rat Brain Following Glycosine Administration. Indian J. Exp. Biol. **8,** 263 (1970).

144. MUNOZ, G. G., and R. MADRONERO: Neue Kondensationsreaktionen mit Imidchloriden. Synthese von Pyrido[2,1-b]chinazolinen. Chem. Ber. **95,** 2182 (1962).

145. NAKASOTO, T., S. ASADA, and K. MARUI: Dehydroevodiamine, Main Alkaloid from the Leaves of *Evodia rutaecarpa.* Yakugaku Zasshi **82,** 619 (1962) [Chem. Abstr. **58,** 3470 b (1963)].

145a. NARITA, H. *et al.*: Bull. Japan Soc. Sci. Fisheries **47,** 935 (1981).

145b. NOGUCHI, T. *et al.*: Bull. Japan Soc. Sci. Fisheries **47,** 909 (1981).

145c. OHNUMA, T., Y. KIMURA, and Y. BAN: Synthetic Studies on Oxindole Spiro-Lactones with Thallium (III) Trinitrate: Formal Synthesis of (±)-Tryptoquivaline G. Tetrahedron Lett. **22,** 4969 (1981).

226 S. JOHNE:

146. ONAKA, T.: A General Three-Step Synthesis of Pyrrolidino[2,1-b]quinazolone Alkaloids via Biogenetically Patterned Path. Tetrahedron Lett. 1971, 4387.
147. OPENSHAW, H. T.: The Quinazoline Alkaloids. In: (133); Vol. III, p. 101 (1953); Vol. VII, p. 247 (1960).
148. ORIPOV, E., L. M. YUN, KH. M. SHAKHIDOYATOV, and CH. SH. KADYROV: Some Reactions of α-Hydroxy- and α-Dimethylaminoformylidene-2,3-polymethylene-3,4-dihydro-4-quinazolinones. Khim. Prir. Soedin. 1978, 603 (Russ).
149. PACHTER, I. J., R. J. MOHRBACHER, and D. E. ZACHARIAS: The Chemistry of Hortiamine and 6-Methoxyrhetsinine. J. Amer. Chem. Soc. 83, 635 (1961).
150. PACHTER, I. J., R. F. RAFFAUF, G. E. ULLYOT, and O. RIBEIRO: Die Trennung und Identifizierung der Alkaloide von Hortia arborea. Angew. Chem. 69, 687 (1957).
151. — — — — The Alkaloids of Hortia arborea Engl. J. Amer. Chem. Soc. 82, 5187 (1960).
152. PACHTER, I. J., and G. SULD: The Structure and Synthesis of Rhetsinine. J. Org. Chem. 25, 1680 (1960).
153. PAKRASHI, S. C., and J. BHATTACHARYYA: Indian Medicinal Plants. IV. Further Alkaloids from Glycosmis arborea. J. Sci. Ind. Res. (India) 21B, (1), 49 (1962).
154. — — Studies on Indian Medicinal Plants. XIV. Interrelationships among the Quinazoline Alkaloids from Glycosmis arborea (Roxb.) DC. Tetrahedron 24, 1 (1968).
155. — — Recent Advances in the Chemistry of Rutaceae Alkaloids. J. Sci. Ind. Res. (India) 24, 293 (1963).
156. — — Studies on Indian Medicinal Plants. Part V: Isolation of Minor Alkaloids from Glycosmis arborea (Roxb.) DC. Ann. Biochem. Exptl. Med. 23, 123 (1963).
157. PAKRASHI, S. C., J. BHATTACHARYYA, L. F. JOHNSON, and H. BUDZIKIEWICZ: Studies of Indian Medicinal Plants — VI. Structures of Glycosmicine, Glycorine and Glycosminine, the Minor Alkaloids from Glycosmis arborea (Roxb.) DC. Tetrahedron 19, 1011 (1963).
158. PAKRASHI, S. C., A. DE, and S. CHATTOPADHYAY: Studies on 4-Quinazolinones: Part I — Convenient Synthesis of Glycosminine and Arborine, the Alkaloids from Glycosmis arborea and Related 4-Quinazolinones. Indian J. Chem. 6, 472 (1968).
159. PALAZZO, S., L. I. GIANNOLA, and S. CARONNA: Reaction of Formation of 2-Substituted 4-Quinazolinones in Polyphosphoric Acid: Synthesis of Glycosminine. Atti Accad. Sci. Lett. Arti Palermo, Parte 1, 34, 339 (1976) [Chem. Abstr. 89, 43309n (1978)].
160. PALLARES, E. S.: The Structure of the Water-Insoluble Pigment of the Bark of "Sangre de Drago". Arch. Biochem. 10, 235 (1946).
161. PETERSEN, S., and E. TIETZE: Die Reaktion cyclischer Lactimäther mit Amino-carbonsäuren. Justus Liebigs Ann. Chem. 623, 166 (1959).
162. PLUGAR, V. N., T. T. GOROVITS, N. TULYAGANOV, and YA. V. RASHKES: Transformations of 4-Quinazolinone Group Alkaloids in Animals. Khim. Prir. Soedin. 1977, 250 (Russ).
163. PLEKHANOVA, N. V., and S. T. AKTANOVA: Alkaloids of Peganum harmala. Issled. Flory Kirgizii na Alkaloidonosnost, Akad. Nauk Kirg. SSR, Inst. Organ. Khim. 1965, 57 (Russ) [Chem. Abstr. 64, 11550f (1966)].
164. PLEKHANOVA, N. V., and G. P. SHEVELEVA: Alkaloids of Linaria transiliensis and Linaria vulgariformis. Issled. Flory Kirgizii na Alkaloidonosnost, Akad. Nauk Kirg. SSR, Inst. Organ. Khim. 1965, 54 (Russ) [Chem. Abstr. 64, 11550e (1966)].
165. PLUGAR, V. N., YA. V. RASHKES, and KH. M. SCHAKHIDOYATOV: Mass Spectra of 2,3-Polymethylene-3,4-dihydro-4-quinazolones with Substituents at C_9. Khim. Prir. Soedin. 1979, 180 (Russ).
166. — — — Comparison of the Mass Spectra of 2,3-Polymethylene-1,2,3,4-tetrahydro-quinazol-4-ones. Khim. Prir. Soedin. 1978, 414 (Russ).

167. PLUGAR, V. N., YA. V. RASHKES, and N. TULYAGANOV: Quantitative Analysis of the Components of Desoxypeganine and Desoxyvasicinone Metabolites. Khim. Prir. Soedin. 1981, 201 (Russ).

168. PRAKASH, A., R. K. VARMA, and S. GHOSAL: Alkaloidal Constituents of *Sida acuta, S. humilis, S. rhombifolia,* and *S. spinosa.* Planta Med. 43, 384 (1981).

169. PRICE, J. R.: Quinazoline Alkaloids. Fortschr. Chem. org. Naturstoffe 13, 330 (1956).

170. RAJAGOPALAN, T. R., S. BHATTACHARJI, and M. L. DHAR: Proceedings, Symposium on Drugs and Antibiotics (Defence Research Laboratory, Kanpur), 1961, 121.

171. RASHKES, YA. V., M. V. TELEZHENETSKAYA, V. N. PLUGAR, and S. YU. YUNUSOV: Mass Spectra of Tetrahydroquinazoline and Tetrahydroquinazolin-4-one Derivatives. Khim. Prir. Soedin. 1977, 378 (Russ).

172. RHFF, R. P., and J. D. WHITE: Synthesis of Cyclopenin and Glycosminine from Phenylpyruvic Acid. J. Org. Chem. 42, 3650 (1977).

173. ROBINSON, R.: The Structural Relations of Natural Products. Oxford: Clarendon Press. 1955.

174. ROSENFELD, A. D., and D. G. KOLESNIKOV: Über l-Peganin aus Blüten und Stengeln von *Peganum harmala* L., Bemerkungen zu den Arbeiten von E. Späth über Peganin. Ber. dtsch. Chem. Ges. 69, 2022 (1936) and references cited therein.

175. SAKAMOTO, S., and K. SAMEJIMA: Preparation of Specific Antibody to 2,3-Trimethylene-4-quinazolone for the Immunoassay of Δ^1-Pyrroline. Chem. Pharm. Bull. 28, 916 (1980).

176. — — Determination of Δ^1-Pyrroline as 2,3-Trimethylene-4-quinazolone. Chem. Pharm. Bull. 27, 2220 (1979).

177. SARKAR, M., and D. P. CHAKRABORTY: Some Minor Constituents from *Glycosmis pentaphylla.* Phytochemistry 16, 2007 (1977).

177a. — — Glycophymoline, a New Minor Quinazoline Alkaloid from *Glycosmis pentaphylla.* Phytochemistry 18, 694 (1979).

178. SCHÄFER, J., and M. STEIN: Untersuchungen über toxische Inhaltsstoffe bei *Galega officinalis* L. Biol. Zentralbl. 88, 755 (1969).

179. — — Über die Variabilität von Inhaltsstoffen in der Geißraute (*Galega officinalis* L.). Naturwissenschaften 54, 205 (1967).

180. SCHILDKNECHT, H., and W. F. WENNEIS: Über Arthropoden-Abwehrstoffe XXV. Anthranilsäure als Precursor der Arthropoden-Alkaloide Glomerin und Homoglomerin. Tetrahedron Lett. 1967, 1815.

181. — — Über Arthropoden-(Insekten-)Abwehrstoffe XX. Strukturaufklärung des Glomerins. Z. Naturforsch. 21b, 552 (1966).

182. SCHILDKNECHT, H., W. F. WENNEIS, K. H. WEIS, and U. MASCHWITZ: Glomerin, ein neues Arthropoden-Alkaloid. Z. Naturforsch. 21b, 121 (1966).

183. SCHINDLER, F., and H. ZÄHNER: Tryptanthrin, ein von Tryptophan abzuleitendes Antibioticum aus *Candida lipolytica.* Arch. Mikrobiol. 79, 187 (1971).

184. SCHÖPF, C., A. KOMZAK, F. BRAUN, and E. JACOBI: Über die Polymeren des Δ^1-Piperideins. Justus Liebigs Ann. Chem. 559, 1 (1948).

185. SCHREIBER, K., O. AURICH, and K. PUFAHL: Isolierung von (+)-Peganin aus der Geißraute, *Galega officinalis* L. Arch. Pharm. 295, 271 (1962).

186. SEN, A. K., S. B. MAHATO, and N. L. DUTTA: Couroupitine A, a New Alkaloid from *Couroupita guianensis.* Tetrahedron Lett. 1974, 609.

187. SHAKHIDOYATOV, KH., A. IRISHBAEVA, and CH. SH. KADYROV: Synthesis of Desoxypeganine and its Derivatives. Khim. Prir. Soedin. 1974, 681 (Russ).

188. SHAKHIDOYATOV, KH., A. IRISHBAEVA, E. ORIPOV, and CH. SH. KADYROV: Quinazolines. VII. Synthesis of 6-Nitro-, Amino-, and 9,9-Dibromodesoxyvasicinone. Khim. Prir. Soedin. 1976, 557 (Russ).

189. SHAKHIDOYATOV, KH. M., and CH. SH. KADYROV: Quinazolines. XII. 4-Thio Analogs of Desoxyvasicinone, its Derivatives and Homologs. Khim. Prir. Soedin. 1977, 668 (Russ).

190. SHAKHIDOYATOV, KH. M., and CH. SH. KADYROV: Quinazolines. X. Synthesis of Methyl-enebis(6,6'-desoxyvasicinone) and its Homologs. Khim. Prir. Soedin. **1977,** 544 (Russ).

191. SHAKHIDOYATOV, KH. M., YA. YAMANKULOV, and CH. SH. KADYROV: Quinazolines. XI. Condensation of Desoxyvasicinone with Aldehydes. Khim. Prir. Soedin. **1977,** 552 (Russ).

192. SHARMA, R. L., R. K. GUPTA, B. K. CHOWDHURY, K. L. DHAR, and C. K. ATAL: Syntheses of Some Vasicine Analogues. Indian J. Chem. **18 B,** 449 (1979).

193. SHARMA, V. K., and M. P. JAIN: Synthesis of Vasicinone Derivatives. Indian J. Chem. **21 B,** 75 (1982).

194. SHEUMACK, D. D., M. E. H. HOWDEN, I. SPENCE, and R. J. QUINN: Maculotoxin: A Neurotoxin from the Venom Glands of the Octopus *Hapalochlaena maculosa* Identified as Tetrodotoxin. Science **199,** 188 (1978).

195. SIDDIQUI, S.: A Reinvestigation of the Alkaloidal Constituents of *Peganum harmala.* Pakistan J. Sci. Ind. Res. **5,** 207 (1962) [Chem. Abstr. **59,** 5213 (1963)].

196. SNIECKUS, V. A.: Quinazoline Alkaloids. Alkaloids **2,** 91 (1972); **3,** 112 (1973); **4,** 124 (1974); **5,** 108 (1975).

197. SOUTHWICK, P. L., and L. CASANOVA: A New Synthesis of dl-Vasicine and a Methoxy Analog. J. Amer. Chem. Soc. **80,** 1168 (1958).

198. SPÄTH, E.: Über das Peganin (Vasicin). Monatsh. Chem. **72,** 115 (1939).

199. SPÄTH, E., and F. KESZTLER-GANDINI: Über ein Nebenalkaloid aus *Adhatoda vasica* Nees. Monatsh. Chem. **91,** 1150 (1960).

200. SPÄTH, E., F. KUFFNER, and N. PLATZER: Synthese und Konstitution des Peganins (Vasicins). Ber. dtsch. Chem. Ges. **68,** 699 (1935).

201. SPÄTH, E., and N. PLATZER: Eine neue Synthese von Pegen-(9) und Peganin (IX. Mitteil. über Peganin). Ber. dtsch. Chem. Ges. **69,** 255 (1936).

202. — — Über Derivate des Peganins und ihre Ring-Homologen (VIII. Mitteil. über Peganin). Ber. dtsch. Chem. Ges. **68,** 2221 (1935).

203. STEPHEN, T., and H. STEPHEN: The Beckmann Rearrangement of cyclo Pentanone Oxime. J. Chem. Soc. **1956,** 4694.

204. STRONG, P. N., and J. F. W. KEANA: Modification of Tetrodotoxin With Succinic Anhydride. Bioorg. Chem. **5,** 255 (1976).

205. SZULZEWSKY, K., E. HÖHNE, S. JOHNE, and D. GRÖGER: Bestimmung der Molekül- und Kristallstruktur sowie der Absolutkonfiguration von Peganin. J. Prakt. Chem. **318,** 463 (1976).

206. TAKEDA CHEMICAL INDUSTRIES, LTD.: Tryptanthrin. Jpn. Kokai Tokkyo Koho 80 47,684 (1980) [Chem. Abstr. **93,** 186 400 d (1980)].

207. TELEZHENETSKAYA, M. V., KH. N. KASHIMOV, and S. YU. YUNUSOV: Peganol, a New Alkaloid from *Peganum harmala.* Khim. Prir. Soedin. **1971,** 849 (Russ).

208. TELEZHENETSKAYA, M. V., and S. YU. YUNUSOV: Alkaloids of *Peganum harmala.* Khim. Prir. Soedin. **1977,** 731 (Russ).

208a. TOTH, G., K. HORVÁTH-DORA, O. CLAUDER, and H. DUDDECK: Alkaloide mit Indolo[2',3':3,4]pyrido[2,1-b]chinazolin-Struktur. VII. Synthese und Untersuchungen des *cis-* und *trans*-Hexahydrorutaecarpins. Justus Liebigs Ann. Chem. **1977,** 529.

209. TSCHECHE, T., and W. WERNER: Evocarpin, ein neues Alkaloid aus *Evodia rutaecarpa.* Tetrahedron **23,** 1873 (1967).

210. TSIEN, R. Y., D. P. L. GREEN, S. R. LEVINSON, B. RUDY, and J. K. M. SANDERS: A Pharmacologically Active Derivative of Tetrodotoxin. Proc. R. Soc. London, Ser. B **191,** 555 (1975).

211. SPRINGER, J. P.: The Absolute Configuration of Nortryptoquivaline. Tetrahedron Lett. **1979,** 339.

212. TSUDA, K.: Über Tetrodotoxin, Giftstoff der Bowlfische. Naturwissenschaften **53,** 171 (1966).

213. TSUDA, K., S. IKUMA, M. KAWAMURA, R. TACHIKAWA, K. SAKAI, C. TAMURA, and O. AMAKASU: The Structures of Tetrodotoxin and its Derivatives. Chem. Pharm. Bull. 12, 1357 (1964) and references cited therein.

214. VASUDEVAN, T. N., and M. LUCKNER: Alkaloide aus Ruta angustifolia Pers., Ruta chalepensis L., Ruta graveolens L. und Ruta montana Mill. Pharmazie 23, 520 (1968).

215. VINCENT, M., J. MAILLARD, and M. BENARD: Structure and Properties of Alkyl Iodides of Alkyldihydro-4-quinazolones. Bull. Soc. Chim. Fr. 1963, 119.

216. WAIBLINGER, K., S. JOHNE, and D. GRÖGER: Zur Biosynthese des Pyrrolidinringes in Peganin. Phytochemistry 11, 2263 (1972).

217. WALETZKY, E., G. BERKELHAMMER, and S. KANTOR: Quinazolinones for Treating Coccidiosis. US Pat. 3,320,124 (1967) [Chem. Abstr. 68, 39 647 v (1968)].

217a. WATERMAN, P. G.: Alkaloids of the Rutaceae. Distribution and Systematic Significance. Biochem. Syst. Ecol. 3, 149 (1975).

218. WILLIAMSON, T. A.: In: Heterocyclic Compounds (R. C. ELDERFIELD, ed.), Vol. VI, p. 351. New York: Wiley. 1957. — LANDQUIST, J. K.: In: Chemistry of Carbon Compounds (E. H. RODD, ed.), Vol. IV B, p. 1308. London: Elsevier. 1959.

219. WOODWARD, R. B.: Structure of Tetrodotoxin. Pure Appl. Chem. 9, 49 (1964).

220. WOODWARD, R. B., and J. Z. GOUGOUTAS: The Structure of Tetrodotoxin. J. Amer. Chem. Soc. 86, 5030 (1964).

221. YAMAZAKI, M., H. FUJIMOTO, and E. OKUYAMA: Structure Determination of Six Metabolites, Tryptoquivaline E, F, G, H, I and J from Aspergillus fumigatus. Chem. Pharm. Bull. 26, 111 (1978).

222. — — — Structure of Tryptoquivaline C (FTC) and D (FTD). Novel Fungal Metabolites from Aspergillus fumigatus. Chem. Pharm. Bull. 25, 2554 (1977).

223. — — — Structure Determination of Six Tryptoquivaline Related Metabolites from Aspergillus fumigatus. Tetrahedron Lett. 1976, 2861.

224. YAMAZAKI, M., and A. IKUTA: Biosynthesis of Evodia Alkaloids. Tetrahedron Lett. 1966, 3221.

225. YAMAZAKI, M., A. IKUTA, T. MORI, and T. KAWANA: Biosynthesis of Evodia Alkaloids. II. The Participation of C_1-Unit to the Formation of Indoloquinazoline Alkaloids. Tetrahedron Lett. 1967, 3317.

226. YAMAZAKI, M., and T. KAWANA: Isolation of Hydroxyevodiamine (Rhetsinine) from the Fruits of Evodia rutaecarpa Hook fil. et Thomson. Yakugaku Zasshi 87, 608 (1967) (Japan) [Chem. Abstr. 67, 64 601 n (1967)].

226a. YAMAZAKI, M., E. OKUYAMA, and Y. MAEBAYASHI: Isolation of Some New Tryptoquivaline — related Metabolites from Aspergillus fumigatus. Chem. Pharm. Bull. 27, 1611 (1979).

227. YVORE, P., N. FOURE, J. AYCARDI, and G. BENNEJEAN: Efficiency of Stenorol in the Chemoprophylaxis of Avian Coccidiosis. Recl. Med. Vet. 150, 495 (1974).

228. ZABOLOTNAYA, E. S., and L. N. SAFRONICH: Alkaloids of Dichroa febrifuga Introduced into the USSR. Lekarstv. Rasteniya 1969 (15), 356 [Chem. Abstr. 76, 32 240 p (1972)].

229. ZHAREKEEV, B. KH., KH. N. KHASHIMOV, M. V. TELEZHENETSKAYA, and S. YU. YUNUSOV: New Alkaloids from Peganum harmala. Khim. Prir. Soedin. 1974, 264 (Russ).

230. ZHAREKEEV, B. KH., M. V. TELEZHENETSKAYA, and S. YU. YUNUSOV: Deoxypeganidine, a New Alkaloid from Peganum harmala. Khim. Prir. Soedin. 1973, 279 (Russ).

231. ZIEGLER, E., W. STEIGER, and TH. KAPPE: Synthesen von Heterocyclen, 130. Mitt.: Über das Glomerin und das Arborin. Monatsh. Chem. 100, 948 (1969).

232. ZUTSCHI, U., P. G. RAO, A. SONI, O. P. GUPTA, and C. K. ATAL: Absorption and Distribution of Vasicine, a Novel Uterotonic. Planta Med. 40, 373 (1980).

(Received December 20, 1983)

Author Index

Page numbers printed in *italics* refer to References

Subject Index

By

A. SIEGEL, Wien

Fortschritte der Chemie organischer Naturstoffe

Progress in the Chemistry of Organic Natural Products

Volume 42:

1982. VII, 323 pages.
Cloth DM 164,—. ISBN 3-211-81706-9

Contents: Y. ASAKAWA: Chemical Constituents of the Hepaticae. — M. HEIDELBERGER: Cross-Reactions of Plant Polysaccharides in Anti-pneumococcal and Other Antisera, an Update.

Volume 41:

1982. 37 figures. VIII, 373 pages.
Cloth DM 196,—. ISBN 3-211-81690-9

Contents: E. HASLAM: The Metabolism of Gallic Acid and Hexahydroxydiphenic Acid in Higher Plants. — D. G. ROUX and D. FERREIRA: The Direct Biomimetic Synthesis, Structure and Absolute Configuration of Angular and Linear Condensed Tannins. — ST. J. GOULD and ST. M. WEINREB: Streptonigrin. — D. J. ROBINS: The Pyrrolizidine Alkaloids. — J. W. DALY: Alkaloids of Neotropical Poison Frogs (Dendrobatidae).

Volume 40:

1981. 21 figures. IX, 295 pages.
Cloth DM 158,—. ISBN 3-211-81624-0

Contents: P. LEFRANCIER and E. LEDERER: Chemistry of Synthetic Immunomodulant Muramyl Peptides. — SUKH DEV: The Chemistry of Longifolene and Its Derivatives. — W. HELLER and CH. TAMM: Homoisoflavanones and Biogenetically Related Compounds. — R. G. COOKE and J. M. EDWARDS: Naturally Occurring Phenalenones and Related Compounds. — C. W. JEFFORD and P. A. CADBY: Molecular Mechanisms of Enzyme-Catalyzed Dioxygenation (An Interdisciplinary Review).

All Volumes and Cumulative Index 1–20 available

Price reduction for subscribers: 10%

Special reduced price (20% reduction) for the complete Series Vols. 1–46 incl. the Cumulative Index to Vols. 1–20

Springer-Verlag Wien New York